Earth Systems Protection and Sustainability

James N. Furze • Saeid Eslamian
Safanah M. Raafat • Kelly Swing
Editors

Earth Systems Protection and Sustainability

Volume 1

Springer

Editors
James N. Furze
Royal Geographical Society
(with the Institute of British Geographers)
London, UK

Saeid Eslamian
Department of Water Engineering
Isfahan University of Technology
Isfahan, Iran

Safanah M. Raafat
Control and Systems Engineering
Department
University of Technology-Iraq
Baghdad, Iraq

Kelly Swing
Tiputini Biodiversity Station
College of Biological
and Environmental Sciences
University of San Francisco de Quito
Quito, Ecuador

ISBN 978-3-030-85828-5 ISBN 978-3-030-85829-2 (eBook)
https://doi.org/10.1007/978-3-030-85829-2

Cover artwork: 'The Sound of the Globe'. Semin Sinem Cacho Duran, Antalya, Turkey. E: semincacho@outlook.com

This Springer imprint is published by the registered company Springer Nature Switzerland AG
The registered company address is: Gewerbestrasse 11, 6330 Cham, Switzerland

This volume is dedicated...
To you, who turns the page to discover the next, for protection and continuation of the beauty on Earth

(James N. Furze)

To my parents: my recently late mother, Fatima, and my late father Mehdi both of whom gave me the foundation of something they never enjoyed – education

(Saeid Eslamian)

To all who seek to enrich their environment, supporting their social and economic development in natural balance with aesthetic glory

(Safanah M. Raafat)

To those who refuse to simply accept the legacy put upon their doorstep, to your growing awareness and mounting creativity that are the essence of hope for the future we all wish for our children

(Kelly Swing)

Foreword

What does Earth Systems Protection (ESP) and Sustainability mean? Well, Earth is the third planet from the Sun; our and that of vast proliferations of diversity and resources home. Systems are sets of connected things or operations which find symphonies of collective operation subject to their variance fields, through time and space parameters. Protection is the instigation of defence against any stress or fluctuation and refers to keeping someone or something safe from harm again within temporal, spatial and material ranges. Sustainability entails being able to continue over a period of time. Each of these terms can be applied in different contexts and perspectives. Indeed the principle of protecting Earth Systems, essentially embodying integrity, contains a seemingly infinite range of diversity and even dimensional qualities which operate in different scenarios. Our definitions lead us to admit that there is a certain paradox if we consider any one set or domain, after all how can we protect multiple, sometimes opposing factors at the same time whilst ensuring that individual elements are sustained? Thus, ESP becomes a matter of being able to predict with intuition gained from insights into converse operations, both subjectively and objectively.

Ironically, the usual connotation of 'ESP' relates to extrasensory perception, clear-sightedness developed independently of known sensing processes. In this context, one might say that ESP entails mysterious precognition, knowledge of the future, without a complete understanding of input ranges or sets. However, the ability to know a systems operation has a probability function, determined by the extent of the subject(s) within which any one element of a set exists. Moreover, the concept for humans is somewhat blurred as surely it requires knowing when and how long into the future one wishes to see.

Through appreciation of the different ways in which we (and other life units) perceive our worlds' geographic distribution of resources and evolutionary development, it is possible to identify patterns of operation which may be seen to converge or oppose each other in different ways. 'Earth Systems Protection and Sustainability' assists us by embracing elements of paradoxical nature to reach alignments. Such alignments, when truly recognized, could indeed give our systems a self-feeding or sustainable nature, whilst protecting the integrity of the systems and maximizing maintenance of the diversity of life. The true promise of environmental sustainability and thus resource management, proffering complementary/cooperative development within life diversity seems to be within our reach. Yet on our approximately five-billion-year-old rocky planet, we must face that our brief existence on Earth and our development currently appear to be driving us towards collapses of life diversity, amidst incredibly unstable flow fields rather than maintenance of life's beauty. The era of mass extinctions is upon us as recognized in the recently coined Anthropocene.

What are the key areas in which we can consider the seemingly impossible regeneration of our beautiful Earth in order to gain sagacity? One might take strategic approaches in practical operations for policy directions. An understanding of how a system balances itself might also be useful to enable sustainability into the future.

Policies are formed by centralized bodies of authority, distributed according to the parameters which separate our complete geographical spread and different cultures. Imperatives stimulate our policy objectives. Thus policy imperatives and operations are of different kinds but all are designed to sustain objectives in a stable manner, potentially representing a self regulating optimally governed world. Our imperatives recognize the key requirements of as many of the components of our demographics as possible into the future. Without policies we lose structure, they are the qualitative representation of our quantitative measurements and help us to mitigate irregularities.

The concentration of greenhouse gases, including carbon dioxide in the terrestrial atmosphere, increased significantly compared to pre-industrial levels, leading to a generalized acceptance of the effect of human activities. In relation to pre-industrial levels, world temperatures were augmented in temperature increases of around 1 °C, though predictions set temperatures with an augmentation of 1.5 °C between 2030 and 2052. The last three decades are commonly regarded as having been the warmest on record, according to our Intergovernmental Panel on Climate Change (IPCC). However, certain regions manifest significant augmentations beyond the world mean.

Supplementary heat output is produced at the most elevated latitudes; as a consequence, the water cycle is affected greatly. This is evident especially in the Arctic, where the pressure is equal to that of the world average, except that the rate is not uniform. The warming of temperatures has led to melting of sea ice. In recent years, global permafrost has melted and snow cover has decreased. In mid-latitude zones, extreme temperatures and intense precipitations have increased with increasing frequency. Lower latitude regions are expected to witness increases in the incidence of extreme hot seasons.

Temperatures of the Earth have led to a number of changes in regional climate, and it is evident that these temperatures are threatening humans, ecosystems and organic life. Sadly, even with a significant reduction in carbon dioxide emissions, studies show that the effects of climate change may not be reversed for at least one thousand years.

Nature has been created in a natural order, in harmonic balance, with extraordinary aesthetic beauty, human beings are members of the community of nature. Therefore, there must be a certain agreement between man and the environment. Nature is a test to mankind, to what extent and in what manner will man execute his task as trustee of the Earth socially and economically?

Mathematics is an essential requirement for the accomplishment of sustainable engineering approaches towards working in real situations. Optimization algorithms, for instance, can provide the most suitable mathematical solution for a sustainable goal with complex constraints. Accordingly, adaptive, robust and optimal control strategies can be formed to reach sustainable requirements. Moreover, the recent progress in deep learning algorithms makes the inclusion of expert knowledge in building a sustainable controlled system far more satisfying and suitable. Advanced algorithms in many effective areas such as robotics, smart buildings, green engineering and renewable energy generation clarify that driving sustainability accentuates the importance of mathematical approaches.

Learning the domains in which systems operate and their relational convergence assists us in developing self-feeding operations in different scenarios. Path planning and deep learning approaches lead us to an understanding of the precise orientation of operational elements and their embedded processes. From the viewpoint of mathematics, an appreciation of opposing subject weights could help to drive systems sustainably through residual differences. The latter leads us to a potentially looped system that could be present through infinite series expansions and contractions, pending the pressures acting on the weights within a loop. Systems learning hence helps us develop supervised learning approaches until the system has sufficient residual variance to enable its natural operation and can function in an unsupervised manner. In mathematic terms, the additive weight of expansions becomes the agents that drive the system potentially indefinitely, even through uncertainties.

Humans, for the most part, are far too preoccupied with the demands of day-to-day life to even have a chance to be worried about mid-century – a scant few decades away – and the notion of nine or ten billion neighbours all looking to grab their piece of the pie. Most of us never see more than dozens to hundreds of fellow human beings at a time, so difficulties in comprehending scenarios beyond our own personal experience are entirely justifiable. Besides, the human mind probably did not evolve to manage truly large numbers; for various reasons, it is no surprise that enormous numbers remain abstract. Whether we might be trying to absorb an estimate of the number of sand grains on a beach, molecules in a cup of coffee or the tonnage of CO_2 emissions from a city of one million inhabitants, any deeper significance unquestionably tends to escape most of us. Certainly, extrapolation from one's own family to regional or global dimensions is further complicated by deficient science education in combination with self-interested political or corporate propaganda. Unfortunately, the numbers connected to anthropogenic impacts on our environment are intimidatingly gigantic and will require participation *en masse* accompanied by proportionate willingness to change mentalities and lifestyles. Further, we must create new technologies and strategies and sincerely seek a functional balance between short-term political or personal economic gains and long-term global benefits, involving significant investments in the distant future. In short, approaches need to become proactive at the macro-scale rather than reactive at the micro-scale.

To understand the implications of having to share space and other vital resources like food and water with another billion or two new arrivals should be a simple matter of summing and dividing. Our brains, and populist governments, however, tend to prefer to oversimplify things and go on using analogies that cannot be scaled up rationally. Where a dozen can eat, so can thirteen. Incidentally, it has been suggested that humankind will require the production of more food in the next three decades than we have produced in all of previous history. Even if those calculations are somewhat awry, the idea is, to say the least, daunting.

Whenever we consider the natural world and our relationship with it, we too often put humankind in a position of lording over the entire planet as though we were not part of the workings of the whole, that we are entitled to every resource out there, that we have been given absolute dominion over everything, that we may consume or make use of all things in our purview, that the mere condition of being human gives us some right to consider everything outside ourselves property to be exploited. While we indeed seem to have control over some aspects of the world we have created, many factors still lie well beyond our powers. It is precisely our arrogance related to the definition of the boundary between the two, the servile versus the indomitable, that gets us into trouble over the long haul. For a couple of centuries, the solution for most perceived problems has been money.

Many problems of the past could be resolved with moderate expenditures, but human nature drives us to hold the purse strings tight until no other alternatives can be employed. The allocation of significant financial outlays, by individuals or governments, often requires a traumatic experience, especially when any portion of the concept approaches something that might be interpreted as being altruistic. The question at present seems to be: just how big will that slap in the face have to be? Looking into the near and not-too-distant future (in relation to the climate crisis, mass extinction, ocean acidification, expanding desertification, plastic saturation, chemical pollution, energy consumption, solid waste and wastewater management, agricultural demand and wild harvest of biotic resources), it appears that humanity will need far greater access to financial assets than ever before in relation to the environment because the challenges have grown far beyond previous dimensions, precisely in proportion to our immensely growing population.

London, UK — James N. Furze
Isfahan, Iran — Saeid Eslamian
Baghdad, Iraq — Safanah M. Raafat
Quito, Ecuador — Kelly Swing
28 May 2021

Preface: From the Coordinating Editor

I am pleased to write the preface for the second book of the Mathematical Advances Towards Sustainable Environmental Systems (MATSES) series. The first book was published in 2017. It recognized the need to understand our planet and learn to manage its resources. Dr James Furze and his associate editors successfully met that need, by presenting pertinent disciplines involved with MATSES, consequently contributing to developing sustainable environmental systems.

A wide range of specialties require a working knowledge of mathematical developments of sustainable systems. Each chapter of MATSES includes a specific field's background information emphasizing a certain sustainable mathematical or engineering problem.

The first volume of *Earth Systems Protection and Sustainability* represents new ideas that have evolved in various fields of science and technology, as well as reflecting age-old fundamental areas and historical communities. New mathematical approaches and social and community perspectives have been tackled and introduced to increase our knowledge of Earth systems, whilst highlighting the overarching themes of protection of both structural integrity and the highly vulnerable components which are essential facets in the dynamical nature of our Earth.

It's our hope and expectation that this book will provide an appealing and effective experience for those that have interest in learning, working or enjoying this increasingly developing area which supports us all into the indefinite future.

Unique points include:

- Frontier research and imperatives for environmental sustainability
- Management and directions for resource usage in both human and naturalized aquatic and terrestrial ecosystems
- An expansive collection of leading authors in both qualitative and quantitative chapters across refined sections
- Community and subject driven 'problems' and 'solutions' which reward ownership for sustainability directly to the individuals/communities the volume is designed to assist

In addition to the 12 author teams from countries across the world, reviewers spread between the authors, community members, governmental officials and organizational unit members (from many of our 'United' core bodies), the editorial collective was built of 4 members spanning the subjective specialisms and disciplines of the volume and the geographical spread of the Earth. We extend our sincere gratitude to the authors, publishers and community members who selflessly gave their time and support in preparation of this book, and with whom we wish to share and reflect our efforts back to. *Earth Systems Protection and Sustainability* is targeted for a global audience (academic, professional, classroom, governmental, unit and community members) in both descriptive and illustrative sections and seeks to include all sectors, thus managing every resource type and usage, ensuring ongoing Earth Systems Protection.

London, UK — James N. Furze
Isfahan, Iran — Saeid Eslamian
Baghdad, Iraq — Safanah M. Raafat
Quito, Ecuador — Kelly Swing
30 May 2021

Introduction

This volume clarifies, classifies and expands areas of Earth Systems Protection and Sustainability. Included are integral subjects with coverage of leading research from subject specialists in different areas, as well as practically driven chapters implementing sustainable developments key to our continued balanced needs.

The four perspectives of the volume are made to reflect different pillars of sustainability that have a reflective and balancing quality, within Earth Systems Protection (ESP). The perspectives are (i) biological/chemical/geographic, (ii) sustainability policy imperatives/social perspectives, (iii) mathematic and applied systems learning and (iv) pictorial – problems and solutions for special focus. With these aspects, we hope to balance both qualitative/quantitative approaches and give reflective ownership of ESP and sustainability to the communities and issues that the volume is designed to assist.

In the first of two volumes, each of the pillars of sustainability in ESP is given in balanced chapters. Groundbreaking mathematic evolutionary patterning with chemical and biological relevance and functions assisting with life's continuation on Earth are given in the first chapter. Soil-borne pathogen management and halophyte uses in agroecology towards sustainability entail the second and third chapters. Industrial extraction and regenerative bioremediation innovation in waste management, in harsh dynamic environments, are covered in the fourth chapter. Extraordinary orchid diversity is included in conservation and sustainable contexts in the fifth chapter. The sixth chapter incorporates technological innovations tackling climate change. Chapter seven documents practical human displacement/management with consideration of learning in an ongoing real crisis situation. Risks from the perspective of the environment, biodiversity, humankind and the planet are seminally discussed in chapter eight. Chapter nine discusses problems of centralized and decentralized approaches in policy formation in the chaotic setting of a country with vast differences in its people and geographical settings amidst huge risks and vulnerabilities. Chapter ten shows sustainable expansion of intelligently driven markets implementing a transition of development state, accentuating development needs and the United Nations Sustainable Development Goals in context. Sustainable consensus in our increasingly uncertain world are given due discussion in chapter eleven, and the twelfth chapter considers robotic path planning in both static and dynamic environments, which inevitably will assist us in increased efficiencies and reduced use of resources as well as in higher mathematic areas applicable to different fields' learning.

Chapters of the first three perspectives are evenly spaced by pictorial contributions made from across the world, highlighting that which we must protect. The pictorial sections refine particularly urgent problems which require attention or additionally provide solutions in different methods of environmental sustainability as operated constructively, in ecology and in communities. The pictorial sections are built from community donated matter and serve to complement the descriptive chapter sections, enabling combined ownership of the volume and synergistic learning to take place between approaches of countering protection and provision of sustainable systems. The first pictorial section takes us to Ecuador amidst the greatest diversity of the planet. The second takes us to Iran and shows a natural solution to extractive industry. The third pictorial section credits Unit inspired world relief organized community approaches, countering huge disturbances and developmental vulnerabilities in a visit to Nepal and India. A heartfelt thanks to those who contributed.

Mathematic advances include both operational/Boolean methods as well as linguistic, logic-based Bayesian approaches and generative mathematics relevant to scenario formation. Mathematic methods include optimization approaches, functional areas, linguistic characterization and mention of areas pertinent for policy makers in their measurements of changing demographics amidst uncertain environments. Inclusion of path planning in specified Cartesian and artificial fields may help provide optimal solutions for different sectors and the circular nature and application of sustainability.

The volume has refined focus and combinatorial function for protection and insurance of sustainability amidst pressures that face us within and beyond current times. Pertinent sections and synergistic elements are covered in order to synthesize key informative nodes to advise academic institutes, governments, international units and community members of the very real dangers facing planet Earth and its diversity. Analytical and scientific chapters are blended with social resilience and socio-economic development consideration, thus enabling the staging of horizons of sustainability within varying scenarios of climatic forces and species dynamics. Variable format is made use of with descriptive chapters and additional illuminating pictorial sections which refine the focus for readers and reward ownership of sustainability and 'ESP' to local communities and issues of most pressing concern. Thus, both 'problems' and potential 'solutions' are given across all levels of life diversity and human development, facilitating timely and future scenario resolution.

The relevance of this volume for development transgresses our present difficulties and assists with the lattice of continued life on Earth through and beyond the current era. A final thanks to those who read or expressed interest in ESP and sustainability, we hope you enjoy your journey as much as the beauty around you benefits from your attention towards the next horizon.

London, UK — James N. Furze
Isfahan, Iran — Saeid Eslamian
Baghdad, Iraq — Safanah M. Raafat
Quito, Ecuador — Kelly Swing
3 June 2021

Contents

Contributors

Aya A. M. Abdellatif Central Laboratory of Organic Agriculture, Agricultural Research Center (ARC), Giza, Egypt

Rizwan Ali Ansari Department of Plant Protection, Faculty of Agricultural Sciences, Aligarh Muslim University, Aligarh, India

Susanta Kumar Biswal School of Applied Sciences, Centurion University of Technology and Management, Bhubaneswar, Odisha, India

Rikina Choudhury Department of Life Sciences, Rama Devi Women's University, Bhubaneswar, Odisha, India

Rajkumari Supriya Devi Biodiversity and Conservation Laboratory, Ambika Prasad Research Foundation, Regional Centre, Imphal, Manipur, India

Semin Sinem Cacho Duran Artist, Antalya, Turkey

Mohamed El Alaoui Department of Production and Industrial Engineering, ENSAM-Meknès, Université Moulay Ismaïl, Meknes, Morocco

École supérieure d'Architecture de Casablanca, Casablanca, Morocco

Saeid Eslamian Department of Water Engineering, College of Agriculture, Isfahan University of Technology, Isfahan, Iran

James N. Furze Royal Geographical Society (with the Institute of British Geographers), London, UK

Laboratory of Biotechnology and Valorization of Natural Resources, Faculty of Sciences-Agadir, Department of Biology, Ibn Zohr University, Agadir, Morocco

Control and Systems Engineering Department, University of Technology-Iraq, Baghdad, Iraq

Tayyaba Hussain Plant Microbe interaction Laboratory, Quaid-i-Azam University, Islamabad, Pakistan

Dina S. S. Ibrahim Department of Nematodes Diseases, Central Laboratory of Biotechnology, Plant Pathology Research Institute, Agricultural Research Center (ARC), Giza, Egypt

Navneet Kaur Department of Life Sciences, Rama Devi Women's University, Bhubaneswar, Odisha, India

Mudassir Khan Department of Healthcare Biotechnology, Atta-Ur-Rahman School of Applied Biosciences (ASAB), National University of Science and Technology (NUST), Islamabad, Pakistan

Sanjeet Kumar Biodiversity and Conservation Laboratory, Ambika Prasad Research Foundation, Regional Centre, Imphal, Manipur, India

Shah Rakesh Kumar Lutheran World Relief, Kathmandu, Nepal

Sudhir Kumar Asian Disaster Preparedness Centre (ADPC), Bangkok, Thailand

Manisha Mahapatra Department of Life Sciences, Rama Devi Women's University, Bhubaneswar, Odisha, India

Shaymaa M. Mahdi Automation and Robotics Research Unit, Control and Systems Engineering Department, University of Technology-Iraq, Baghdad, Iraq

El Hassan Mayad Laboratory of Biotechnology and Valorization of Natural Resources, Faculty of Sciences-Agadir, Ibn Zohr University, Agadir, Morocco

Manal Mostafa Department of Nematodes Diseases, Central Laboratory of Biotechnology, Plant Pathology Research Institute, Agricultural Research Center (ARC), Giza, Egypt

A. A. Muftau Baniyas East, Abu Dhabi, United Arab Emirates

Gyawali Narayan Agriculture and Forestry University, Rampur, Chitwan, Nepal

Farai Nhakwi Faculty of Finance and Economics, Zhejiang Gongshang University, Zhejiang, China

School of Business Sciences, Chinhoyi Universitu of Technology, Chinhoyi, Zimbabwe

Abdullatif A. Olanrewaju Department of Construction and Green Engineering, Universiti Tunku Abdul Rahman, Kampar, Malaysia

Amin Parnian Young Researchers Club, Masjed Soleyman Branch, Islamic Azad University, Masjed Soleyman, Iran

Amir Parnian National Salinity Research Center (NSRC), Agricultural Research Education and Extension Organization (AREEO), Yazd, Iran

Safanah M. Raafat Automation and Robotics Research Unit, Control and Systems Engineering Department, University of Technology-Iraq, Baghdad, Iraq

Firas A. Raheem Automation and Robotics Research Unit, Control and Systems Engineering Department, University of Technology-Iraq, Baghdad, Iraq

Sakti Kanta Rath Department of Life Sciences, Rama Devi Women's University, Bhubaneswar, Odisha, India

Kafayat O. Shobowale Department of Mechatronics Engineering, Air Force Institute of Technology, Kaduna, Nigeria

Kelly Swing Tiputini Biodiversity Station, College of Biological and Environmental Sciences, University of San Francisco de Quito, Quito, Ecuador

Tropical Ecology Program, Boston University, Boston, MA, USA

Jamuna Tudu Department of Life Sciences, Rama Devi Women's University, Bhubaneswar, Odisha, India

Part I
Yasuní Biosphere Reserve, Ecuador – Beauty, Extractive and Consequent Industries in the Earth's Most Biodiverse Location

Fig. 1 Aerial image of rainforest along the Tiputini River just west of Tiputini Biodiversity Station, Yasuní, Orellana Province, November 2004

Fig. 2 A million kinds of insects may occupy the Yasuní region of westernmost Amazonia. Here, a predatory katydid known as the "spiny devil" (*Panacanthus cuspidatus*), November 2011

Fig. 3 The Yasuní Biosphere Reserve, encompassing an area similar to the country of Belgium in extension, has over 145 species of frogs, including the Map Treefrog (*Boana geographica*), June 2006

Fig. 4 Schneider's dwarf caiman (*Paleosuchus trigonatus*) is just one of more than 120 species of reptiles in the park, June 2008

Fig. 5 Yasuní includes approximately 600 avian species, such as the Blue-headed Parrot (*Pionus menstruus*), May 2010

Fig. 6 Within Ecuador's eastern lowland rainforest, an area no larger than Austria, there are 5 feline species, 13 marsupial species, 36 rodents (up to >50kg) and a dozen primate species. The endemic Golden-mantled Tamarin (*Saguinus tripartitus*) is present at Tiputini Biodiversity Station and across Yasuní, April 2013

Fig. 7 View upstream of the Tiputini River overlooked from Tiputini Biodiversity Station, April 2018

Fig. 8 Submerged forest, classically known as igapó, Tiputini Biodiversity Station, April 2011

Fig. 9 Oña Tega, a Waorani man using a blowgun, a traditional part of indigenous life before the arrival of the oil industry in their ancestral territory, Orellana Province, eastern Ecuador, April 2006

Fig. 10 As oil access roads impact remote areas like eastern Ecuador, traditional small-scale slash-and-burn agriculture often converts to massive clear-cuts accompanied by extensive forest fires, November 2018

Fig. 11 In operations where the natural gas is poor quality or in relatively low quantities, it is burned off as a precautionary measure. Oropendola nests here in close proximity to a gas flare within the Yasuní area of Ecuador, November 2012

Fig. 12 The outdated, common practice of gas flaring attracts innumerable insects to their deaths from across expansive areas of the forest. The scorched and killed are keystone pollinators, insectivorous birds and bats. Gas flares here against the rainforest backdrop, Joya de los Sachas, Amazonian Ecuador, March 2006

Fig. 13 Cleanup of an oil spill in Yasuní, Ecuador. The image represents progress after about six weeks of dedicated activity. This phase of remediation continued for most of a year. Due to oil seeping into the ground, this section of forest was eventually removed by bulldozers along with a layer of substrate before allowing natural succession to take over the slow recovery process, March 2008

Fig. 14 Workover rig in Block 16, eastern Ecuador. This heavy machinery is used to recover wells that have declined in daily production. Each barrel of crude oil extracted requires the handling of about 20 additional barrels of formation water laden with salts and heavy metals, August 2007

Fig. 1

Fig. 2

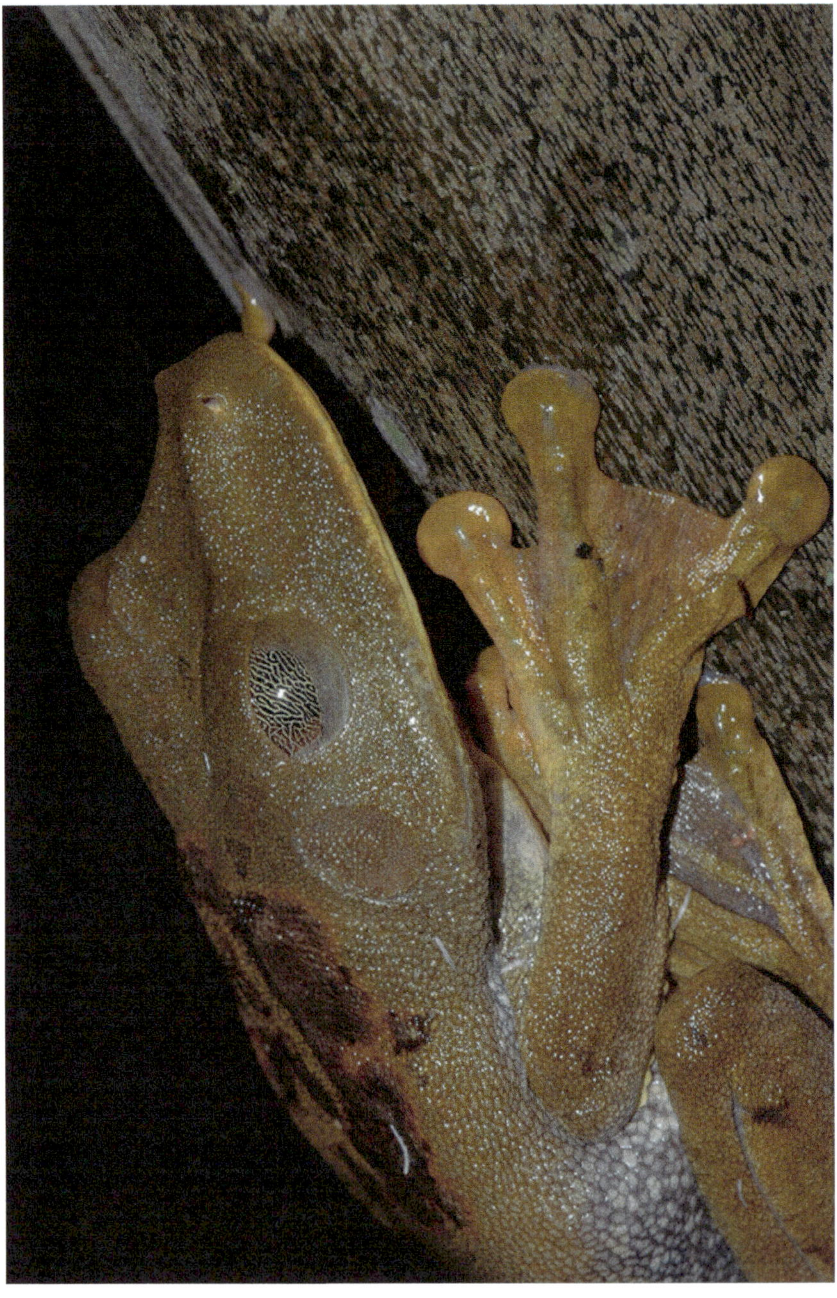

Fig. 3

Fig. 4

Fig. 5

Fig. 6

Fig. 7

Fig. 8

Fig. 9

Fig. 10

Fig. 11

Fig. 12

Fig. 13

Fig. 14

Chapter 1
Generators, Harmonics and Evolutionary Emergence

James N. Furze and El Hassan Mayad

Abstract Planet Earth was formed around five billion years ago; evolutionary proliferations reach over four billion years. Studies of evolution are made up of different fields of study with alternative perspectives. The current work gives a brief review of these areas, along with modelling techniques applied to date. This chapter demonstrates the mathematic unification of the different fields of study in order to further an engineering-based structure for advancement in our understanding of the pathways which lead to different constructs in life. Computational, biological and chemical studies lead us to propose that there are two main patterns formed in evolution of life – cellular and molecular.

We used a laboratory simulation to clarify the formation of polymers from monomers under dynamic conditions to set a basis for future simulation. Methods discussed are of both a Bayesian and Boolean nature. Application of logistic regression requires partitioning of variance within the systems; after discussing the background required, we made use of application of higher mathematic technique (multi-objective genetic algorithm) to generate variance within the different scales of evolution, the result of which was analogous with the Fisher equation model of gene distribution within populations. The results of the distribution were subjected to least squares polynomial regression in order to reinforce the rational truth of the models formed. Each copula distribution was noted in univariate terms and applied to Sklar's theorem in order to give a concise description of the pathways as alternative functions. Laboratory simulated data was also subjected to a robust linear

J. N. Furze (✉)
Royal Geographical Society (with the Institute of British Geographers), London, UK

Laboratory of Biotechnology and Valorization of Natural Resources,
Faculty of Sciences-Agadir, Department of Biology, Ibn Zohr University, Agadir, Morocco

Control and Systems Engineering Department, University of Technology-Iraq, Baghdad, Iraq
e-mail: james.n.furze@gmail.com; jamesfurze@hotmail.com

E. H. Mayad
Laboratory of Biotechnology and Valorization of Natural Resources,
Faculty of Sciences-Agadir, Ibn Zohr University, Agadir, Morocco
e-mail: e.mayad@uiz.ac.ma

J. N. Furze et al. (eds.), *Earth Systems Protection and Sustainability*,
https://doi.org/10.1007/978-3-030-85829-2_1

regression and the resultant variance plotted. Field and laboratory trials validate the existence of triangular relationships within communities; further harmonic constants between interacting species may be found. Variance in the trials showed sine/cosine contraction and expansion, similar to that seen in mathematical and genetic results covered. Extensive use of copular frameworks, Mamdani and refined Takagi-Sugeno-Kang algorithms and additional novel DANCE (differential algorithmic network centred emergence), application of robotic techniques and functional expressions are recommended as primary constructs covering different categories of life in passive and active terms pending their interaction.

Keywords Evolution · Dynamic · Cellular · Molecular · Copula · Simulation · DANCE

1.1 Introduction

Evolution of life's kingdoms spans the last four billion years on Earth. Studies of the creation of life have involved chemistry, physics, biology, biochemistry and astrobiology. Mixtures of inorganic molecules originating in interstellar ices were seen to combine and perform self-replication with respect to their properties (Dworkin et al. 2001); ultimately these molecules formed simple organic structures seen to be capable of performing self-organization due to their chemical and physical properties in dynamic conditions of 'volcanic' Earth. In 'small ponds' assumed to form on early terrestrial landforms, dehydration and rehydration drove molecules far from equilibrium; lipids formed and captured systems of polymers through multiple cycles. This increases the chance that collections of molecules will emerge having one or more functions required for the origin of life, and crucially selection of vesicles encapsulating these polymers led to stepwise increments towards the emergence of functional systems capable of growth, reproduction and evolution (Gordon 1993; Damer and Deamer 2015).

Early life was seen to be 'communal', based purely on chemical/physical properties of molecules in the dynamic conditions of the young Earth (Deamer et al. 2006; Marzban et al. 2014). We can make use of the proposed scenario of volcanic early Earth and draw similarities with processes that occur today in locations such as in the Kamchatka peninsula in Russia or indeed any other volcanic sites.

Starting from a simple, astrophysically relevant, ice mixture (water, methanol, ammonia and carbon monoxide), a complex mixture of compounds, including amphiphiles and fluorescent molecules, is generated in low temperatures by photolysis. Adding additional sources of energy thought to be present on the pre-biotic Earth (such as temperature extremes) similar to geothermal sites enables the build-up of membranous structures from amphiphilic molecules and fats (Lu et al. 2021) and results in formation of simple nucleotide structures and ribose. Hence simple protocells capable of self-replication via RNA-directed processes have been hypothesized and proved to exist in laboratory conditions which simulate those on early Earth (Schrum et al. 2010; Mungi and Rajamani 2015).

Reactive dynamics of the early Earth were (and still are) present in different dimensions (such as rock formation geodynamic combinations; 3:1 concentrations of formaldehyde-hydrogen cyanide; hydrated/dehydrated cycling). In the orchestra of chaotic molecular collisions, biological evolution formed as a result of a process similar to the formation of geometric fractals. Model formation in three different forms gave rise to different forms of self-replicating 'cells' in different phylogenies (Alicea 2014). However there are several points which are uncertain in the hypothesis of molecular/cellular radiations, and application of a systems approach is the leading current thinking to resolve the difference between cellular and observable/molecular scales of postulated biogenic processes (Grover et al. 2015). Initial evolutionary processes included chemical construction, implying monomer formation, polymerization reactions and formation of complex assemblies. Each of the stages is subjected to selection and differential evolution in variable scenarios. Further modelling is required to resolve evolutionary processes which occurred under alternate environmental scenarios. The future of this work is to expand the complex range of polymerization and molecules formed through experimentation in external dynamic conditions. Logically, variable dynamic conditions lead to different proportional combinations of emergent 'organic' systems (Monnard 2016).

Many models have been used to simplify and operationalize the subtle but complex mechanisms of biological evolution. The so-called toy models are gross simplifications that nevertheless attempt to retain major essential features of evolution, bridging the gap between empirical reality and formal theoretical understanding. Thirteen models which describe evolution qualify as toy models, including the tree of life, branching processes, adaptive ratchets, fitness landscapes and the role of nonlinear avalanches in evolutionary dynamics. Such toy models are intended to capture features such as evolutionary trends, coupled evolutionary dynamics of phenotype and genotype, adaptive change, branching and evolutionary transience (Alicea 2014).

The engineered design of biological processes leads our systematic understanding of evolution, though is at an early stage. Such perspective will develop as a greater volume of previous research is factored, together with increasingly advanced combinatorial Bayesian analysis. A functional-copular approach is proposed using mathematic method, resolving formation of evolutionary pathway functions (Furze et al. 2013a; Schölzel and Friedrichs 2008). Such an approach links together phylogenetic diversification through functional evolutionary adaptation both temporally and spatially over the course of evolution (Manceau et al. 2017). The first of these approaches links phylogenies through common trait structures between lineages, though does not consider the expansive functional range of the traits due to the interactions between the traits themselves. Indeed further research indicates that conserving phylogenetic diversity may often be seen as a poor strategy for conserving functional diversity with evolutionary value (Mazel et al. 2017). However, the link between functional and strategical variation may be resolved by monitoring systems at alternate time scales and in micro-, cellular and macro-ranges. Recent studies have shown that we may enable common structures which link modelling of life forms and processes in select environments at the macro-scale (Cabral et al. 2016). Considering relations of strategical and functional diversity in Earth systems of high vulnerability accentuates sustainability within ecosystems which continue to evolve (Furze et al. 2017; Basener and Sanford 2018; Lal et al. 2021); such approaches should resolve the variance between genetic trait evolutions at different levels and may be used to feed back and give prediction of functional scenarios in the future. Retaining resilient functional and phylogenetic diversity is a major challenge in conservation today, as pressures on Earth systems result in huge losses to biodiversity amidst heavily polluted backgrounds. It is essential that we align our activities in cooperation with the needs of natural systems, or risk perversion and irretrievable changes to evolutionary pathways through cascades of feedback mechanisms operating ecologically, chemically and as a result of species interactions (Swing 2017). In order to give the appropriate weight of each subject area to the process, combinatorial mathematic method (Furze et al. 2013b), computational theory (Prusinkiewicz and Lindenmayer 1990) and simulations of evolution itself (Damer et al. 2012) have recently been formed to allow for differential scenarios across time periods.

The main motivation of this chapter is to expand/generate the variance which we may use to consider the relationships of how genetic material is formed and evolution takes place in different scales of life and within ecosystems. The use of this mathematics has been criticised due to our lack of knowledge about how the patterns vary and their causes in the first place; a motivation of this study is to clarify the origins of the generative process of evolution and project proliferation pertinent to organisms of all kingdoms.

Life's evolution is proposed to follow two main pathways: cellular and molecular systems operated simultaneously over the same time period of around four billion years. The objectives of this study are to show functions for cellular and molecular evolution, to show a discrete molecular evolutionary function and to show differences between proliferations of the split pathways of evolution (with separate chemical and biological laboratory data). We also show emergence of life within extreme environments from data collected in the field. Finally we form functional mathematics, giving a basis for expansion of evolutionary elements and their pathways, thus forming different constructs. As well as in 'exobiological' systems and explorations (Fortney et al. 2016), this will be relevant in some of the vulnerable ecosystems we are currently failing to protect, including those subject to extreme dynamics present on Earth.

1.2 Methods/Results

1.2.1 The Use of Logistic Regression as a Basis for Simulations of Evolution

Typically, the generalized logistic (Richard's) curve is used to summarize growth models; this takes the following form:

$$Yt = A + \frac{K - A}{\left(C + Qe^{-Bt}\right)^{1/v}} \tag{1.1}$$

In (1.1), Y is the dependent variable and t is time. A = the lower asymptote; K = the upper asymptote (carrying capacity when $A = 1$). Effectively these are 'plateau' stages within evolution which may be determined in extreme parameters, including those of volcanic thermal (oceanic) vents (Mullineaux et al. 2018; Deamer and Georgiou 2015). Cycling towards K is mathematically, chemically and biologically related as discussed in the final section of this work; B = growth rate; $v > 0$ affects near which asymptote maximum growth occurs; Q = component related values of $Y(0)$; and C = the maximum distribution of Y (usually 1). The logistic curve may be used as a simplified model for evolution as when displayed graphically it shows a sigmoid growth curve with a Poisson/Gumbel copula route. However, due to the unknown number of multiple elements (non-differentiated) which are in effect in evolutionary terms, its use is highly controversial as the logistic approach does not allow for a clear statement of error due to its non-differentiated form. It is important therefore to make a statement of the precise number of chemical elements dispersing, in order to clarify the error parameters which occur across the period of evolution; secondly it is important to clarify the length of the (approximately 4 billion years) period of evolution without which the logistic curve may not be used (Furze and Zhu 2014). Without the former two sets of variables being clarified, the use of logistic regression is a theoretical model due to Q above and time being undefined.

1.2.2 *Translating Evolutionary Data Generation into Biological Terms*

1.2.2.1 Evolutionary Data

Evolution operates along two pathways, evolutionary factors or genes are subject to opposing mechanisms of dispersal: conservation and mutation. In the current chapter, these pathways are referred to as cellular and molecular. The conservative cellular process gives a slower evolutionary rate over unit time, though molecular evolution gives a steeper rate over unit time. Gaining from conservation achieves a gain of frequency, though conversely mutating into a new variety results in a loss of frequency. Theoretically evolution consists of both molecular and cellular pathways simultaneously.

In the current study, we simulated the cyclical basis of hydrated/rehydrated, two-phase cycles and following Moore's law of exponential growth in the communities of molecules obtained two curves for cellular and molecular evolutionary pathways as shown in Fig. 1.1.

Figure 1.1 shows a three-dimensional structure which was obtained following Poisson simulation of vectors representing those in cellular [x] and molecular [y]

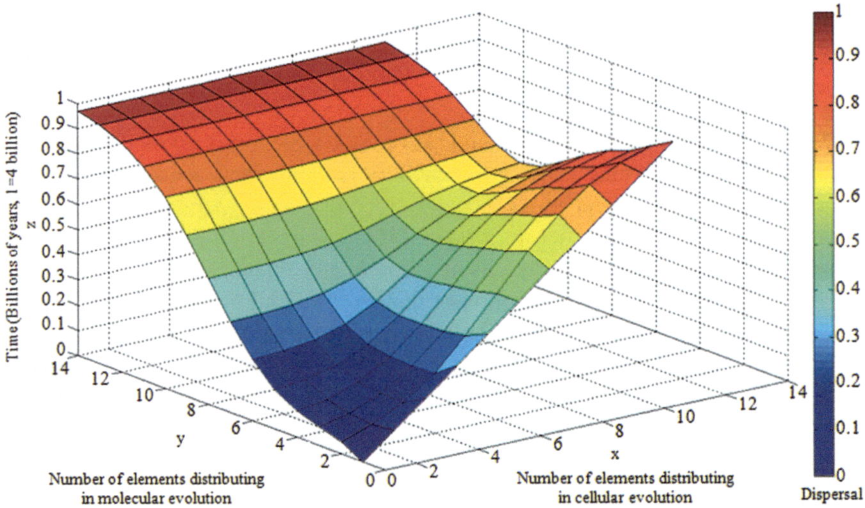

Fig. 1.1 Three-dimensional surface plot of binomial/Poisson distribution of cellular and molecular evolutionary pathways. (Furze and Mayad 2021)

evolution over time [z]. The data for these curves was simulated using a multiple objective genetic algorithm (MOGA) approach (Read et al. 2016; Furze et al. 2017) using a population size of 14 elements. The spread of each curve emulates that which is found following the Fisher equations (Edwards 1994) of natural selection and was also shown in a recent publication (Furze and Mayad 2021).

1.2.2.2 Fisher Expansion of Evolution

Let the distribution frequency of any gene in a population be x_i, where $0 \leq x_i \leq 1$. Then take the fitness that the gene confers upon the organism as w_i, where $0 \leq w_i \leq 1$. Regarding the locus where x_i acts, all genes competing for that spot have an average fitness of $\overline{w}$ such that $0 \leq \overline{w} \leq 1$. In the case that the new gene is fitter than average, then $w_i > \overline{w}$. We can now denote the change in frequency x_i, x_i', x_i'' in each generation approximately as follows:

$$\text{First generation } x_i = x_i \tag{1.2}$$

$$\text{Second generation } x_i' = x_i w_i / \overline{w} \tag{1.3}$$

$$\text{Third generation } x_i'' = x_i' w_i / \overline{w} \tag{1.4}$$

This is converted to a Fisher equation, by the frequency, Δx_i, as:

$$\Delta x_i = x_i' - x_i \tag{1.5}$$

$$= x_i w_i / \overline{w} - x_i \tag{1.6}$$

$$= x_i \left(w_i / \overline{w} - 1 \right) \tag{1.7}$$

This shows that the fitter w_i makes an individual above $\overline{w}$, the greater is $w_i - \overline{w}$, so the faster x_i spreads until $\overline{w}$ rises to w_i.

Suppose a population of any given species has 1000 individuals. One individual has an allele, 'x_2', resistant to high acidity ($w_2 = 1$), and the other 999 'x_1' individuals have only 50% resistance ($w_1 = 0.5$). With low acidity, each generation ($w_i - w$) = 0, so x_2 does not increase ($\Delta x_2 = 0$). Yet once increased acidity is present, x_1 halves each generation, while x_2 increases its frequency from $x_2 = 0.001$ to $x_2 = 1.0$ by about the 18th generation.

The equation for the front curve of Fig. 1.1 is established. This shows a conserved sequence. Both curves are logistic, although the Fisher equation is iterative.

$$\Delta x_i = x_i \left(w_i / \overline{w} - 1 \right) \tag{1.8}$$

Illustrating a standard logistic convention, as in the following equation:

$$X_i = \varepsilon_i / \left(1 + \varepsilon_i \right) \tag{1.9}$$

1.2.2.3 Cellular and Molecular Evolution

Each component spread of Fig. 1.1 is shown in Figs 1.2 and 1.3 along with their least squares regression coordinates (through the 7th degree polynomial), as processed through Matlab (Version 2018a).

In Figs. 1.2 and 1.3, 'e' is notation for times 10 to the power. Shown in Figs. 1.2 and 1.3, the elements were dispersed over unit time. The difference between the curves is due to the coefficient factor ($\partial 1$) being lower in cellular evolution (larger scale, fewer elements, lower fitness), achieving a lower independent 'generator' of the curve; these are defined further in Table 1.1 which gives the rules for linear, quadratic and cubic distributions.

In Table 1.1, C represents cellular; M represents molecular; and rules 1, 2 and 3 represent the distributions shown in the case of cellular and molecular distributions, as such they may be shown in concise terms (Furze and Mayad 2021) by the expression:

$$F_i \leq \sum_{j=1}^{n} M_j B_{ji}, \left(i = 1, \ldots n \right) \tag{1.10}$$

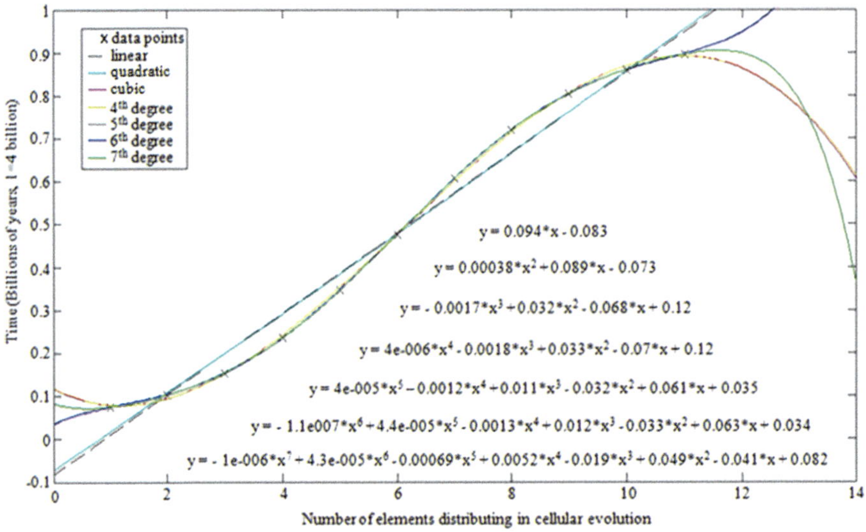

Fig. 1.2 Two-dimensional cellular evolution binomial distribution. (Furze and Mayad 2021)

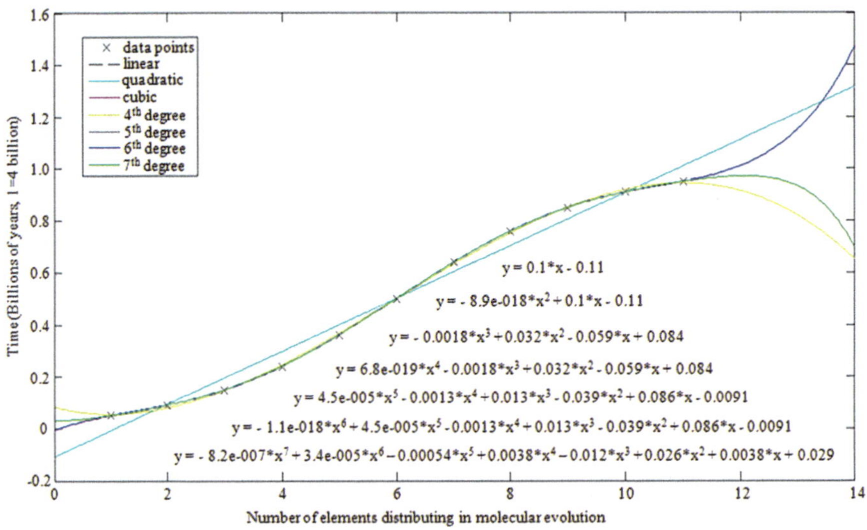

Fig. 1.3 Two-dimensional molecular evolution binomial distribution. (Furze and Mayad 2021)

Table 1.1 Rational polynomials give rules for elements dispersing through evolution (Furze and Mayad 2021)

Rule	Variables									
	$\partial 1$		$\partial 2$		$\partial 3$		$\partial 4$		ε	
	C	M	C	M	C	M	C	M	C	M
1. $Z = \partial 1x + \partial 2 \pm \varepsilon$	0.094	0.1	−0.083	0.11					0.13898	0.14225
2. $Z = \partial 1x^2 + \partial 2x + \partial 3 \pm \varepsilon$	0.00038	−8.9e−18	0.089	0.1	0.073	−0.11			0.13854	0.14225
3. $Z = \partial 1x^3 + \partial 2x^2 + \partial 3x + \partial 4 \pm \varepsilon$	−0.0017	−0.0018	0.032	0.032	−0.068	0.059	0.12	0.084	0.020352	0.02251

In (1.10) F represents the function of data i in the space of Fig. 1.2 and Fig. 1.3, n represents the number of elements, j is the spread of data i, M represents the mid-point of the linear regression and B represents the Z matrix space. Increasing the dynamic in the ratio of opposing conditions (such as hydrolysis-condensation reactions; volatile agents; amount of heat/light) on the elements dispersing in the above Poisson distributions would result in translation of the copula distributions. This may be seen as a generator function for the cellular/molecular 'switch'. The difference, subject to error (variance) in both cellular and molecular pathways, represents the fuzzy integral and consequently defines the membership functions which may be used in the following Laplace equation:

$$A = \int_u \mu A \ldots (y) / y \tag{1.11}$$

In (1.11) $\int_u$ stands for the union of fuzzy singletons (elements dispersing) and μA is the grade of membership in A (or y). '…' is used to represent the iterative process which may be repeated through additional dispersed elements.

The copula type we apply is referred to as a Gumbel, represented by the following:

$$\phi G_u = \left(-\log G_u\right)^{\theta G_u} \tag{1.12}$$

ϕG is the set generator of the union equal to the negative logarithm of the generator union, G_u raised to the power of the symmetrical difference, Θ in the generator union itself (Wang et al. 2012; Alhadlaq and Alzaid 2020; Aldhufairi et al. 2020; Ota and Kimura 2021). Further copula exploration should reveal mathematic relations around the error of the dispersed elements of cellular and molecular evolution, which gives consistent exploration of the Bayesian (discrete) nature of the distributions. Carrying out these steps via hybrid genetic algorithm and functional expansion elucidates the variance between distributions fills in the mathematic gaps to make use of logistic relations.

Although we have not clarified enough of the structural error to make use of the logistic regression expression (1.1) shown earlier, we may link the distributions with use of Sklar's theorem (Sklar 1959; Durante et al. 2013), which states that multivariate distributions may be expressed as copula functions and evaluated as a two-dimensional distribution function. Univariate distribution functions (F_1, … Fn) can be linked to a multivariate distribution function, H as shown:

$$H\left(x_1,\ldots,x_n\right)=C\left(F_1\left(x_1\right),\ldots,F_n\left(x_n\right)\right) \tag{1.13}$$

given that H is a unifying term that summarizes Hilbert space harmonics of discrete univariate distribution functions, which may be given in Laplacian rule-based systems (with sinusoidal variations).

In biological context we may refer to equation (1.11) as an expression of evolutionary harmonics used to summarize differential algorithmic network centred emergence (DANCE). The Gumbel Copula has been used to represent data of climatic and dynamical systems; thus (1.12) may be discretely applied to represent individual expressions of categories of life and their generation (such as that of unicellular life, microbial populations and macro-ecological populations). The elements which represent the dynamic conditions which form the union in the copula give a Poisson/skewed or sinusoidal pattern. This pattern is represented at all scales throughout life, from genetic bottlenecks imposed in vulnerable locations, island landmasses or indeed by the levels of volatile agents and extremes imposed on populations of life-units as a result of modern anthropic 'extractive' approaches. Harsh changes have been present throughout the changing conditions on planet Earth (Judson 2017). Harmonic expressions summarize complex life constructs in order that we can gain greater understanding of the parameters of life (Brack 2004; Furze and Mayad 2021) and their effects on the emergence of novel life systems.

Fundamental evolutionary patterning has been represented through Moore's law of exponential growth, the Fisher equation of natural selection (respecting genetic factors) and by using an expansive mathematical basis.

1.2.3 Laboratory Simulated Polymer Formation

Laboratory studies were carried out in order to concisely define the resolution of molecular evolution; polymers resembling nucleic acids can be synthesized from monomers in simulated hydrothermal conditions undergoing wet-dry cycles. In order to establish a formula that can describe the yield of polymers, variables to be included are as follows:

1. Fraction of monomers becoming polymers in each cycle by simple ester bond synthesis (Fe)
2. Fraction of polymers being hydrolysed back to monomers in each cycle (Fh)
3. Fraction of polymers being synthesized on a template in each cycle (Ft)
4. Fraction of monomers being degraded in each cycle (Fd)
5. Number of cycles

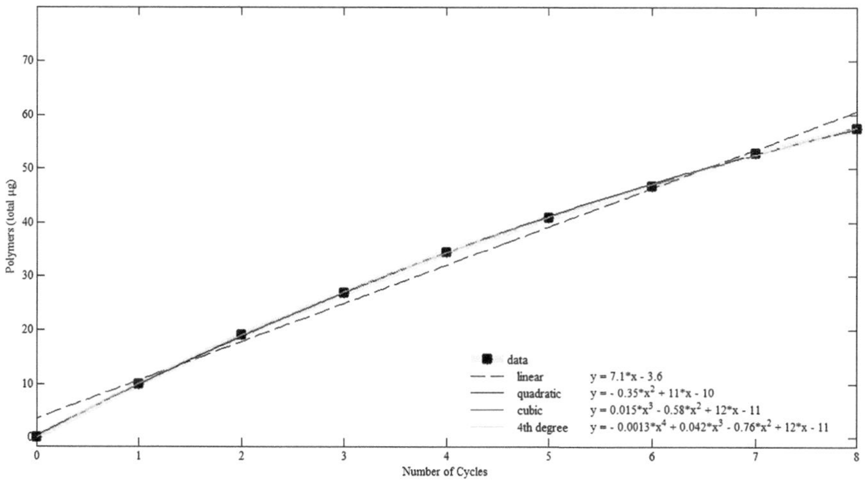

Fig. 1.4 An established framework for laboratory simulated data of polymer formation, assuming hydrolysis, degradation and template refeeding are absent

In the above, 1–4 are constants that may be chosen.

For instance, in the simplest case, assume that no polymers are hydrolysed. Now suppose that we start out with 100 μg of monomers and that 10% of them become polymerized in the first cycle. At the end of the first cycle, we will have 10 μg of polymers and 90 μg of monomers remaining.

Given the assumption that there is a 10% yield of polymers and that we use 100 micrograms of monomers at the start, data obtained is shown in Fig 1.4.

The reduction in residual variance (error) resulting from higher polynomial orders accounts for the increase in accuracy closer to the theoretical model formed from a genetic evolutionary basis such as the Fisher equation, the distribution of which is analogous to that of the data shown in Figs. 1.1, 1.2 and 1.3. The residual variance follows a sine wave pattern, the parameters of which are shown in Fig. 1.5 and indicated in the earlier steps of 1–5 which establish the conditions for the cycle. Plotting linear, quadratic, cubic and fourth-degree polynomials through Fig. 1.4 data gives a result which is consistent with the simulated data pattern shown in Figs. 1.2 and 1.3 (contact for further details). Differentiating via the Taylor expansion to a higher polynomial (up to 7th order) will account for prediction of unknown variables of hydrolysis, degradation and template refeeding; however these are not shown in Fig. 1.5 due to the instability of higher-order polynomials. More laboratory and simulated studies are required to establish this point in reality. Research in this area works towards establishing the parameters for rule bases and scenarios for alternative scales of cellular and molecular evolution (Bains 2020) in variable settings of life systems. The latter may well lead to sets of data (Tugrul et al. 2019; Taylor et al. 2020; van den Hoogen et al. 2020; Echeverría-Londoño et al. 2018; Albrecht et al. 2018), which translate the distributions of Fig 1.1 to satisfy a full cone, respecting biological, chemical and geological research.

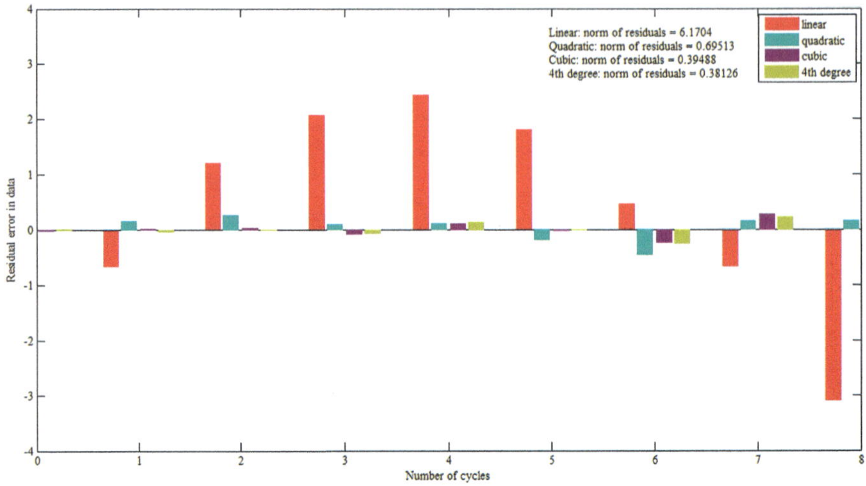

Fig. 1.5 Residual error of laboratory simulated data for molecular evolution

Research of extreme dynamics and gradient effects is essential to determine not only the formation of life but factors which determine its variance. Considering terrestrial systems enables the recognition of emergent populations, essentially triangular relations in life constructs and the sinusoidal waves, which represent the greater copula harmonics (multiple functions or strings of life present) in evolution.

1.3 Discussion: Evolutionary and Numerical Connotations

We have shown the bidirectional nature of molecular and cellular evolution using theoretical and laboratory-based data; the expansion of evolution is also consistent with that which takes place in extreme dynamics in the field (Furze and Mayad 2021) where ancient ternary groupings of life's constructs were seen. The laboratory-based data brings us closer to a concise definition of the resolution on which we must consider chemical and biological reactions and interactions in order to construct a concise algorithmic basis for evolution. A suggested future direction is to first construct algorithms in accordance with differential scenarios using a Mamdani Fuzzy basis (Mamdani 1974) and then with the more refined parameters of a Takagi-Sugeno-Kang Fuzzy basis (Takugi and Sugeno 1985). These higher mathematic techniques enable us to build scenarios, which cater for the evolution of the different categories of life that exist today. Such methods also illustrate the expanding and contracting nature of evolution which occurs as illustrated through the pattern of variance which is followed in dynamic evolution, also documented in the proposed use of Markov chains to describe evolution (Djordjevic 2015). The pattern followed may be reflected through sine/cosine relations such as those which are present in the

Fourier transform. Research in this area is ongoing, although differential algorithmic network centred emergence (DANCE) and harmonic evolutionary algorithmic resilience (HEAR) have recently been identified to give a basis for passive and active interactions between individuals, species and even trophic proportionality (Furze and Mayad 2021). Copula (distribution) expansion identifies similarity in biological/theoretical patterns with laboratory simulated patterns (all of which share the same variance patterns).

We have met our original objectives of linking the functions for molecular and cellular evolution together through the use of Sklar's theorem and combinatorial approaches of MOGA, laboratory and field studies (Furze and Mayad 2021; Filali et al. 2021). Continued use of these methods is proposed to confirm the categories of life present within different kingdoms in existence today. We have shown the mathematical basis on which different categories of copulas may be formed. Excitingly the use of the generator Gumbel equation (1.12) enables us to numerically observe the state of evolution in the conditions of any given time period.

Regarding chemical/genetic formations and the potential laboratory scenarios shown in Fig. 1.5, the special cases of hydrolysis, template refeeding and damaging processes should be considered in future research. A combination of laboratory, computer simulated and mathematic exploration produces rational rule structures which feed back to experimental and field results in order to reflect the dynamics operating. Processes such as template refeeding give an exponential increase in polymer formation and may be detected beyond the third polynomial level. Depurination (loss of adenine from monomers) simulation should result in the presence of shorter peaks and troughs given the correct resolution is observed for molecular systems. The number of times that we differentiate to get the polynomials logically indicates the number of processes which are occurring in order that we see the effects. This can be tested in laboratory/field conditions. Given that we establish the main sets of variables operating within a systematic approach to evolution, there is great potential for expansion of evolutionary copulas to show the formation of different constructs. Hence concise statements can be used to show different nodes of evolution throughout the categories of life. Variance and indeed rational truth of the evolutionary operating systems can be generated mathematically. Advancing through complex polynomials and validating their presence within copulas holds great promise for our understanding of evolution as decreased variance in the evolution of cellular and molecular systems is shown. Extensive use of higher mathematics, dispersal methods, rational polynomials and an algorithmic basis enables use of Markov chains across various categories of life; additionally this will produce Bayesian vectors for use in logistic regression. Integrating fieldwork and different laboratory methods gives practical indication of the variance, which may be shown mathematically.

Determining variation in life and its processes certainly requires use of multi-objective genetic algorithms and subsequent functional scenario generation; however with use of the higher mathematics we work towards discretely qualifying life parameters, the overstepping of which of course lead to changes in life processes and hence evolution. Evolution of processes and species proliferations is indirectly guided by feedback and feed-forward of the same life systems as well as the environments within which they are found. Naturally it is imperative that the process is not destroyed at the 'preliminary' stages (e.g. by chemical imbalances). In order to establish discrete parameters across different scales of evolution, we suggest further exploration in extreme and dynamic environments in both terrestrial and aquatic systems.

Evolutionary emergence of soil community diversity may be 'forced' (induced), in the extremes of pH and drought. This leads us to suggest that within extreme environments there are activators (effectors) and actuators (causal) of diversity across the vast range of macro-ecological species and ecosystems found on Earth (Novikov and Copley 2013; Adams et al. 2016, Khanal et al. 2015; De Marchi et al. 2017; Furze and Mayad 2021). Expansion of the evolutionary generator (Gumbel copula), Eq. (1.12), for cellular/molecular systems enables calculation of differential algorithmic network centred emergence (DANCE), also expressed as a stochastic integral Laplacian expression, Eq. (1.11) – stating evolution at any given set of differential parameters. Harmonic evolutionary algorithmic resilience (HEAR) represents an accelerator of evolution between differing levels of suppression in communities. In the case that the difference is great (overlapping), suppression is effective and the organism (or element) is termed as active [1]. In the case that the difference is minimal (smoother or aligned), there is no suppression and the organism (or element) is classified as passive [0]. HEAR thus enables application of binary terms of evolution.

1.4 Conclusion

The current chapter shows that copula distributions may be used to account for sets of variables, which occur within evolution. Rules on which the curves can be reconstructed may be stated in both cellular and molecular terms. The key linking processes of Earth's diversity are pathogenicity and virulence complexes found within the soil and other growth environments across evolution. Microbial and macro-species contain orthologs with both normal and promiscuous activities, which are triggered by genetic mutation events over time. Further experimentation is required to identify members of edaphic, terrestrial, aquatic and atmospheric environments within ternary groups. It may be interpolated that the microbial community begets the macro-ecological community and thus enables levels of primary ecological production. Investigation of the soil environment and other microbiomes may be of assistance in making the best use of resources (nutrients) in terms of their availability in even the most extreme environments.

The ranges of extremes within which cascades of evolution may be observed require further investigation, in order that threshold values can be identified around which general increases or decreases in diversity are seen. It follows that directional trends should be observed in which clades of species assemblages adapted to different sets of extremes will form. The formation of such monotypic conditions is detrimental to maximum diversity throughout life as it leads to dramatic species losses as conditions change. The relations, which expand throughout evolution, require urgent investigation, as when concentrations of diversity are present, discrete chemical pathway changes may result in lack of evolutionary potential. The additive effect of species assemblages and chemical pathways being 'misdirected' by harsh changes in their environment has drastic consequences on species numbers across greater time scales.

The use of remote monitoring and robotic techniques is suggested to sample extreme environments, which are analogous with the state of Earth at the beginning of evolution. We envisage that an increased range of samples would help us to define the complete (Fourier) transform series for evolution. From the latter, we may also make use of alternative functions (such as the cone function and its inverse) to show contraction and expansion of evolutionary patterns. The use of radial functions holds great promise for the mathematic categorization of life. This chapter is the first to suggest the use of generators of diversity; further we identify the presence of harmonic relationships in biological and chemical systems. Considering the harmonics of life leads us to observe emergence in all environmental settings, it has been identified that virulence and pathogenicity are key factors warranting further investigation in evolutionary constructs. We propose future work should overlap the formation of model structures in linguistic terms with biochemical (laboratory) and field studies to further emphasize the complexity of life and its constructs.

Given that within life constructs the combined harmonics of life operate in cooperative/synergistic terms, perhaps future work may query the very nature of triangular pirouettes in Hilbert space itself; do extremes exist in Hilbert space or are they just our biological perspectives of frequency, oscillation and density, which parameterize novel mathematic expressions such as extremum seeking algorithms? Combinatorial mathematic, topological and unified field theory explorations are imminent in this exciting field to bring evolutionary emergence and sustainability of life constructs to the fore for the benefit of both human and other life systems. Given that we acknowledge our interdependence, role and position in the array of Earth's beauty, we must understand and categorize biological diversity in order to impose appropriate protection of the highest order.

Acknowledgements Special thanks go to David Deamer and Bruce Damer for their contributions towards experimental studies of monomer formation, polymer formation and further evolutionary processes. Thanks also must go to numerous field workers, theoretical biologists and mathematicians who helped inspire this work.

References

Adams MS, Adams RB, Wessman CA et al (2016) Nutritional cues tie living organisms to their environment and its sustainability. Front Nutr. https://doi.org/10.3389/fnut.2016.00028

Albrecht J, Classen A, Vollstädt MGR et al (2018) Plant and animal functional diversity drive mutualistic network assembly across an elevational gradient. Nat Commun. https://doi.org/10.1038/s41467-018-05610-w

Aldhufairi FA-A, Samanthi RGM, Sepanski JH (2020) New families of bivariate copulas via unit Lomax distortion. Risks 8:106. https://doi.org/10.3390/risks8040106
Alhadlaq W, Alzaid A (2020) Distribution function, probability generating function and Archimedean generator. Symmetry 12:2108. https://doi.org/10.3390/sym12122108
Alicea B (2014) Toy models for macroevolutionary patterns and trends. Biosystems 122:25–37. https://doi.org/10.1016/j.biosystems.2014.06.012
Bains W (2020) Getting beyond the toy domain. Meditations on David Deamer's "Assembling Life". Life. https://doi.org/10.3390/life10020018
Basener WF, Sanford JC (2018) The fundamental theorem of natural selection with mutations. J Math Biol 76:1589–1622. https://doi.org/10.1007/s00285-017-1190-x
Brack A (2004) The chemistry of life's origins. In: Seckbach J (eds) Origins. cellular origin, life in extreme habitats and astrobiology, vol 6. Springer, Dordrecht. https://doi.org/10.1007/1-4020-2522-X_5
Cabral JS, Valente L, Hartig F (2016) Mechanistic simulation models in macroecology and biogeography: state-of-art and prospects. Ecography. https://doi.org/10.1111/ecog.02480
Damer B, Deamer D (2015) Coupled phases and combinatorial selection in fluctuating hydrothermal pools: A scenario to guide experimental approaches to the origin of cellular life. Life 5(1):827–887. https://doi.org/10.3390/life5010872
Damer B, Newman P, Norkus R et al (2012) Cyberbiogenesis and the EvoGrid: a twenty-first century grand challenge. In: Seckbach J. (eds) Genesis – In The Beginning. Cellular Origin, Life in Extreme Habitats and Astrobiology, vol 22. Springer, Dordrecht. https://doi.org/10.1007/978-94-007-2941-4_16
De Marchi M, Sengar B, Furze J (2017) Prospects for sustainability in human environment patterns: dynamic management of common resources. In: Furze J, Swing K, Gupta A et al (eds) Mathematical advances towards sustainable environmental systems. Springer, Cham. https://doi.org/10.1007/978-3-319-43901-3_14
Deamer DW, Georgiou CD (2015) Hydrothermal conditions and the origin of cellular life. Astrobiology 15(12):1091–1095. https://doi.org/10.1089/ast.2015.1338
Deamer D, Singaram S, Rajamani S et al (2006) Self assembly processes in the prebiotic environment. Philos Trans R Soc Lond B 361:1809–1818. https://doi.org/10.1098/rstb.2006.1905
Djordjevic IB (2015) Markov chain-like quantum biological modeling of mutations, aging, and evolution. Life 5(3):1518–1538. https://doi.org/10.3390/life5031518
Durante F, Fernández-Sánchez J, Sempi C (2013) A topological proof of Sklar's theorem. Appl Math Lett 26(9):945–948. https://doi.org/10.1016/j.aml.2013.04.005
Dworkin JP, Deamer DW, Sandford SA et al (2001) Self assembling amphiphilic molecules synthesis in simulated interstellar / precometary ices. PNAS 98(3):815–819
Echeverría-Londoño S, Enquist BJ, Neves DM et al (2018) Plant functional diversity and the biogeography of biomes in North and South America. Front Ecol Evol. https://doi.org/10.3389/fevo.2018.00219
Edwards AWF (1994) The fundamental theorem of natural selection. Biol Rev 69:443–474. https://doi.org/10.1111/j.1469-185X.1994.tb01247.x
Filali Alaoui I, Ait Hamza M, Benjlil H et al (2021) Phytoparasitic nematodes of organic vegetables in the Argan Biosphere of Souss-Massa (Southern Morocco). Environ Sci Pollut Res. https://doi.org/10.1007/s11356-021-12986-8
Fortney JJ, Marley MS, Laughlin G et al (2016) The hunt for planet nine: atmosphere, spectre, evolution, and detectability. Astrophys J Lett 824:L25. https://doi.org/10.3847/2041-8205/824/2/L25
Furze JN, Mayad EH (2021) Harmonics, evolutionary generators, DANCE and HEAR – functional dimensions. Environ Sci Pollut Res. https://doi.org/10.1007/s11356-021-13159-3
Furze JN, Zhu Q (2014) Orthogonal relations coupling renewable energy and sustainable plant systems. In: Proceedings of 2014 International Conference on Modelling, Identification and Control, Melbourne, VIC, Australia, 3-5 Dec 2014. https://doi.org/10.1109/ICMIC.2014.7020719
Furze JN, Zhu Q, Qiao F et al (2013a) Functional enrichment of utopian distribution of plant life-forms. Am J Plant Sci 4(12A):37–48. https://doi.org/10.4236/ajps.2013.412A1006

Furze JN, Zhu Q, Qiao F et al (2013b) Mathematical methods to quantify and characterise the primary elements of trophic systems. Int J Comput Appl Technol 47(4):314–325
Furze JN, Zhu Q, Hill J et al (2017) Biological modelling for sustainable ecosystems. In: Furze J, Swing K, Gupta A et al (eds) Mathematical advances towards sustainable environmental systems. Springer, Cham. https://doi.org/10.1007/978-3-319-43901-3_2
Gordon R (1993) The fractal physics of biological evolution. In: Dynamical phenomena at interfaces, surfaces and membranes. Nova Science Publishers, New York
Grover MA, He CY, Hsieh MC et al (2015) A chemical engineering perspective on the origins of life. Processes 3(2):309–338. https://doi.org/10.3390/pr3020309
Judson OP (2017) The energy expansions of evolution. Nature Ecol Evol 1:1–9. https://doi.org/10.1038/s41559-017-0138
Khanal A, McLoughlin SY, Kershner JP et al (2015) Differential effects of a mutation on the normal and promiscuous activities of orthologs: implications for natural and directed evolution. Mol Biol Evol 32(1):100–108. https://doi.org/10.1093/molbev/msu271
Lal DM, Sreekanth GB, Shivakrishna A et al (2021) Ecosystem health status and trophic modeling of an anthropogenically impacted small tropical estuary along India's west coast. Environ Sci Pollut Res. https://doi.org/10.1007/s11356-021-12857-2
Lu J, Liu M, Fan Y et al (2021) TCDD induced lipid accumulation by impairment of autophagic flux in THP-1 macrophages. Environ Sci Pollut Res. https://doi.org/10.1007/s11356-021-13258-1
Mamdani EH (1974) Applications of fuzzy algorithms for simple dynamic plant. Proc IEEE 121(12):1585–1588. https://doi.org/10.1049/piee.1974.0328
Manceau M, Lambert A, Morlon H (2017) A unifying comparative phylogenetic framework including traits coevolving across interacting lineages. Syst Biol 66(4):551–568. https://doi.org/10.1093/sysbio/syw115
Marzban C, Viswanathan R, Yurtsever U (2014) Earth before life. Biol Direct 9(1). https://doi.org/10.1186/1745-6150-9-1
Mazel F, Mooers A, Riva GVD et al (2017) Conserving phylogenetic diversity can be a poor strategy for conserving functional diversity. Syst Biol 66(6):1019–1027. https://doi.org/10.1093/sysbio/syx054
Monnard PA (2016) Taming prebiotic chemistry: the role of heterogeneous and interfacial catalysis in the emergence of a prebiotic catalytic/information polymer system. Life. https://doi.org/10.3390/life6040040
Mullineaux LS, Metaxas A, Beaulieu SE et al (2018) Exploring the ecology of deep-sea hydrothermal vents in a metacommunity framework. Front Mar Sci. https://doi.org/10.3389/fmars.2018.00049
Mungi CV, Rajamani S (2015) Characterization of RNA-like oligomers from lipid-assisted nonenzymatic synthesis: implications for origin of informational molecules on early Earth. Life 5(1):65–84. https://doi.org/10.3390/life5010065
Novikov Y, Copley SD (2013) Reactivity landscape of pyruvate under simulated hydrothermal vent conditions. PNAS 110(33):13283–13288. https://doi.org/10.1073/pnas.1304923110
Ota S, Kimura M (2021) Effective estimation algorithm for parameters of multivariate Farlie–Gumbel–Morgenstern copula. Jpn J Stat Data Sci. https://doi.org/10.1007/s42081-021-00118-y
Prusinkiewicz P, Lindenmayer A (1990) The algorithmic beauty of plants. Springer, New York. https://doi.org/10.1007/978-1-4613-8476-2
Read MN, Alden K, Rose LM et al (2016) Automated multi-objective calibration of biological agent-based simulations. J R Soc Interface. https://doi.org/10.1098/rsif.2016.0543
Schölzel C, Friedrichs P (2008) Multivariate non-normally distributed random variables in climate research –introduction to the copular approach. Nonlinear Process Geophys 15:761–772. https://doi.org/10.5194/npg-15-761-2008
Schrum JP, Zhu TF, Szostak J (2010) The origins of cellular life. Cold Spring Harb Perspect Biol. https://doi.org/10.1101/cshperspect.a002212
Sklar A (1959) Fonctions de repartition á n dimensions et leurs marges. Publ L'institut Statistique l'Université Paris 8:229–231

Swing K (2017) Challenges to conservation. In: Furze J, Swing K, Gupta A et al (eds) Mathematical advances towards sustainable environmental systems. Springer, Cham. https://doi.org/10.1007/978-3-319-43901-3_6
Takagi T, Sugeno M (1985) Fuzzy identification of systems and its applications to modeling and control. IEEE Trans Syst Man Cybernet 15(1):116–132. https://doi.org/10.1109/TSMC.1985.6313399
Taylor LS, Phillips G, Bernard EC et al (2020) Soil nematode functional diversity, successional patterns, and indicator taxa associated with vertebrate decomposition hotspots. PLoS One 15(11):e0241777. https://doi.org/10.1371/journal.pone.0241777
Tugrul S, Ozhan K, Akcay I (2019) Assessment of trophic status of the northeastern Mediterranean coastal waters: eutrophication classification tools revisited. Environ Sci Pollut Res 26:14742–14754. https://doi.org/10.1007/s11356-018-2529-6
van den Hoogen J, Geisen S, Wall DH et al (2020) A global database of soil nematode abundance and functional group composition. Sci Data. https://doi.org/10.1038/s41597-020-0437-3
Wang L, Guo X, Zeng J et al (2012) Copula estimation of distribution algorithms based on exchangeable Archimedean copula. Int J Comput Appl Technol 43(1):13–20. https://doi.org/10.1504/IJCAT.2012.045836

Chapter 2
Sustainable Management of Soil-Borne Plant Pathogens

Dina S. S. Ibrahim, Manal Mostafa, Aya A. M. Abdellatif, and Rizwan Ali Ansari

Abstract In an attempt to fulfill the increased food demands of an explosive population, synthetic fertilizer and pesticide-ridden food production have steadily increased, considerably affecting the agroecosystem. Consumers throughout the world have not been informed of the detrimental effects of these chemicals. Many soil-borne pathogens including phytonematodes are aggressively managed in the presence of non-judicious chemical fertilizers. Resultantly, many developed as well as developing countries have embraced organic cultivation efforts and experienced outstanding results. In sustainable management, a wide range of biocontrol microorganisms including fungi, bacteria, and actinomycetes are available for use at a commercial scale without causing any perturbation to the natural biota. Recently, biocontrol and microbial-based biopesticides have provided great promise in soil and plant health improvement. Mechanisms such as antibiosis, hyperparasitism, food and space competition, and induced systemic resistance (ISR) induction are implicated in the reduction of nematode/pathogen populations. Organic matter and beneficial microorganisms improve plant growth and yield performance and also curtail the attack of a wide spectrum of pests and pathogens. However, there are some minor lacunas encountered in the use of these microorganisms at a large scale which require addressing in future studies. Genomics to metagenomics studies are required to obtain amicable solutions for producers.

D. S. S. Ibrahim (✉) · M. Mostafa
Department of Nematodes Diseases, Central Laboratory of Biotechnology, Plant Pathology Research Institute, Agricultural Research Center (ARC), Giza, Egypt
e-mail: Dina.Serag@arc.sci.eg; mickeyocean@yahoo.com; manal.mustafa93@gmail.com

A. A. M. Abdellatif
Central Laboratory of Organic Agriculture, Agricultural Research Center (ARC), Giza, Egypt
e-mail: aya_aloraby90@yahoo.com

R. A. Ansari
Department of Plant Protection, Faculty of Agricultural Sciences, Aligarh Muslim University, Aligarh, India
e-mail: rizwans.ansari@gmail.com

J. N. Furze et al. (eds.), *Earth Systems Protection and Sustainability*,
https://doi.org/10.1007/978-3-030-85829-2_2

Keywords Sustainable management · Soil health · Plant protection · Diseases · Management

2.1 Introduction

Plant health and agroecosystem potentiality are dependent on beneficial phytobiomes. Ecosystem services such as storage of organic matter and decomposition, biogeochemical cycling, and reduction of plant pathogens propagules are carried out in various ecological settings containing interacting organisms, including plants themselves (Janvier et al. 2007; Ansari and Mahmood 2019a, b). Today plant diseases are responsible for tremendous losses in different crops in both arable (Raaijmakers et al. 2009) and pastoral agriculture (Dignam et al. 2016). The nature of the complexity and how and to what extent good soil health affects disease progression continue to be deliberated.

Pastoral farming systems support the production of grazing livestock and cover more than 25% of the Earth's ice-free land surface (FAOSTAT 2011). Controlling plant disease is complicated by the multi-pathogen complexes that appear with annual, biennial, and perennial plants and the expansive nature and potentially challenging topography of pastoral-based agriculture systems, which impede delivery of external inputs (Ansari and Khan 2012a, b; Dignam et al. 2016). Naturally, disease-suppressive soils include consortia of beneficial microbes such as plant growth-promoting rhizobacteria (PGPR) involved in the protection of susceptible plant hosts from soil-borne disease (Mendes et al. 2011; Penton et al. 2014; Cha et al. 2016; Raaijmakers and Mazzola 2016; Ansari et al. 2017).

The phenomenon of general suppression is related to the competitive potential of the total soil microbial community; specific suppression is driven by the antagonistic activity of an individual or specific group of microorganisms (Weller et al. 2002; Raaijmakers and Mazzola 2016). The wealth of understanding correlated with the well-characterized models of soil suppressiveness in arable systems (Cha et al. 2016; Carrión et al. 2018) provides opportunities to explore and exploit mechanisms in pastoral agricultural systems. The latter improves our understanding of the processes that underlie release from pathogen pressure in natural grassland systems (Maron et al. 2011; Schnitzer et al. 2011; Latz et al. 2012, 2016; Mommer et al. 2018). Surprisingly, only a few studies have focused on the distribution of disease-suppressive microbiota in agricultural grasslands and the mechanisms by which these communities relate or respond to soil management practices (Dignam et al. 2016; Wakelin 2018).

Understanding soil physicochemical properties that affect microbial communities and their activities accentuates the design of management practices which steer ecosystem services (Nielsen et al. 2015). All plants and animals are subject to infection from one or more species of parasitic nematodes. Plant-parasitic nematodes cause heavy annual losses to major crops. Economic losses related to nematode infection would be greater without the application of effective strategies of nematode management and tactics that decrease losses.

A phylogenetically diverse range of soil microbial communities have been correlated with the suppression of plant diseases (Raaijmakers and Mazzola 2012). For instance, *Pseudomonas* spp. have been repeatedly implicated and are responsive to varying management practices across agricultural systems (Garbeva et al. 2004; van Overbeek et al. 2012; Walters et al. 2018; Mahmood et al. 2019; Ansari et al. 2020a, b).

Dignam et al. (2018) conducted in-depth molecular studies across pastoral soils to identify mechanisms by which indigenous soil microbes may be manipulated to improve the capacity of soil to suppress plant disease agents. Notably, variation in soil organic matter (SOM) quality has been positively correlated with both taxonomic (*Pseudomonas* community composition) and functional indicators of disease-suppressive activity in soils. Controlling practices that lead to significant alteration of soil organic matter content and quality (including chemical composition and decomposability) could devastatingly increase soil suppressiveness. Contributions of management-induced changes in biotic and abiotic soil properties to soil suppression have been studied. Previous measurements of soil organic matter quantity and consistency have been shown to vary with plant residue management (Simpson et al. 2012; Adair et al. 2013).

Due to environmental concerns, researchers have focused on finding suitable alternatives to chemical pesticides for controlling soil-borne pathogens and plant parasitic nematodes (Larkin et al. 1998; Yimer et al. 2018). In this context, alternative strategies including crop rotation, solarization, biofumigation, grafting, and application of biocontrol agents or organic amendments, such as composts, are of considerable interest among scientists and agricultural producers (Bailey and Lazarovits 2003; Louws et al. 2010a, b). The objective of the current chapter is to summarize the current knowledge of the most effective approaches used in the control of soil-borne diseases.

2.2 Management of Soil-Borne Pathogens

Biotic stressors including microbial pathogens, nematodes, and weeds attack different crops causing huge yield losses of 20 and 40% to global agricultural productivity. As such, these impediments are considered as the main obstacle in successful crop cultivation (Teng 1987; Oerke 2006). Soil-borne pathogens including fungi, bacteria, and nematodes are affected by physical, chemical, and biological properties of the soil and agricultural practices including irrigation, fertilization, and tillage regimes (Katan and Gamliel 2011). It is challenging to detect and diagnose a variety of soil-borne pathogens as responsible for serious plant diseases. Symptoms of soil-borne diseases, effected by different pathogens, are similar. Symptoms include root rot, root blackening, wilt, yellowing, stunting or seedling damping-off, bark cracking, and twig or branch dieback. Pathogens are well known as damaging factors due to extensive persistence even in the absence of host plants or suitable environmental conditions. Resistant structures such as cysts, sclerotia, chlamydospores, or oospores may be freely formed (Mihajlović et al. 2017). Yet the vitality of soil-borne pathogens varies and depends on prevailing environmental conditions. Thus, the majority of soil-borne pathogens cannot be controlled by a single approach. Many physical, chemical, and biological control strategies, such as that of use of host plant resistance, crop rotation, sanitation, and destruction of residual crop roots, nematicides, organic amendments, and use of eco-friendly fungi, bacteria, and other biological control agents have been reported to effectively control soil-borne pathogens (Fortuner 1991; Mihajlović et al. 2017).

One of the most effective approaches to disinfest soil is the application of steam or flooding the soil with hot water. The main disadvantage of these approaches is their high cost as they consume a high rate of fuel and involve expensive and sophisticated machinery (McGovern and McSorley 1997). Soil solarization, which involves heating the soil by solar energy, has proved to be a satisfactory alternative, bringing soil populations of pathogens to unharmful levels in areas with appropriate weather conditions (Basallote-Ureba and Melero-Vara 1993; Katan et al. 2012). Further, fumigants can be considered a major tool for soil disinfestation. However, certain fumigants are not readily degraded in soil and cause pollution of underground water and the environment (Mihajlović et al. 2017).

Disease suppression by biocontrol agents is the sustained manifestation of interactions among plant, pathogen, the biocontrol agent, the microbial community around the plant, and the physical conditions of the environment. Biological control of soil-borne diseases is complex because they occur in dynamic environments at the interface of root and growth media known as the rhizosphere. Rhizospheric microorganisms interact beneficially with their host plant via several mechanisms. They can promote plant growth directly by the improvement of nutrient acquisition or hormonal stimulation or indirectly affect plant health by reducing the severity of phytopathogens (Berg and Smalla 2009). The rhizosphere is subject to dramatic changes on a short temporal scale – including rain events and daytime drought. Changes result in fluctuations in salt concentration, pH, osmotic potential, water potential, and soil particle structure. The dynamic nature of the rhizosphere makes it an interesting setting for the interactions that lead to disease and biocontrol of disease (Waisel et al. 2002; Handelsman and Stabb 1996). Most plants exhibit inhibitory and stimulatory biochemical interactions with other plants and microorganisms, referred to as allelopathy. Root exudates of higher plants have the ability to affect microflora in the rhizosphere. Plants may secrete different bioactive compounds as root exudates that prevent phytopathogens from infecting crops. Medical plants play a vital role in controlling disease and negate the need for undesirable hazardous chemicals, thereby protecting the environment. Volatile essential oils extracted from medicinal plants have been reported to possess antimicrobial activity against a wide range of plant pathogens (Tanović et al. 2014). Further, oregano, fennel, and laurel oils demonstrate antimicrobial activity against soil-borne fungi of beans under laboratory conditions (Turkolmez and Soylu 2014).

2.3 Organic Additives Used with Crop Rotations

Organic matter and its replenishment have become the core of soil health management programs. Characteristics of the soil including physical, chemical, and biological properties are a function of organic matter content and quality. Adding organic matter to soil induces diverse and important biological activities (Widmer et al. 2002). Crop rotation practices and organic matter applications have the potentiality for the restoration of soil health and increased productivity of degraded highland crop fields.

Crop rotation undoubtedly provides multiple benefits during crop production. To minimize the severity of root-knot nematodes, the effectiveness of the application of botanical toxicants or plant product derivatives has been confirmed (Al-Askar 2012; Khalil et al. 2012; Ansari et al. 2019). Organic additives help preserve, sustain, or replenish soil resources, including organic matter, nitrogen, phosphorous, and nutrient inputs, as well accentuate physical and chemical properties (Ball et al. 2005; Ladygina and Hedlund 2010; Sasse et al. 2018). Crop rotation has a positive impact on soil fertility, condition, and aggregate stability. Small grains, especially barley, are highly recommended for improvement of organic matter content and reduction of problems from pink root and *Fusarium* basal plate rot of onion (Schwartz 2011). The most beneficial rotations in the family of *Brassicaceae* include that of broccoli, cabbage, cauliflower, turnip, radish, canola, rapeseed, and numerous mustards, which produce sulfur compounds that break down to produce isothiocyanates. The latter is toxic to a wide range of soil organisms and cleanse the soil as part of a process referred to as biofumigation (Youssef and Lashein 2013). Further, crop rotation strategy improves soil water management and reduces erosion (Ball et al. 2005). However, crop rotation is not useful in the management of diseases caused by soil-borne pathogens that possess a wide host range and those that form long-living survival structures, such as sclerotia, cysts, or oospores (Umaerus et al. 1989). In addition, crop rotation does not affect disease organisms that survive on or in the seed, such as cereal smuts. Equally, crop rotation does not affect disease organisms that blow in, such as cereal rusts (Kheyrodin 2011). Application of crop rotation used in combination with other controlling strategies is highly recommended.

2.4 Biological Control

Biological control is defined as the antagonistic effect of organisms which act as biocontrol agents against soil-borne pathogens (Afzal et al. 2013; Ansari et al. 2017; Berendsen et al. 2018; Ansari et al. 2020a, b). Usually, biocontrol agents are applied individually to control the proliferation of different plant pathogens. While some studies reported the potential benefits of a single application of biocontrol agents, in many other cases studies report insignificant results because a single biocontrol agent may not be active in all types of soil environments and agricultural ecosystems (Raupach and Kloepper 1998). These result in insufficient colonization of the agents, low tolerance to changes in environmental conditions, and variability in the production of antifungal metabolites (Weller et al. 2002). Mixtures of biocontrol agents have the advantage of activity with broad-spectrum properties.

Mixed applications significantly enhance the efficacy and reliability of biocontrol agents and maximize the induction of defense enzymes in plants (Latha et al. 2009). The main character of an effective biocontrol strain is its ability to compete and persist in the environment and to colonize, proliferate, and establish itself on plant parts. Furthermore it should be affordable to produce strains at a large scale and maintain good vitality without specialized storage systems (Harman 1996; Lamovsek et al. 2013). Soil application of biocontrol agents, viz., *Trichoderma viride*, *T. harzianum*, fluorescent *Pseudomonas*, *Serratia marcescens*, and *Bacillus subtilis*, effectively decreases the severity of root rot diseases caused by soil-borne pathogens in numerous economic crops (Loganathan et al. 2010; Shafique et al. 2015b). *Trichoderma* spp. are endophytic fungi which grow in plant tissue without causing disease and are well known to secrete large quantities of toxic metabolites such as gliotoxins which have antifungal activity. The antagonistic activity of *Trichoderma* spp. is related to direct mycoparasitism in addition to competition for nutrients and space (Sharon et al. 2001; Afzal et al. 2013). Thus, *Trichoderma* spp. are mycoparasites that have been considered as powerful biocontrol agents for foliar and soil-borne pathogens as well as species of plant-parasitic nematodes (Kowsari et al. 2014). Species of *Trichoderma* such as *T. harzianum*, *T. asperelloides*, and *T. hamatum* have proven nematicidal potentiality against root-knot nematodes (Sharon et al. 2001; Sayed et al. 2019) and can be developed for introduction as strong biocontrol agents. Increasingly, *Trichoderma* spp. are evaluated for their activity against root-knot nematodes on a wide range of crops, such as okra, tomato, mung bean, cucumber, bell pepper, and sugar beet (Meyer et al. 2001). Afzal et al. (2013) reported that *T. viride* is effective in inhibiting *F. solani*, *F. oxysporum*, and root-knot nematodes on okra, used alone or in combination with *Pseudomonas aeruginosa*. However, variations in efficacy of isolates, biocontrol ability, and reproducibility of consistent results and effects under variable environmental conditions hinder their development and application at a large scale (Sharon et al. 2001).

Different species of bacteria and actinobacteria can be applied more easily and have proven satisfactory in the control of soil-borne diseases (Ramarathnam et al. 2011). The bioactivity of the pigment extracted from *Serratia marcescens* was shown to inhibit the vitality of nematodes at their juvenile stage (Rahul et al. 2014). Furthermore, *S. marcescens* secrete hydrolytic enzymes (proteases, chitinases, lipases, and cellulases), which result in severe deformities in fungal mycelia and play a role in controlling root rot disease in tea after field application (Purkayastha et al. 2018).

Actinobacteria are important saprophytes which have the ability to degrade a wide range of plant and animal debris in the process of decomposition. Certain genera, such as *Streptomyces* and *Micromonospora*, are known for producing bioactive metabolites including enzymes and antibiotics with broad-spectrum properties. *Streptomyces asterosporus*, an endophytic actinobacteria, produces large quantities

Table 2.1 Common microorganisms used as biocontrol agents against soil-borne diseases

Plant pathogen	Host	Biocontrol agent	References
Sclerotium rolfsii	Sugar beet	*Trichoderma harzianum*	Paramasivan et al. (2014)
Fusarium oxysporum f. sp. ciceris	Chickpea	*Pseudomonas; Bacillus* spp.	Karimi et al. (2012)
Rhizoctonia solani	Tomato	*Trichoderma harzianum*; *Serratia proteamaculans*	Youssef et al. (2016)
Ralstonia solanacearum	Tomato	*Streptomyces microflavus*	Shen et al. (2020)
Meloidogyne spp.	Tomato, Cucumbers	*T. harzianum*, *T. asperelloides* and *T. hamatum*, *T. viride* *Paecilomyces lilacinus*	Sayed et al. (2019) and Yankova et al. (2014)
Heterodera schachtii	Sugar beet	*T. harzianum* and *T. virens*	Moghadam et al. (2009)

of hydrogen cyanide, siderophores, chitinases, and β-1,3-glucanases. These actinobacteria significantly inhibit *Fusarium* root rot disease severity in tomato seedlings (Goudjal et al. 2016). Furthermore *S. antibioticus* showed nematicidal activity against *Meloidogyne incognita* with the culture supernatant of the strain inducing 100% juvenile mortality. This significant nematicidal activity may be related to actinomycins secreted by *Streptomyces* antibiotics as part of an important secondary metabolite production (Sharma et al. 2019). Biological control can be considered as a safe and effective alternative strategy to reduce the heavy use of harmful chemical pesticides. Common biocontrol agents are laid out in Table 2.1.

2.5 Nanomaterials as a New Approach for the Management of Soil-Borne Pathogens

Nanotechnology is an intriguing and rapidly advancing science and has the potential to revolutionize many scientific, technological, medical, and agricultural disciplines (Khan and Rizvi 2014). Nanotechnology has potential use in the management of plant diseases. The most simple and obvious way to protect plants from pathogen invasion is the direct application of nanoparticles in the soil, on seeds, or on foliage. Their effects on non-target organisms, particularly mineral fixing/solubilizing microorganisms, will be of great importance in the direct application of nanoparticles in soil.

Different types of nanomaterials including carbon tubes and cups can be utilized as carriers for valuable chemicals such as pheromones, systemic acquired resistance (SAR) inducing chemicals, polyamine synthesis inhibitors, or even concentrated active ingredients of pesticides, due to their controlled release under flooded conditions (Khan et al. 2014a). Impacts should be addressed to determine the scope and use of nanoparticles (NPs) in the control of plant diseases from two main perspectives: the direct effect of NPs on pathogens and the use of nanomaterials in pesticide formulation, namely, "nanopesticides."

NPs may prove very useful in the diagnosis of plant pathogens/diseases and pesticide residue analysis, considering the ultra-small size of the particles and their high degree of reactivity/sensitivity. Moreover, one of the nanomaterial techniques – nanoencapsulation – applies to antioxidants and antimicrobials, identified as colloid-based nano-incorporation collaboration. Lipid-based nanoencapsulation techniques of encapsulation are based on biologically derived polymeric nanocarriers. Further encapsulation techniques are based on non-biological polymeric nanocarriers, incorporation of cyclodextrin, electrospraying, electrospinning, carbon nanotubes, and nanocomposite encapsulation.

In the control of soil-borne pathogens, several researchers address the effect of nanoencapsulation techniques on antioxidant/antimicrobial function. Pisoschi et al. (2018) emphasizes the importance of selecting the right encapsulation form. Bioactive compound safety and controlled release are accomplished, but consideration should be given to the effect of nanomaterials on human health and the environment. The rise, retention, or decrease of bioactivity depends on relationships formed between the encapsulated compound functional groups and the encapsulating nanomaterial.

2.6 The Mechanisms of Nanomaterials for Management of Soil-Borne Pathogens

The main mechanism of action resulted from nanoparticles in antimicrobials and pesticides is that of reactive oxygen species (ROS), which induce oxidative stress and release superoxide, free radicals, and particles that can react and upset peptide interactions in the cell wall of microscopic organisms (Makhluf et al. 2005). ATP is synthesized by the reduction of molecular oxygen to water in the mitochondria of cells by a series of coupled proton and electron transfer reactions. A small percentage of oxygen is not entirely reduced during this process, resulting in the formation of superoxide anion radicals and consequently other radicals containing oxygen. Therefore, ROS require results from alleging oxidative metabolism for cell division, most of which happens in the mitochondria. Superoxide anion radicals, hydroxyl radicals, singlet oxygen, and hydrogen peroxide (H_2O_2) comprise biologically important ROS (Yin et al. 2012; Prasad et al. 2017). Via intense oxidative stress, the blast of ROS causes damage to all the macromolecules of the cell causing lipid peroxidation, protein modification, enzyme interruption, and degradation of ribonucleic acids (RNA) and deoxyribonucleic acids (DNA). ROS contributes to cell death at high concentrations and induces severe DNA disruption and mutations at low concentrations, in bacteria and other cellular systems present in soil environments (Wang et al. 2011; Matějka and Tokarský 2014).

Nanoparticles cannot cross the nuclear membrane and so accumulate in the cytoplasm where, as the nuclear membrane breaks down, they obtain entry to the nucleus during mitosis (Singh et al. 2009). The direct interaction of nanoparticles with proteins associated with RNA and DNA can result in the genetic material being physically affected. Another cause for DNA disruption may be interaction with the structure or action of the DNA repair enzymes in cell nuclei.

Nanoparticles are unable to reach the nuclear membrane and thus aggregate in the cytoplasm, where, when the nuclear membrane divides, they can access the heart amid mitosis (Singh et al. 2009). Physical degradation of nucleic acids can result from the direct interaction of nanoparticles with DNA and DNA-related proteins. An additional cause for DNA disruption may be interaction with the structure or role of the DNA repair enzymes in the nucleus (Huang et al. 2015; Mostafa et al. 2018).

In Fig. 2.1 nanoparticles enter the cell via endocytosis and release ions that induce the formation of ROS. ROS products cause many damages within the cell and eventually cause cell death: (a) DNA denaturation and damage, (b) unfolded protein and damaging, (c) mitochondria dysfunction, and (d) lipid peroxidation, cell membrane disruption, and intracellular content leakage.

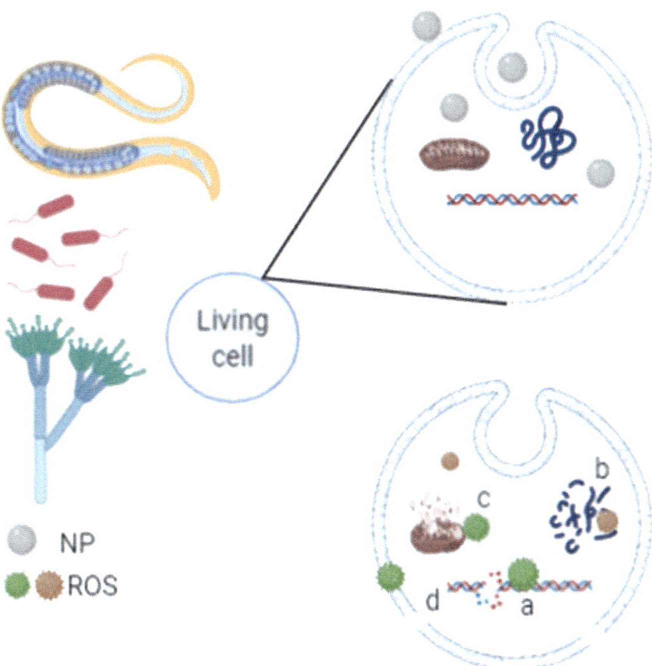

Fig. 2.1 Possible mechanisms of nanoparticle pesticides

2.7 Effect of Nanoparticles on Plant Pathogens and Microorganisms

The physiochemical properties of nanoforms differ greatly from their macro-forms; it is important to analyze the impact of NPs on microorganisms in order to take advantage of this technology in plant protection, particularly for phytopathogens. The behavior of microorganisms may be impaired by nanoparticles due to their ultra-small size and reactivity.

Inhibition of colonization of *Staphylococcus aureus*, *P. aeruginosa*, *Escherichia coli*, and *Klebsiella pneumoniae* has been achieved using silver NPs (Logeswari et al. 2015). Silver nanoparticles (30 nm) synthesized with *Solanum trilobatum* and *Ocimum tenuiflorum* leaf extracts have high antimicrobial activity against *S. aureus* and *E. coli*, respectively (Logeswari et al. 2015). The available evidence so far on this aspect has demonstrated that the nanoparticles have a significant impact on bacteria and fungal colonization. Such effects, however, are suppressive and also stimulatory and should not be generalized.

2.7.1 *Effect of Nanoparticles on Bacteria*

The antibacterial effect of zinc nanoparticles has been extensively studied by Jayaseelan et al. (2012) against *P. aeruginosa*. The maximum inhibition zone for bacteria colonization (22 ± 1.8 mm) was reported at 25 ng mL^{-1}ZnO NPs. So, a new antimicrobial compound was revealed by ZnO NPs action. Bryaskova et al. (2011) examined the antibacterial action of synthesized Ag NPs/PVP (polyvinylpyrrolidone-based hybrid materials with silver nanoparticles) against three distinct classes of bacteria, *Staphylococcus aureus* (gram-positive bacteria), *E. coli* (gram-negative bacteria), and *P. aeruginosa* (gram-negative non-fermentable bacteria), and also against *B. subtilis* spores.

The efficacy of CuO NPs as antimicrobials has been identified by Azam et al. (2012) against *S. aureus*, *B. subtilis*, *P. aeruginosa*, and *E. coli*. Guzman et al. (2009) noted that silver nanoparticles displayed elevated antimicrobial and bactericidal behavior against extremely methicillin-resistant strains of bacteria such as *E. coli*, *P. aeruginosa*, and *S. aureus*. In general, it has been observed that the antibacterial behavior of nanoparticles is dependent on the concentration, morphology, metabolism, membrane intracellular selective permeability, and the form of the microbial cell.

2.7.2 *Effect of Nanoparticles on Plant Pathogenic Fungi*

In the processing of food products, plant pathogens, bacteria, fungi, viruses, and nematodes, provide major limitations (Khan et al. 2011, 2012; Khan 2012a; Khan and Jairajpuri 2010a, b, 2012). The exploitation of nanotechnology for the treatment of plant pathogens has great potential. Deepak et al. (2013) report that $CuSO_4$ and $Na_2B_4O_7$ were found to be most successful in controlling rust disease of field peas among nanoforms of 15 micronutrients. Sunflower damping-off and charcoal rot diseases were suppressed by microelements including manganese and zinc (Abd El-Hai et al. 2009).

Nanoparticles of silver have a fungicidal effect against various yeasts and molds, different strains of *Candida*, and *Aspergillus brasiliensis* (Bryaskova et al. 2011). The fungicidal effectiveness of ZnO NPs has been reported against post-harvest pathogenic fungi *Botrytis cinerea* and *Penicillium expansum* (He et al. 2011). To investigate antifungal activities of ZnO NPs and to classify variations in morphology and cellular compositions of fungal hyphae, conventional microbiological plating, scanning electron microscopy (SEM), and Raman spectroscopy have been used. At concentrations greater than 3 mmol L^{-1}, ZnO NPs (70 ± 15 nm) inhibited the growth of both *B. cinerea* and *P. expansum*. In the hyphae of *B. cinerea*, the NP treatments induce deformation and stopped the growth of conidiophores as well as conidia in *P. expansum*, which inevitably contributes toward the death of fungal hyphae.

Plant pathogenic fungi (*Alternaria alternata*, *Sclerotinia sclerotiorum*, *Macrophomina phaseolina*, *Rhizoctonia solani*, *B. cinerea*, and *Curvularia lunata*) have been shown to be significantly inhibited by a concentration of 15 mg L^{-1} of silver NPs (Krishnaraj et al. 2012). Further, suppression of colonization of *A. flavus* has been achieved with the use of zinc nanoparticles at 25 mg mL^{-1} (Jayaseelan et al. 2012).

2.7.3 Effect of Nanoparticles on Plant Pathogenic Nematodes

The emerging branch of bionanotechnology combines biological principles with chemical and physical approaches for the production of nano-sized particles with specific functions. Bionanotechnology is an economic substitute for physical and chemical methods of nanoparticle formation (Ahmed et al. 2006). The synthesis of metal nanoparticles with greener methods and its application in biological fields are flourishing fields of research (Narayanan and Park 2014). Three types of nanoparticles, silicon oxide (SiO_2NP, 11–14 nm), silver (AgNP, 20 nm), and titanium oxide (TiO_2NP, 20 nm), have been shown to be toxic to *M. incognita*, in in vitro and in vivo experimentation on tomato (Ardakani 2013). The smaller size of the nanoparticles with a large surface area increases interactions with microbial cells to implement a broad range of potential antimicrobial activities (Martinez et al. 2010).

The effect of AgNPs has been evaluated for their potential nematicidal effects on *M. incognita* infecting tomato (Ahmed El-Deen and Bahig 2018; El-Batal et al. 2019; Kalaiselvi et al. 2019). Further, ZnO NPs have an inverse effect on the cuticle and hypodermis of nematodes by affecting lipid, glycogen, and mucopolysaccharides (Siddiqui et al. 2018). Further, gold nanoparticles show promise in the management of root-knot nematodes (Thakur et al. 2018; Hu et al. 2018).

2.7.4 Nanopesticides

The presence of a formulation containing an active ingredient applied around the plant root at the initial crop growth stage helps in protecting the plant from the invasion of pathogens and bringing its population down below economic threshold levels (Khan et al. 2014a, b). As all the propagules/spores of a pathogen do not invade the host at one time, within their intermittent attack, consequent persistence or gradual release of an active ingredient in the root zone improves effectiveness of formulations (Khan et al. 2011). Furthermore, timely and slow release of an active element decreases the amount of pesticide required for disease control. Such controlled release has the added benefit of prior minimization of the effects of the pesticide on man and the environment (Khan and Jairajpuri 2012).

Controlled release of active ingredients may be achieved through a nanotechnological method in which nanomaterials act as a carrier for the chemicals. Hence refined formulations potentially reduce pesticide inputs associated with environmental hazards, namely, disease vectors and parasitic organisms. Nanopesticides decrease application rates as chemical quantities required are effective in the order of 10–15 times smaller than those applied through classical formulations. Hence higher efficiency is achieved for sufficient control of diseases. With a smaller size, better kinetic stabilization, low viscosity, and optical transparency, nanoemulsions improve pesticide delivery systems (Xu et al. 2010). As a carrier for pesticide delivery, nanoemulsions enhance bioavailability and solubility of the active ingredients of chemicals. Thus nanopesticides have very small particles of active ingredients or other small engineered structures with pesticide properties (Bergeson 2010b). Furthermore, nanopesticides enhance the release and wettability of agricultural formulations and the movement of unwanted pesticides (Bergeson 2010a).

Nanomaterials and biocomposites show beneficial properties such as permeability, stiffness, crystallinity, solubility, thermal stability, and biodegradability (Bouwmeester et al. 2009; Bordes et al. 2009), necessary for the formation of nanopesticides. Nanopesticide formulations present an enormous specific surface area and accordingly have an increased affinity to target molecules (Yan et al. 2015). Nanopesticide delivery techniques such as nanoencapsulation, nanoemulsions, nanocages, and nanocontainers show effectiveness in plant protection programs (Bouwmeester et al. 2009; Lyons and Scrinis 2009; Bergeson 2010b). Corradini et al. (2010) examined the prospect of using chitosan nanoparticles, an extremely degradable antibacterial, for the slow release of NPK fertilizer. Further, kaolin clay-based nanolayers have been developed to be used as cementing and coating material for the controlled release of fertilizers (Liu et al. 2006). Principally, nano-clay materials provide interacting surfaces with high aspect ratios for encapsulation which facilitates their use as agrochemicals such as fertilizers, plant growth promoters, and pesticides (Ghormade et al. 2011). In general, there are three types of controlled release systems (CRS): zero-order, first-order, and square-root time release, each of which may be tailored to environmental conditions and pest/pathogen biology. Nano-formulations are released faster in the soil, yet slowly in plants with residue levels below regulatory criteria in foodstuffs.

2.8 Integrated Approaches of Soil Management

After a century of incremental research, technological advances are connecting with a need for sustainable crop growth, leading to yield increases. Severe diseases in many crops are caused by soil-borne pathogens which have combined lineaments based on their close connections with the soil. Interactions between the pathogen and the host, in turn, interact between both biotic and abiotic environmental components.

Basic management strategies employed to condition and improve soil include disruption of one or more of the disease components, at any stage of disease development, to achieve an economic depression in diseases with minimal disturbance to the environment (Katan 2017; Mihajlović et al. 2017). Soil management is achieved through physical, biological, cultural, physiological, chemical, and genetic approaches. Further, management may utilize soil disinfestations via biofumigation, fumigation, anaerobic soil disinfestation, or soil solarization. Interestingly, application of fungicides, organic amendments, biocontrol, crop rotation, resistant cultivars, grafting, induced resistance, and cultural practices may be combined in integrated pest management programs.

Integrated pest management (IPM) of soil-borne pathogens aims to combine control approaches in an environmentally optimal manner for rational and sustainable disease reduction (Porter et al. 2010). IPM aims to achieve sustainable increases in yields and income development in terms of plant diseases with low negative effects on environmental and natural resources while facilitating a reduction of pesticide use. A systematic approach to sustainable pest management has been illustrated. Lewis et al. (1997) mention that long-term resolutions for pest management may only be carried out by restructuring and managing methods that maximize an array of built-in preventative strengths, with therapeutic strategies. Further, Chellemi et al. (2016) indicate four pillars in the management of soil-borne pathogens which seek to prevent the introduction and separation of pathogens into the crop systems and reduce pathogen populations to manageable levels.

Soil disinfestation should incorporate an effective reduction of soil pest populations with minimal damage to soil microbial and beneficial activities, such as mycorrhizae. Importantly, disinfestation should not leave phytotoxic residues (Katan 2017). Soil solarization effects mild soil heating (45–55 °C) in upper soil layers at depths of 5–20 cm and 35–40 °C in lower (30–45 cm) layers. Solarization works best on heavy soils containing clay, loam, or mixtures of the two. These soils hold more water than light soils, enabling steam production every day. Steam is required to kill nematodes, weed seeds, and insect eggs in the soil. Solarization may be less effective on sandy soil, which drains faster and produces less steam. To maximize the benefit of solarization in sandy soils, drip irrigation lines should be laid under clear plastic covers with water added regularly. The warming of the soil has no radical effect on resident biotic components. However, it results in the physical thermal killing of pathogens in upper, hotter soil layers and stimulates a beneficial microbial shift in the less heated soil layers, which contributes to pathogen control (Culman et al. 2006; Gelsomino and Cacco 2006; Ozylmaz et al. 2016).

Breeding potential plant hosts for resistance to pests/pathogens is a very effective method for controlling pathogens with no negative effects on the environment. Further, grafting scions of commercially desirable but susceptible cultivars on rootstocks resistant to soil-borne pathogens supplies plants with functional resistance that is equal to non-grafted cultivars with genes for resistance. The grafting approach provides pliability because it is relatively easier and faster to replace a rootstock, when a new physiological race appears than to breed a new cultivar (Louws et al. 2010a, b; Mihajlović et al. 2016).

Novel approaches of transferring genes across plant breeding barriers, particularly including wild species, and inclusion of resistance controlled by multiple genes offer tremendous resources for soil-borne pathogen resistance. Isolating and cloning genes of resistance can facilitate direct gene transfer within and across crop species. Our limited understanding of soil-borne pathogen genetics relates to relative disease responses overtime on resistant crops and advances only through research at the molecular, organismal, and population levels. A better understanding of these processes is fundamental for the wise deployment of resistant cultivars in cropping systems. Furthermore, the application of organic amendments has been suggested for management of soil-borne disease strategies (Bonanomi et al. 2010).

Organic amendments, such as animal and green manure, peats, composts, and organic wastes, are viable propositions to control soil-borne diseases (Colla et al. 2012; Arnault et al. 2013; Mehta et al. 2014). Crop rotation gives various benefits to crop production. As mentioned in Sect. 2.3, rotations are associated with enhanced soil productivity, increased soil tilth, reduced erosion, improved soil water management, and aggregate stability and textural improvement (Li et al. 2020). Crop rotation is without a doubt a valuable method for plant disease management. Although ineffective when used singularly, reductions in disease caused by soil-borne pathogens that have a wide host range or produce long-living survival structures occur (Tillmann et al. 2016). Soil pesticides may be used with seedling diseases because of the need to protect plants for relatively short periods (Chase 2012).

Different approaches of management of soil-borne pathogens are summarized in Fig. 2.2.

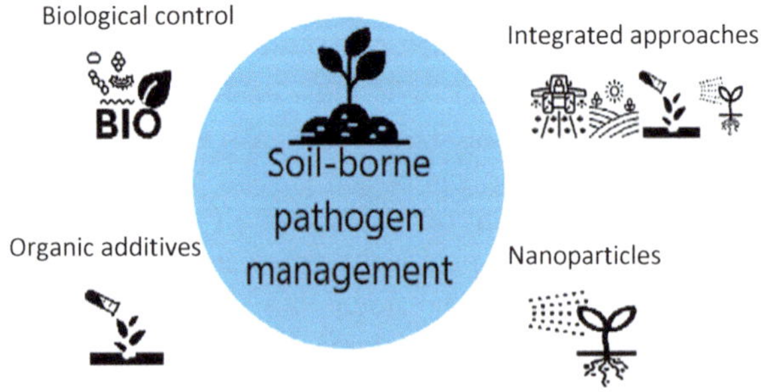

Fig. 2.2 Sustainable management of soil-borne plant pathogens

2.9 Using Soil-Borne Pathogen Biodiversity to Contribute to Sustainable Agriculture

Research is in demand to identify, select, and adopt cropping systems (including cover crops, antagonistic crops, green manure crops, inter-planting, rotations, organic amendments, and minimal tillage) that improve the soil diversity of pathogens and other fauna and microflora and repress known species of plant-parasites in agroecosystems. Much could be learned from the "biological balance" in the natural ecosystem, which affects fewer changes in the physical and biotic environment. Future soil-borne pathogen management must utilize sustainable agricultural practices that take into account beneficial, detrimental, and other soil-borne pathogens species in the rhizosphere and soil.

2.10 Conclusions and Future Prospects

Nowadays, the main impediment in crop production is the lack of successful and safe opportunities for controlling soil-borne diseases. The application of single approaches to control soil-borne diseases is not sufficient. Non-chemical options, such as soil solarization, crop rotation, soil amendments, or even biological control, may be ineffective when applied alone. However, all of these when used in combination become viable as components of an integrated pest management strategy, although they do not completely eliminate pathogens from the soil. Initial results obtained by combining different methods for the control of soil-borne diseases imply a necessity to continue research in this area in order to ensure long-lasting sustainability of crop protection. This chapter described a group of eco-friendly approaches that may be effective when used in combination to control a wide range of soil-borne diseases.

In conclusion, the introduction of chemical fertilizers and pesticides for the management of the soil-borne disease can result in good yield. However, a wide range of human diseases and ecological perturbations may be encountered. The biological control of soil-borne disease is an alternative approach to counter further biodiversity loss. The application of biocontrol agents and organic additives not only improves soil fertility but also controls disease. Recently, nanoparticles have been introduced in the intensification of agricultural crop production. Judicious application of nanomaterials along with biocontrol agents and organic additives is likely to exhibit better results in terms of plant disease management and crop produce intensification in sustainable cropping systems.

References

Abd El-Hai KM, El-Metwally MA, El-Baz SM et al (2009) The use of antioxidants and microelements for controlling damping-off caused by *Rhizoctonia solani* and charcoal rot caused by *Macrophomina phasoliana* on sunflower. Plant Pathol J 8:79–89

Adair KL, Wratten S, Lear G (2013) Soil phosphorus depletion and shifts in plant communities change bacterial community structure in a long-term grassland management trial. Environ Microbiol Rep 5(3):404–413

Afzal SA, Tariq S, Sultana V et al (2013) Managing the root diseases of okra with endo-root plant growth promoting Pseudomonas and *Trichoderma viride* associated with healthy okra roots. Pak J Bot 45(4):1455–1460

Ahmad Z, Pandey R, Sharma S et al (2006) Alginate nanoparticles as antituberculosis drug carriers: formulation development, pharmacokinetics and therapeutic potential. Indian J Chest Dis Allied Sci 48(3):171–176

Ahmed Hammad NE, Bahig Ahmed ED (2018) Effectiveness of silver nanoparticles against root–knot nematode, *Meloidogyne incognita* infecting tomato under greenhouse conditions. J Agric Sci 10(2):148–156

Al-Askar A (2012) In vitro antifungal activity of three Saudi plant extracts against some phytopathogenic fungi. J Plant Protect Res 52(4):458–462

Ansari RA, Khan TA (2012a) Parasitic association of root-knot nematode, Meloidogyne incognita on guava. e-Journal Sci Technol (e-JST) 5:65–67

Ansari RA, Khan TA (2012b) Diversity and community structure of phytonematodes associated with guava in and around Aligarh, Uttar Pradesh, India. Trends Biosci 5(3):202–204

Ansari RA, Mahmood I (eds) (2019a) Plant health under biotic stress – volume i, organic strategies. Springer, Singapore

Ansari RA, Mahmood I (eds) (2019b) Plant health under biotic stress: volume 2: microbial interactions. Springer, Singapore

Ansari RA, Rizvi R, Sumbul A et al (2017) PGPR: current vogue in sustainable crop production. In: Probiotics and plant health. Springer, Singapore, pp 455–472

Ansari RA, Sumbul A, Rizvi R et al (2019) Organic soil amendments: potential tool for soil and plant health management. In: Plant health under biotic stress. Springer, Singapore, pp 1–35

Ansari RA, Rizvi R, Sumbul A et al (2020a) Management of phytonematodes: recent advances and future challenges. Springer, Singapore

Ansari RA, Rizvi R, Sumbul A et al (2020b) Plant-growth-promoting rhizobacteria (PGPR)-based sustainable management of phytoparasitic nematodes: current understandings and future challenges. In Management of phytonematodes: recent advances and future challenges. Springer, Singapore, pp 51–71

Ardakani AS (2013) Toxicity of silver, titanium and silicon nanoparticles on the root-knot nematode, *Meloidogyne incognita*, and growth parameters of tomato. Nematology 15(6):671–677

Arnault I, Fleurance C, Vey F et al (2013) Use of Alliaceae residues to control soil-borne pathogens. Ind Crop Prod 49:265–272

Azam A, Ahmed AS, Oves M et al (2012) Size-dependent antimicrobial properties of CuO nanoparticles against Gram-positive and-negative bacterial strains. Int J Nanomedicine 7:3527

Bailey KL, Lazarovits G (2003) Suppressing soilborne diseases with residue management and organic amendments. Soil Tillage Res 72(2):169–180

Ball BC, Bingham I, Rees RM et al (2005) The role of crop rotations in determining soil structure and crop growth conditions. Can J Soil Sci 85(5):557–577

Basallote-Ureba MJ, Melero-Vara JM (1993) Control of garlic white rot by soil solarization. Crop Prot 12:219–223

Berendsen RL, Vismans G, Yu K et al (2018) Disease-induced assemblage of a plant-beneficial bacterial consortium. ISME J 12(6):1496–1507

Berg G, Smalla K (2009) Plant species and soil type cooperatively shape the structure and function of microbial communities in the rhizosphere. FEMS Microbiol Ecol 68:1–13

Bergeson LL (2010a) Nanosilver pesticide products: what does the future hold? Environ Qual Manag 19(4):73–82
Bergeson LL (2010b) Nanosilver: US EPA's pesticide office considers how best to proceed. Environ Qual Manag 19:79–85
Bonanomi G, Antignani V, Capodilupo M et al (2010) Identifying the characteristics of organic soil amendments that suppress soilborne plant diseases. Soil Biol Biochem 42:136–144
Bordes P, Pollet E, Avérous L (2009) Nano-biocomposites: biodegradable polyester/nanoclay systems. Prog Polym Sci 34(2):125–155
Bouwmeester H, Dekkers S, Noordam MY et al (2009) Review of health safety aspects of nanotechnologies in food production. Regul Toxicol Pharmacol 53(1):52–62
Bryaskova R, Pencheva D, Nikolov S et al (2011) Synthesis and comparative study on the antimicrobial activity of hybrid materials based on silver nanoparticles (AgNps) stabilized by polyvinylpyrrolidone (PVP). J Chem Biol 4(4):185
Carrión VJ, Cordovez V, Tyc O et al (2018) Involvement of Burkholderiaceae and sulfurous volatiles in disease-suppressive soils. ISME J 12(9):2307
Cha JY, Han S, Hong HJ et al (2016) Microbial and biochemical basis of a Fusarium wilt-suppressive soil. ISME J 10(1):119
Chase AR (2012) Chemical control of Fusarium wilt in greenhouse crops. In: Gullino ML, Katan J, Garibaldi A (eds) *Fusarium* wilts of greenhouse vegetable and ornamental crops. APS Press, USA, pp 117–120
Chellemi DO, Gamliel A, Katan J et al (2016) Development and deployment of system-based approaches for the management of soilborne plant pathogens. Phytopathology 106:216–225
Colla P, Gilardi G, Gullino ML (2012) A review and critical analysis of the European situation of soilborne disease management in the vegetable sector. Phytoparasitica 40(5):515–523
Corradini E, De Moura MR, Mattoso LH (2010) A preliminary study of the incorporation of NPK fertilizer into chitosan nanoparticles. Express Polym Lett 4(8)
Culman S, Duxbury J, Lauren J et al (2006) Microbial community response to soil solarization in Nepal's rice wheat cropping system. Soil Biol Biochem 38:3359–3371
Deepak S, Ashish K, Singh AK et al (2013) Induction of resistance in field pea against rust disease through various chemicals/micronutrients and their impact on growth and yield. Plant Pathol J (Faisalabad) 12(2):36–49
Dignam BE, O'Callaghan M, Condron LM et al (2016) Challenges and opportunities in harnessing soil disease suppressiveness for sustainable pasture production. Soil Biol Biochem 95:100–111
Dignam BE, O'Callaghan M, Condron LM et al (2018) Effect of land use and soil organic matter quality on the structure and function of microbial communities in pastoral soils: implications for disease suppression. PLoS One 13(5):e0196581
El-Batal AI, Attia MS, Nofel MM et al (2019) Potential nematicidal properties of silver boron nanoparticles: synthesis, characterization, in vitro and in vivo root-knot nematode (*Meloidogyne incognita*) treatments. J Clust Sci 30(3):687–705
FAOSTAT (2011) Food and agriculture organization statistics Database. In: Food and Agriculture Organization of the United Nations, Statistics Database
Fortuner R (1991) The Hoplolaiminae. In: Nickle WR (ed) Manual of agricultural nematology. Marcel Dekker, New York, pp 669–720
Garbeva P, van Veen JA, van Elsas JD (2004) Assessment of the diversity, and antagonism towards *Rhizoctonia solani* AG3, of Pseudomonas species in soil from different agricultural regimes. FEMS Microbiol Ecol 47(1):51–64
Gelsomino A, Cacco G (2006) Compositional shifts of bacterial groups in a solarized and amended soil as determined by denaturating gel electrophoresis. Soil Biol Biochem 38:91–102
Ghormade V, Deshpande MV, Paknikar KM (2011) Perspectives for nano-biotechnology enabled protection and nutrition of plants. Biotechnol Adv 29(6):792–803
Goudjal Y, Zamoum M, Sabaou N et al (2016) Potential of endophytic Streptomyces spp. for biocontrol of Fusarium root rot disease and growth promotion of tomato seedlings. Biocontrol Sci Technol Taylor & Francis 26(12):1691–1705

Guzmán MG, Dille J, Godet S (2009) Synthesis of silver nanoparticles by chemical reduction method and their antibacterial activity. Int J Mater Metallur Eng 2(7):91–98

Handelsman J, Stabb EV (1996) Biocontrol of soilborne plant pathogens. Plant Cell 8(10):1855

Harman GE (1996) *Trichoderma* for biocontrol of plant pathogens: from basic research to commercialized products. In: Proceedings of the Cornell Community Conference on Biological Control, Ithaca, vol 1113. Available via Cornell University. http://web.entomology.cornell.edu/shelton/cornell-biocontrol-conf/talks/harman.html. Accessed 5 May 2021

He L, Liu Y, Mustapha A et al (2011) Antifungal activity of zinc oxide nanoparticles against *Botrytis cinerea* and *Penicillium expansum*. Microbiol Res 166(3):207–215. https://doi.org/10.1016/j.micres.2010.03.003

Hu C, Wu GH, Lai SF et al (2018) Toxic effects of size-tunable gold nanoparticles on *Caenorhabditis elegans* development and gene regulation. Sci Rep 8:15245

Huang S, Wang L, Liu L et al (2015) Nanotechnology in agriculture, livestock, and aquaculture in China. A review. Agron Sustain Dev 35(2):369–400

Janvier C, Villeneuve F, Alabouvette C et al (2007) Soil health through soil disease suppression: which strategy from descriptors to indicators? Soil Biol Biochem 39(1):1–23

Jayaseelan C, Rahuman AA, Kirthi AV et al (2012) Novel microbial route to synthesize ZnO nanoparticles using *Aeromonas hydrophila* and their activity against pathogenic bacteria and fungi. Spectrochim Acta A Mol Biomol Spectrosc 90:78–84

Kalaiselvia D, Mohankumara A, Shanmugama G et al (2019) Green synthesis of silver nanoparticles using latex extract of *Euphorbia tirucalli*: a novel approach for the management of root knot nematode, *Meloidogyne incognita*. Crop Prot 117:108–114

Karimi K, Amini J, Harighi B et al (2012) Evaluation of biocontrol potential of *Pseudomonas* and *Bacillus* spp. against Fusarium wilt of chickpea. Aust J Crop Sci 6(4):695–703

Katan J (2017) Diseases caused by soilborne pathogens: biology, management and challenges. J Plant Pathol 99(2):305–315

Katan J, Gamliel A (2011) Soilborne diseases, control by physical methods. In: Gliński J, Horabik J, Lipiec J (eds) Encyclopedia of agrophysics, Encyclopedia of Earth sciences series. Springer, Dordrecht

Katan J, Shtienberg D, Gamliel A (2012) The integrated management concept in the context of soilborne pathogens and soil disinfestation. In: Gamliel A, Katan J (eds) Soil solarization: theory and practice. APS Press, USA, pp 91–97

Khalil ME, Allam A, Barakat AT (2012) Nematicidal activity of some biopesticide agents and microorganisms against root-knot nematode on tomato plants under greenhouse conditions. J Plant Protect Res 52(1):47–52

Khan MR (2012a) Nematodes, an emerging threat to global forests: assessment and management. Plant Pathol J 11:99–113

Khan MR (2012b) Nematodes, an emerging threat to global forests: assessment and management. Plant Pathol J 11:99–113

Khan MM, Khan MR, Mohiddin FA (2012) The relative performance of different inoculation methods with *Alternaria brassicae* and *A. brassicicola* on Indian mustard. Plant Pathol J 11:93–98

Khan MR, Ashraf S, Rasool F et al (2014a) Field performance of Trichoderma species against wilt disease complex of chickpea caused by *Fusarium oxysporum f. sp. ciceri* and *Rhizoctonia solani*. Turk J Agric For 38(4):447–454

Khan MR, Haque Z, Kausar N (2014b) Management of the root-knot nematode *Meloidogyne graminicola* infesting rice in the nursery and crop field by integrating seed priming and soil application treatments of pesticides. Crop Prot 63:15–25

Khan MR, Jairajpuri MS (2010a) Nematode infestation in food crops-national scenario. In: Khan MR, Jairajpuri MS (eds) Nematode infestations part I: food crops. National Academy of Sciences, India, pp 1–16

Khan MR, Jairajpuri MS (2010b) Nematode infestation in industrial crops-national scenario. In: Khan MR, Jairajpuri MS (eds) Nematode infestations part II: industrial crops. National Academy of Sciences, India, pp 1–19

Khan MR, Jairajpuri MS (2012) Nematode infestation in horticultural crops, national scenario. In: Khan MR, Jairajpuri MS (eds) Nematode infestation part III: horticultural crops. National Academy of Sciences, India, pp 1–30
Khan MR, Majid S, Mohidin FA et al (2011) A new bioprocess to produce low-cost powder formulations of biocontrol bacteria and fungi to control fusarial wilt and root-knot nematode of pulses. Biol Control 59(2):130–140
Khan MR, Rizvi TF (2014) Nanotechnology: scope and application in plant disease management. Plant Pathol J 13(3):214–231
Kheyrodin H (2011) Crop rotations for managing soil-borne plant diseases. Afr J Food Sci Technol 2(1):001–009
Kowsari M, Motallebi M, Zamani RM (2014) Construction of new GFP-tagged fusants *for Trichoderma harzianum* with enhanced biocontrol activity. J Plant Protect Res 54(2):122–131
Krishnaraj C, Ramachandran R, Mohan K et al (2012) Optimization for rapid synthesis of silver nanoparticles and its effect on phytopathogenic fungi. Spectrochim Acta A Mol Biomol Spectrosc 93:95–99
Ladygina N, Hedlund K (2010) Plant species influence microbial diversity and carbon allocation in the rhizosphere. Soil Biol Biochem 42(2):162–168
Lamovsek J, Urek G, Trdan S (2013) Biological control of Root-Knot Nematodes (*Meloidogyne* spp.): microbes against the pests. Acta Agric Slovenica 101(2):263–275
Larkin RP, Roberts DP, Gracia-Garza JA (1998) Biological control of fungal diseases. In Fungicidal activity,chemical and biological approaches. (pp 141–191). New York, NY: Wiley
Latha P, Anand T, Ragupathi N et al (2009) Antimicrobial activity of plant extracts and induction of systemic resistance in tomato plants by mixtures of PGPR strains and Zimmu leaf extract against *Alternaria solani*. Biol Control 50(2):85–93
Latz E, Eisenhauer N, Rall BC et al (2012) Plant diversity improves protection against soil-borne pathogens by fostering antagonistic bacterial communities. J Ecol 100(3):597–604
Latz E, Eisenhauer N, Rall BC et al (2016) Unravelling linkages between plant community composition and the pathogen-suppressive potential of soils. Sci Rep 6:23584
Lewis WJ, van Lenteren JC, Phatak SC et al (1997) A total system approach to sustainable pest management. Proc Natl Acad Sci U S A 94:12243–12248
Li F, Liang X, Li H et al (2020) Enhanced soil aggregate stability limits colloidal phosphorus loss potentials in agricultural systems. Environ Sci Eur 32(17). https://doi.org/10.1186/s12302-020-0299-5
Liu XM, Feng ZB, Zhang FD et al (2006) Preparation and testing of cementing and coating nano-subnanocomposites of slow/controlled-release fertilizer. Agric Sci China 5(9):700–706. https://doi.org/10.1016/S1671-2927(06)60113-2
Loganathan M, Sible GV, Maruthasalam S et al (2010) *Trichoderma* and chitin mixture based bioformulation for the management of head rot (*Sclerotinia sclerotiorum* (Lib.) deBary) –root-knot (*Meloidogyne incognita* Kofoid and White; Chitwood) complex diseases of cabbage. Arch Phytopathol Plant Protect 43(10):1011–1024. https://doi.org/10.1080/03235400802214885
Logeswari P, Silambarasan S, Abraham J (2015) Synthesis of silver nanoparticles using plants extract and analysis of their antimicrobial property. J Saudi Chem Soc 19(3):311–317
Louws FJ, Rivard CL, Kubota C (2010a) Grafting fruiting vegetables to manage soilborne pathogens, foliar pathogens, arthropods and weeds. Sci Hortic 127(2):127–146
Louws J, Rivard CL, Kubota C (2010b) Grafting fruiting vegetables to manage soilborne pathogens, foliar pathogens, arthropods and weeds. Sci Hortic 127(2):127–146
Lyons K, Scrinis G (2009) Under the regulatory radar nanotechnologies and their impacts for rural Australia. In: Merlan F (ed) Tracking rural change: community, policy and technology in Australia, New Zealand and Europe. Australian National University E Press, Canberra, pp 151–171
Mahmood I, Rizvi R, Sumbul A et al (2019) Potential Role of Plant Growth Promoting Rhizobacteria in Alleviation of Biotic Stress. In Plant Health Under Biotic Stress (pp. 177-188) Springer, Singapore
Makhluf S, Dror R, Nitzan Y et al (2005) Microwave-assisted synthesis of nanocrystalline MgO and its use as a bacteriocide. Adv Funct Mater 15(10):1708–1715

Maron JL, Marler M, Klironomos JN et al (2011) Soil fungal pathogens and the relationship between plant diversity and productivity. Ecol Lett 14(1):36–41

Martinez F, Peggy G, Olive L et al (2010) Synthesis, characterization and evaluation of antimicrobial and cytotoxic effect of silver and titanium nanoparticles. Nanomedicine 6:681–688

Matějka V, Tokarský J (2014) Photocatalytical nanocomposites: a review. J Nanosci Nanotechnol 14(2):1597–1616

McGovern RJ, McSorley R (1997) Physical methods of soil sterilization for disease management including soil solarization. In: Rechcigel NA, Rechcigel JE (eds) Environmentally safe approaches to disease control. CRC Lewis, Boca Raton

Mehta CM, Palni U, Franke-Whittle IH et al (2014) Compost: its role, mechanism and impact on reducing soil-borne plant diseases. Waste Manag 34(3):607–622

Mendes R, Kruijt M, De Bruijn I et al (2011) Deciphering the rhizosphere microbiome for disease-suppressive bacteria. Science 332(6033):1097–1100

Meyer SL, Roberts DP, Chitwood DJ et al (2001) Application of *Burkholderia cepacia* and *Trichoderma virens*, alone and in combinations, against *Meloidogyne incognita* on bell pepper. Nematropica:75–86

Mihajlović M, Rekanović E, Hrustić J et al (2016) Ulogaagrotehničkihmera usuzbijanjupatogenaizzemljišta u konceptuintegralnezaštitepaprike. Biljnilekar 44(4):333–342

Mihajlović M, Rekanović E, Hrustić J et al (2017) Methods for management of soilborne plant pathogens. Pesticidi i fitomedicina 32(1):9–24

Moghadam E, Rouhani H, Flahi Rastegar M (2009) Biological control of sugar beet cyst forming nematode with trichoderma under in vitro and green house condition. Isfahan Univ Technol-J Crop Prod Proc 13(48):301–312

Mommer L, Cotton TA, Raaijmakers JM et al (2018) Lost in diversity: the interactions between soil-borne fungi, biodiversity and plant productivity. New Phytol 218(2):542–553

Mostafa M, Almoammar H, Abd-Elsalam KA (2018) Nanoantimicrobials mechanism of action. In: Nanobiotechnology applications in plant protection. Springer, Cham, pp 281–322

Narayanan KB, Park HH (2014) Antifungal activity of silver nanoparticles synthesized using turnip leaf extract (*Brassica rapa* L.) against wood rotting pathogens. Eur J Plant Pathol 140(2):185–192

Nielsen UN, Wall DH, Six J (2015) Soil biodiversity and the environment. Annu Rev Environ Resour 40:63–90

Oerke EC (2006) Crop losses to pests. J Agric Sci 144:31–43

Ozylmaz U, Benlioglu K, Yildis A et al (2016) Effects of soil amendments combined with solarization on the soil microbial community in strawberry cultivation using quantitative real-time PCR. Phytoparasitica 44:661–680

Paramasivan M, Chandrasekaran A, Mohan S et al (2014) Ecological management of tropical sugar beet (TSB) root rot (*Sclerotium rolfsii* Sacc.) by rhizosphere *Trichoderma* species. Arch Phytopathol Plant Protect 47(13):1629–1644

Penton CR, Gupta VV, Tiedje JM et al (2014) Fungal community structure in disease suppressive soils assessed by 28S LSU gene sequencing. PLoS One 9(4):e93893

Pisoschi AM, Pop A, Cimpeanu C et al (2018) Nanoencapsulation techniques for compounds and products with antioxidant and antimicrobial activity – a critical view. Eur J Med Chem 157:1326–1345

Porter I, Banks J, Mattner S et al (2010) Global phase out of methyl bromide under the Montreal Protocol: implications for bio-protection, biosecurity and the ozone layer. In: Gisi U, Chet I, Gullino ML (eds) Recent developments in management of plant disease. Springer, New York, pp 293–309

Prasad R, Gupta N, Kumar M et al (2017) Nanomaterials act as plant defense mechanism. In: Nanotechnology. Springer, Singapore, pp 253–269

Purkayastha GD, Mangar P, Saha A et al (2018) Evaluation of the biocontrol efficacy of a Serratia marcescens strain indigenous to tea rhizosphere for the management of root rot disease in tea. PLoS One 13(2):e0191761

Raaijmakers JM, Mazzola M (2012) Diversity and natural functions of antibiotics produced by beneficial and plant pathogenic bacteria. Annu Rev Phytopathol 50:403–424

Raaijmakers JM, Mazzola M (2016) Soil immune responses. Science 352(6292):1392–1393

Raaijmakers JM, Paulitz TC, Steinberg C et al (2009) The rhizosphere: a playground and battlefield for soilborne pathogens and beneficial microorganisms. Plant Soil 321(1–2):341–361

Rahul S, Chandrashekhar P, Hemant B et al (2014) Nematicidal activity of microbial pigment from Serratia marcescens. Nat Prod Res 28(17):1399–1404

Ramarathnam R, Dilantha Fernando WG, de Kievit T (2011) The role of antibiosis and induced systemic resistance, mediated by strains of *Pseudomonas chlororaphis*, *Bacillus cereus* and *B. amyloliquefaciens*, in controlling blackleg disease of canola. BioControl 56:225–235

Raupach GS, Kloepper JW (1998) Mixtures of plant growth-promoting rhizobacteria enhance biological control of multiple cucumber pathogens. Phytopathology 88(11):1158–1164

Sasse J, Martinoia E, Northen T (2018) Feed your friends: do plant exudates shape the root microbiome? Trends Plant Sci 23:25–41

Sayed M, Abdel-Rahman T, Ragab A, Abdellatif A (2019) Biocontrol of Root-Knot Nematode *Meloidogyne incognita* by Chitinolytic *Trichoderma* spp. Egypt J Agron 18(1):30–47

Schnitzer SA, Klironomos JN, HilleRisLambers J et al (2011) Soil microbes drive the classic plant diversity–productivity pattern. Ecology 92(2):296–303

Schwartz HF (2011) Soil-borne diseases of onion. Crop Ser Dis 2(940)

Shafique HA, Sultana V, Ara J, Haque SH, Athar M (2015b) Role of antagonistic microorganisms and organicamendment in stimulating the defense System of okra against root rotting fungi. Polish J. Microbiol, 64(2), 157–162

Sharma M, Jasrotia S, Ohri P, Manhas RK (2019) Nematicidal potential of *Streptomyces antibioticus* strain M7 against *Meloidogyne incognita*. AMB Express 9(1):168

Sharon E, Bar-Eyal M, Chet I et al (2001) Biological control of the root-knot nematode *Meloidogyne javanica* by *Trichoderma harzianum*. Phytopathology 91(7):687–693

Shen T, Lei Y, Pu X et al (2020) Identification and application of *Streptomyces microflavus* G33 in compost to suppress tomato bacterial wilt disease. Appl Soil Ecol 157:1037242. https://doi.org/10.1016/j.apsoil.2020.103724

Siddiqui ZA, Khan A, Khan MR et al (2018) Effects of zinc oxide nanoparticles (ZnO NPs) and some plant pathogens on the growth and nodulation of Lentil (*Lens culinaris* Medik.). Acta Phytopathol Entomol Hungarica 53(2):195–212

Simpson M, McLenaghen RD, Chirino-Valle I et al (2012) Effects of long-term grassland management on the chemical nature and bioavailability of soil phosphorus. Biol Fertil Soils 48(5):607–611

Singh N, Manshian B, Jenkins GJ et al (2009) NanoGenotoxicology: the DNA damaging potential of engineered nanomaterials. Biomaterials 30(23–24):3891–3914

Tanović B, Gašić S, Hrustić J et al (2014) Effect of emulsifiable concentrate (EC) of oregano essential oil on Monilinia spp. In: Book of Abstracts, 5th CASEE Conference: Healthy Food Production and Environmental Preservation – the Role of Agriculture, Forestry and Applied Biology. University of Novi Sad, Faculty of Agriculture, Novi Sad, p 48

Teng PS (ed) (1987) Crop loss assessment and pest management. APS Press, St Paul

Thakur RK, Dhirta B, Shirkot P (2018) Studies on effect of gold nanoparticles on *Meloidogyne incognita* and tomato plants growth and development. BioRxiv 428144

Tillmann M, von Tiedemann A, Winte M (2016) Crop rotation effects on incidence and diversity of *Fusarium* species colonizing stem bases and grains of winter wheat. J Plant Dis Protect 124(2):121–130

Turkolmez S, Soylu EM (2014) Antifungal efficacies of plant essential oils and main constituents against soilborne fungal disease agents of bean. J Essent Oil Bearing Plants 17(2):203–211

Umaerus VR, Scholte K, Turkensteen LJ (1989) Crop rotation and the occurrence of fungal diseases in potatoes. In: Vos J, Van Loon CD, Bollen GJ (eds) Effects of crop rotation on potato production in the temperate zones, Developments in plant and soil sciences, vol 40. Springer, Dordrecht. https://doi.org/10.1007/978-94-009-2474-1_15

van Overbeek LS, Senechkin IV, van Bruggen AH (2012) Variation in microbial responses and *Rhizoctonia solani* AG2-2IIIB growth in soil under different organic amendment regimes. Can J Plant Pathol 34(2):268–276. https://doi.org/10.1080/07060661.2012.679622

Waisel Y, Eshel A, Kafkafi U (2002) Plant roots – the hidden half, 3rd eds. New York,Basel, 1136pp.

Wakelin SA (2018) Managing soil microbiology: realising opportunities for the productive land-based sectors. N Z J Agric Res 61(3):358–376

Walters WA, Jin Z, Youngblut N et al (2018) Large-scale replicated field study of maize rhizosphere identifies heritable microbes. Proc Natl Acad Sci 115(28):7368–7373

Wang S, Lawson R, Ray PC et al (2011) Toxic effects of gold nanoparticles on *Salmonella typhimurium* bacteria. Toxicol Ind Health 27(6):547–554

Weller DM, Raaijmakers JM, Gardener BBM et al (2002) Microbial populations responsible for specific soil suppressiveness to plant pathogens. Annu Rev Phytopathol 40(1):309–348

Widmer TL, Mitkowski NA, Abawi GS (2002) Soil organic matter and management of plant-parasitic nematodes. J Nematol 34(4):289

Xu L, Liu Y, Bai R et al (2010) Applications and toxicological issues surrounding nanotechnology in the food industry. Pure Appl Chem 82(2):349–372

Yan Y, Kuramae EE, Klinkhamer PG, van Veen JA (2015) Revisiting the dilution procedure used to manipulatemicrobial biodiversity in terrestrial systems. Appl Environ Microbiol 81:4246–4252

YankoVa, V, Valchev, N, Markova D (2014). Effectiveness of phytopesticide neem azal T/S ® against tomato leafminer (Tuta absoluta meyrick) in greenhouse tomato. Bulgarian J. Agricultural Science 20(5):1116–1118.

Yimer HZ, Nahar K, Kyndt T et al (2018) Gibberellin antagonizes jasmonate-induced defense against *Meloidogyne graminicola* in rice. New Phytol:1–15

Yin JJ, Liu J, Ehrenshaft M, Roberts JE et al (2012) Phototoxicity of nano titanium dioxides in HaCaT keratinocytes – generation of reactive oxygen species and cell damage. Toxicol Appl Pharmacol 263(1):81–88

Youssef MM, Lashein AM (2013) Effect of cabbage (Brassica oleracea) leaf residue as a biofumigant, on root knot nematode, *Meloidogyne incognita* infecting tomato. J Plant Protect Res 53(3):271–274

Youssef SA, Tartoura KA, Abdelraouf GA (2016) Evaluation of *Trichoderma harzianum* and *Serratia proteamaculans* effect on disease suppression, stimulation of ROS-scavenging enzymes and improving tomato growth infected by *Rhizoctonia solani*. Biol Control 100:79–86

Chapter 3
Empirical Values of Halophytes in Agro-ecology and Sustainability

Tayyaba Hussain and Mudassir Khan

Abstract Salinity is an increasingly urgent problem causing tremendous yield losses on a global scale. The problem is a marked imperative in arid and semiarid regions. To maximize crop productivity, and alleviate environmental stress, these areas require either reduction of salinity or the use of salt-tolerant crops. Halophytes are plants capable of normal growth in saline habitats and are able to thrive on "ordinary" soil, though these plants have a capacity to tolerate concentrations over 0.5% throughout their life cycle. As a consequence of rapid climate change, the proportion of saline areas is increasing daily, providing motivations for development of salt-tolerant crops to cope with the adverse conditions and contribute to long-term sustainability goals.

Research efforts are directed toward studying phytoremediation of saline environments in order to efficiently ameliorate salts from both soil and water. Challenges of attaining sustainable environments need to be addressed through mitigating global climate change while enabling a cooperatively sustained food industry. Many features of halophytes are highlighted in this chapter, easing the improvement of salt tolerance in crops in the future. Genetic and physiological screening of halophytes facilitates the contribution of halophytes with respect to long-term environmental sustainability.

Keywords Halophytes · Salt-responsive genes · Salinity · Phytoremediation · Environmental sustainability · Crop modification

---

Tayyaba Hussain and Mudassir Khan contributed equally.

T. Hussain (✉)
Plant Microbe Interaction Laboratory, Quaid-i-Azam University, Islamabad, Pakistan
e-mail: tayyaba.hussain@bs.qau.edu.pk

M. Khan
Department of Healthcare Biotechnology, Atta-Ur-Rahman School of Applied Biosciences (ASAB), National University of Science and Technology (NUST), Islamabad, Pakistan
e-mail: kmudassir71@yahoo.com

J. N. Furze et al. (eds.), *Earth Systems Protection and Sustainability*,
https://doi.org/10.1007/978-3-030-85829-2_3

3.1 Introduction

The word "halophytes" is derived from the Greek prefixes *halo* meaning salt and *phyte* meaning plants. Halophytes are plants which have the ability to survive in habitats containing salt concentrations over 0.5% (Stocker 1928). These plants acclimated growth in natural saline environments, via different mechanistic approaches, and are thus "salt tolerant." More recently, Greenway and Munns (1980) depicted that these plants have the ability to thrive in salt-rich environments in which soil solutions exhibit a minimum of 3.3 bars, which is equal to the presence of approximately 70 mM monovalent salt in the soil. Plants unable to sustain growth in these conditions are coined as non-halophytes (Dansereau 1957). Although only 2% of the world's flora is represented by halophytes, this figure exhibits a limited knowledge of plant diversity. In the course of evolution, plants develop a plentitude of morphological, structural, and physiological changes to flourish in salt-rich environments (Mishra and Tanna 2017). It has been estimated that salinity has a tremendous effect on overall soil biogeography, 10% of the world's land surface is affected, and indeed fertile texture suffers greatly with almost 50% of irrigated land being contaminated (Shah 2018).

Implementations of bio-saline practices in the agro-industry sector have been instigated by the scientific community to mitigate salinity naturally via plants with high salt tolerance adaptability. Glenn et al. (1999) reported different practices which showed the significant impacts of halophytes on agriculture. Principally, it would be conducive to analyze the mechanistic approach of halophyte growth in saline environments, which may be integrated in normal wild plants to develop salt-tolerant cultivars (Serrano 1996; Rausch et al. 1996; Zhu et al. 1997). Consequently, halophytic plants can be used as a standard for evaluation of plants growing in agronomic settings (Glenn et al. 1999). Metagenomic approaches involving genomic alteration and integration of genetic factors of halophytes into other plants pave the way for domestication of transgenic crops (Llerena 1994; Squires 1994; Ashraf et al. 2010).

3.1.1 Why Study Halophytes?

Increased salinization is a major motivation to develop our knowledge of halophytes. Salinization is the process of excessive accumulation of salts. Primarily, salts are sodium, potassium, calcium, and magnesium. These salts slowly leach into the soil and affect the rhizosphere and overall water table. Poor drainage, improper crop rotation, and inadequate farming practices are significant factors which contribute to salinization. High levels of salts occur in droughted areas, especially in arid and semi-arid regions with low annual rainfall (Jolly et al. 2008). There are two categories of salinization, natural and artificially induced. The former is caused by physical/chemical weathering, acid rain, high temperature, and presence of natural geological salt deposits (Xia et al. 2020). The latter is due to inappropriate man-made practices such as improper crop rotation, poor drainage, and supply of salty water to crops. Waterlogging also induces salinization. Ironically, supply of excessive water to respective areas seems the best practice to ameliorate salt from the rhizosphere and flush the soil. However, this practice ultimately causes the leaching of water-soluble salts from the rhizosphere into the groundwater table, rendering agronomic practices difficult to manage and production of food limited in the face of increasing human populations.

Arable area is decreasing gradually due to progressive soil salinization, which ultimately depletes crop productivity to dangerously low levels (Kefu et al. 2002). Munns and Tester (2008) estimated several crop decreases in dry matter production in response to salinity, revealing 50% decreases at 80 mM NaCl for rice (*Oryza sativa* L.); 100 mM NaCl for durum wheat (*Triticum turgidum* ssp. durum); and 120 mM NaCl for barley (*Hordeum vulgare* L).

In northern China salinity poses a serious soil problem, affecting cultivated and native plants (Liu et al. 2002); a widely distributed halophytic species, seepweed (*Suaeda salsa* L.), alleviates salinization. Seeds of seepweed have 30–40% edible oil content, which is considered an ideal source of unsaturated fatty acids. Fresh branches of seepweed are edible as a healthy and fresh vegetable (Kefu et al. 2002; Wang et al. 2001). Hence, environmentally degraded soil may be remediated to an extent through halophyte growth. It is important to cultivate a large number of halophytes to improve environments:in areas of concern. Further, kallar grass (*Leptochloa fusca* L.), a "salt grass," is able to accumulate an average of 20 t ha^{-1} dry matter after 4–5 cuts per year and is considered a useful plant in improving saline-sodic soil conditions to successfully sustain vegetation growth after 5-year periods (Mahmood et al. 1994). Correspondingly, quinoa (*Chenopodium quinoa* L.), grown in Bolivia under suboptimal soil conditions, exhibits low yield potential (< 0.5 t ha^{-1}), but under optimal conditions its yielding potential can be substantially increased up to 3–4 t ha^{-1} (Adolf et al. 2013). Overall natural and artificially induced salinization compound and endanger the environment. The introduction of salt-tolerant crops provides hope as these plants can take up excess salts from the soil and assist soil rehabilitation. Using halophytes in remediation approaches may also be complemented with production of transgenic salt-tolerant cultivars (Aslam et al. 2011).

The remaining sections of the chapter consider the future prospects and goals of halophytic integration. In Sect. 3.2 objectives of halophytic research are succinctly stated. Strategies implementing salinity treatment are detailed in Sect. 3.3. Section 3.4 discusses distribution of halophytes. Section 3.5 proposes the use of halophyte habitats as natural "laboratories." In Sect. 3.6 adaptation mechanisms and the physiology of salinity avoidance are covered. In Sect. 3.7, ion compartmentalization is detailed. In Sect. 3.8 halophyte screening methods are given. In Sect. 3.9 salt tolerance mechanisms are summarized. Section 3.10 discusses salt tolerance genes. Section 3.11 outlays effective treatment of salinity by halophytes. Section 3.12 mentions the contributions of halophytes in environmental sustainability. Section 3.13 indicates future prospects and functions for halophyte species. Potential techniques for genetic integration of halophytes are covered in Sect. 3.14 and concluding remarks are given in Sect. 3.15.

3.2 Mitigation of Salinization

The principle objective of future halophyte research is to introduce salt-resistant cultivars to mitigate salinization. To understand and realize this goal, two research directions are proposed. Firstly, to breed salt-tolerant relatives with wild varieties to induce salt resistance via conventional breeding principles (Flowers 1989); secondly, naturally salt-tolerant plants may be cultivated at domestic levels to ameliorate salinity. Using both of these means, physiological traits and responsible gene functioning against salinity can be identified, which assists in the development of transgenic crops (Epstein et al. 1980).

3.3 Strategies of Salinity Treatment Implementing Technological Applications

Treatment of highly saline soils involves different approaches and techniques, which can mainly be separated into two basic categories. Ex situ techniques remediate excavated contaminated soil by different means and potentially includes chemical extraction, thermal treatment, and solidification. In situ techniques are preferred where remediation is carried out with exception of excavation of the contaminated soil. However, ex situ treatment of the soil is more expensive as it requires returning the soil to the restored site after treatment. In situ techniques are favored over ex situ with low cost benefits and minimum impact on the ecosystem (Sauer et al. 1996).

South-west Haryana in India, parts of Rajasthan, and other adjacent areas that comprise the "Thar Desert" found at 27.4695° N, 70.6217° E, in predominantly an arid zone interspersed with saline domains, will benefit from in situ techniques. Saline soil and its native flora have unique biology, enlisted in Table 3.1.

Saline stress is relieved via phytoremediation, which remediates the soil with the use of beneficial microbes and potential halophytes. Biotechnological interventions which enhance phytoremediation play a key role in crop sustainability; among these marker-assisted technology, quantitative trait loci (QTL) mapping, and gene tagging techniques are being adopted in screening and selection of desirable traits in plant genomes (Vera-Estrella et al. 2005; Devi et al. 2017).

Table 3.1 The native flora of Thar Desert

Plant	Common name	Family	References
Salsola baryosma	Dandy	Amaranthaceae	Sharma and Ramawat (2014)
Zygophyllum simplex	Zygophyllum	Zygophyllaceae	
Trianthema triquetra	Horse-Purslane	Aizoaceae	
Tamarix aphylla	Athel tree	Tamaricaceae	Charan and Sharma (2016)
Portulaca meridiana	Chickenweed	Portulacaceae	
Zygophyllum simplex	Zygophyllum	Zygophyllaceae	
Haloxylon recurvum	Saxaul	Amaranthaceae	
Haloxylon salicornicum	Rimth saltbush	Amaranthaceae	
Suaeda fruticosa	Shrubby seablite	Amaranthaceae	
Salsola baryosma,	Dandy	Chenopodiaceae	
Sesuvium sesuvioides	Desert pink	Aizoaceae	
Chenopodium murale	Goosefoot	Amaranthaceae	Rajaram et al. (2006)

3.4 Halophyte Distribution

Diverse halophyte distribution covers 1% of the world's flora, enabling responses to abiotic stress, such as drought stress, saline stress, osmotic stress, habitat, and even stress distribution among taxa of flowering plants (Flowers et al. 2010). Plants which can complete their life cycle at 200 mM NaCl are categorized as halophytes (Flowers and Colmer 2008).

Most Cyperaceae, Poaceae, and Brassicaceae species and also a large number of dicotyledons such as *Aster tripolium*, *Glaux maritima*, and *Plantago maritima* are halophytic. These plants can cope with salty soils in nature when required, though may be indifferent to their habitat in their patterns of distribution. Examples of highly adaptive species are *Myosurus minimus*, *Potentilla anserina*, and *Chenopodium glaucum*, which can grow in any habitat. Many species, such as *Agrostis stolonifera*, *Festuca rubra*, and *Juncus bufonius* populations, live on salty soils with those on salt-free soils varying genetically. The latter indicates that strategically grouped competitive stress-tolerant species have highly adaptable genetic factors which are differentially triggered according to their suitability with the habitat. Halophytes can be classified in terms of ecological factors. Classifications of halophytes are obligate, facultative, habitat-indifferent, and glycophytic (Von Sengbusch 2003) and are shown and defined in Table 3.2.

Saline conditions alter growth patterns. Obligate halophytes require a continuous supply of salt for optimum growth, while facultative halophytes have no such issue and can grow in saline as well as in non-saline condition. Quinoa is a potential facultative halophyte and is able to survive in extreme saline stress. Indeed the species tolerates soil electrical conductivity greater than 40 dS m^{-1}. This species has been cultivated since around 3500 years ago for a dietary source (Jacobsen et al. 2003; Razzaghi et al. 2011; Bonales-Alatorre et al. 2013). Obligate halophytes grow in salty habitats only.

Table 3.2 Types of halophytes and their habitats

Types of halophyte	Habitat	Examples
Obligate	Growth in high saline environment Salt concentration greater than 200 mM NaCl	*Arthrocnemum macrostachyum* *Frankenia salina*
Facultative	Moderate saline environment, salt concentration approximately less than 200 mM NaCl	*Aster tripolium* L. *Plantago lanceolata*
Habitat-indifferent	Can grow in salt-free environment but can grow better as compare to glycophytes	*Salsola imbricate* *Agrostis stolonifera*
Glycophyte	Preferably grow in low salt concentrations, less than 100 mM NaCl	*Zea mays* *Vicia faba* *Oryza sativa*

Domestication of halophytes will inevitably lead to establishment of completely new and artificial agro-ecosystems with cooperative benefits such as the production of highly productive yields of food, fodder, fiber, and fuel while also giving phytoremediation impact, for example, *Vetiveria zizanioides* (Chen et al. 2004). Such an approach may fulfill the needs of rapidly increasing human populations. The major factor behind soil salinization are anthropogenic activities and climatic changes. Halophytes are preferred by governmental bodies to remediate damage caused by salinization of soil and fresh water (Seydehmet et al. 2018; Parnian and Furze 2021).

3.5 Habitats of Halophytes as Natural Laboratories

Saline ecosystems remain a research spotlight and their consideration is of ongoing importance for humans. Indeed, many archeologists theorize that the first colonization of the new world was due to pre-Aleut fisher-gatherers swimming along the Pacific "kelp highway" during the last ice age (Pringle 2008). Studies of brown algae beds and its ecology aided our understanding of the key mechanisms of food webs and top-down trophic processes of regulation mechanisms (Welch and Graham 2004).

Bottom-up regulation has been studied in salt deserts in ephemeral alkaline lakes. Allelopathy, its traditional breeding and biotechnological integration, leads us toward understanding of the salinization and desalinization process. It is basically a chemical interaction between living organisms like plants. To exploit allelopathy, we need to study the ecology of competitive bushes in salt semi-desert areas (Woodell et al. 1969; Charley and West 1975). Manipulations of allelochemicals can be managed by root exudation. High concentrations of chemicals are leached out by well-structured irrigation systems (Jabran et al. 2015).

3.6 Adaptation of Halophytes to Resist Salinity

Different mechanisms of adaptation have been recognized by studies relating to plant salt tolerance. Adaptations include compartmentalization of ions, production of osmolytes, responses of germination, osmotic adaptation, succulence, enzyme responses, salt excretion, and genetic control (Koyro et al. 2011). Halophytes adapt themselves by establishment of compartments at a cellular level, specifically in vacuoles, where sodium and chloride ions play key role in osmotic adjustment (Flowers et al. 1986). Halophytes secrete a myriad of osmotically compatible solutions which are responsible for osmotic adjustments. Likewise, production of monosaccharides and disaccharides assists halophytes to absorb more water under saline stress conditions (Weber 2009). Potential xerophytic halophytes such as *Haloxylon ammodendron* and *Zygophyllum xanthoxylum* are reported to absorb of high quantities of ions and retain a great amount of water (Wang et al. 2004). Abiotic stress directly interacts with the soil and affects germination rates, which ultimately determine the fate of seed to germinate and the health of the plants. Further rainfall significantly assists in triggering the growth of seedlings through dilution of soil salinity (El-Keblawy et al. 2020). Plant systemic immune systems develop regulatory networks to mitigate salinity stress, seen in ROS production in response to cytotoxicity. In defense processes plants synthesize vital enzymes such as peroxidase, superoxide dismutase, and catalase in turn regulating ROS signaling pathways (Brito et al. 2021). Salt excretion is well adapted by halophytes to balance salinity stress. Similarly, halophytes prefer to excrete salts by glandular tissues present on green leaf. These glands excrete excessive sodium and chloride ions and heavy metals (Wang et al. 2014). Many genes are activated by the influence of salinity stress resulting in product expression of zeaxanthin oxidase, ABA-aldehyde, and 9-cis-epoxycarotenoid dioxygenase (Gorham 1995).

Eco-physiological processes used by tissues and cells of halophytes are common in vascular plants. Halophytes' versatility and phylogeny incorporate novel organisms, salt pan-based agriculture, and phytoremediation (Flowers et al. 2010).

Morphological research and genetic analysis enable exploration of "hidden" species diversity of the sub-halophiles, coastal halophytes, and inland grasslands (Hassan et al. 2016). Understanding diversity leads us to make use of specific genetic factors and background. Biotechnological application furthers development of transgenics. Production of transgenic crops has different applications which potentially enhance both human productivity and ecosystem value. Halophytic transgenic species include developed gourmet vegetables while retaining ornamental qualities or even biofuel uses. Agricultural plants are sensitive to a low amount of sodium chloride. Maas and Grattan (1999) showed soil salinity of lower than 2 dS m^{-1} resulted in a reduction of yield of vegetable species productivity, examples of which are enlisted in Table 3.3.

Basic osmotic adjustment has evolved, despite its polyphyletic origins, through inorganic salt accumulation, NaCl in the vacuole, and organic solutes in cell cytoplasm. Glycophytes and halophytes have variable function ion-transport systems.

Table 3.3 Reduction of vegetable yield in moderately saline environments

Plant	Yield reduction
Phaseolus vulgaris L.	19%
Capsicum annuum L.	14%
Zea mays L.	12%
Solanum tuberosum L.	12%

Glycophytes evolved under natural selection pressure adapting to retain low sodium concentration in aerial parts of plants (Cheeseman 2015). In halophytes Na/H^+ antiporters are required for Na^+ and H^+ ion uptake; Na^+ ion leakage is prevented by specialized lipid vacuoles (Glenn et al. 1999). *Suaeda maritima* exhibit a large vacuole among halophytes, which occupies 77% of mesophyll cells (Hajibagheri et al. 1984). This feature enables tolerance in higher concentrations of salts with up to 500 mM potential accumulation (Dracup and Greenway 1985). *S. maritima* can bear a concentration of the Na+ in the cell sap that is up to 800 mM.

Salt accumulation varies from species to species. However, a prominent feature of halophytes is their capacity of salt accumulation through a range of strategies (Dajic 1996). Based on different adaptive mechanisms, halophytes are classified as salt excluding, salt excreting, and salt accumulating.

Salt stress can be efficiently relieved by salt exclusion. A low uptake of Na^+ at root cortex leads to Na+ ion exclusion (Davenport et al. 2005). Ultra-filtration mechanisms are possessed by such plants; likewise, *Bruguiera gymnorrhiza*, *Kandelia candel*, *Ceriops candolleana*, and *Rhizophora mucronata* have specific characteristics to establish exclusion.

Internal salt levels can be regulated by salt-excreting plants through their foliar glands. Examples of salt-excreting species are *Acanthus ilicifolius*, *Avicennia marina*, *Avicennia officinalis*, *Aegiceras corniculatum*, and *Avicennia alba*. Such plants release salts via salt bladders. These modified cells release salt on the surface of leaves in a liquid form which subsequently becomes crystallized.

Salt accumulators accumulate high concentration of salt and hence overcome toxicity of salt by succulence development. Examples of this are shown by *Sonneratia acida*, *Lumnitzera racemosa*, *Salvadora persica*, *Sonneratia apetala*, *Sonneratia alba*, *Suaeda nudiflora*, *Sesuvium portulacastrum*, and *Excoecaria agallocha*.

3.7 Halophytes Are Ion Compartmentalization Specialists

Ion compartmentalization mechanisms are used for salt tolerance and have been explored at the sub-cellular level. Fructose-1,6-bisphosphatase is an enzyme confined to chloroplasts which works more effectively in halophytes compared to glycophytes (Bose et al. 2017). Retention of K^+ is achieved without plasma membrane activation. Activation of H^+ ATPase in mesophyll cells of halophytes has been demonstrated by ion influx studies. This strategy is very energy efficient among halophytes (Percey et al. 2016). Na^+ can be extruded effectively via salt overly sensitive (SOS1) Na^+/H^+ antiporters present on plasma membranes. Further Na^+ compartmentalized inside vacuoles via Na+/H+ exchangers (NHX) assists in the movement of excessive Na+ ions into vacuoles (Tuteja 2007). NHX is an antiporter of NA^+/H^+ present on the tonoplast (Yamaguchi et al. 2013).

Z. xanthoxylum is a succulent xerophyte in which the AKT1 (*Arabidopsis* K^+ transporter 1) gene has been identified, which is responsible for effective K^+ uptake in roots. This plant modulates selective uptake of K and Na^+ from roots (Ma et al. 2017). Under salt stress halophytes possess specialized transporters for the homeostasis of ions. In glycophytes salt tolerance can be enhanced by integration of these transporter transgenes (SeNHX1 and PutNHX1 from *Salicornia europaea* and *Puccinellia tenuiflora*). Further cytosolic K^+ retention occurs; Na^+ sequestration takes place in *Arabidopsis thaliana* (Liu et al. 2017).

Halophytes have the ability to tolerate different ionic concentrations of salts. Ionic stress is reduced by the amount of Na^+ which accumulates in the cytosol of cells, though this strategy is confined to plants whose leaves are able to carry out transpiration (Woodrow et al. 2011). True halophytes have developed transport systems which facilitate accumulation of salt in their aerial parts (Dajic 2006). In high concentrations of soil salinity, salt exclusion is achieved by lower permeability of roots (Zhu 2001; Flowers and Hajiagheri 2001). Exclusion of sodium is associated with salt tolerance in glycophytes including barley, rice, and wheat (James 2011). There are a range of different factors which account for exclusion of salts or ions. K^+ loading is of greater priority to that of Na^+ in the xylem, effectively removing salt from upper parts of xylem, leaf sheaths, and stem (Munns 2002).

Sensation of Na^+ ions is detected by receptors of membranes extracellularly, while intracellular sensation is either via enzymes in the cytoplasm which are sensitive to Na^+ ions or by membrane proteins (Woodrow et al. 2011). Survival of halophytes is based on refined strategies. Accumulation of compatible solutes under high salt stress occurs (Lee et al. 2008). Moreover, leaf tissues of halophytes are adopted to accumulate high levels of salt ions. This adaptation is crucial for the generation of a water gradient potential along roots and shoots for the maintenance of water flux throughout plants (Silveira et al. 2009; Herbst 2001).

3.8 Halophyte Screening

Halophyte planting under similar environmental conditions enables effective comparison (Pasternak 1990). Observations screened over 3 sequential years allow visual observations to be made. Hence 78 plant species showed very good growth even when irrigated with 100% seawater, whereas a further 22 species found to have best growth at 15% seawater were identified (Ventura et al. 2015).

A database named HALOPH was initialized and published by Aronson in 1989. HALOPH included around 1560 species of plants based on the capacity of plant to tolerate salt concentrations of 80 mM NaCl with electrical conductivity 7.8 dSm^{-1}.

Thus plants whose growth is inhibited by salt concentration more than 80 mM are termed as glycophytes (Aronson 1989). Halophyte species were able to grow when exposed to irrigation water with an electrical conductivity of 7–8 dS m^{-1}. HALOPH was dedicated to provide economic uses of different plant species. The database was modified from the code developed by G.E Wickens and coworkers for the survey of economic plants for semi-arid and arid lands (Wickens 2013). Aronson's data was extended (Menzel and Lieth 2003). An interactive version of the HALOPH database has been compiled recently and can be found at http://www.sussex.ac.uk/affiliates/halophytes (accessed 19 September 2019).

Multiple applications of halophytes are enlisted in Table 3.4.

Scientists collaborate in research efforts on cultivation of halophyte crops, from areas of extremities, with ample rainfall or from dry regions, encountering the inland salt pans and salinity problems in coastal areas.

Different research groups around the globe particularly the Rozema group in the Netherlands (Katschnig et al. 2013; de Vos et al. 2013 ; Rozema and Schat 2013; Koyro et al. 2011), Papenbrock in Germany, Abdelly in Tunisia (Buhmann and Papenbrock 2013), Ksouri et al. (2012), and the Khan research group from Pakistan (Gul and Khan 2003) present halophytes as cash crops. New species are presented with novel growing techniques. Ornamental cultivation of halophyte plants for floriculture or landscaping is still in its infancy. Efforts are directed for improved performance toward testing of glycophytes under salt stress (Cassaniti et al. 2013). The explicit use of halophytes in chemical, food, and ornamental plant-growing industries has also been explored (Koyro et al. 2011). Plants are best suited for raw material in industries; the foremost application of halophytes is bioremediation. In context, selection of improved varieties of halophytes is required for agricultural production, though they are barely domesticated.

"COST Action" has played significant role in utilization of halophytes with novel features for forthcoming agricultural growers and documented online at http://www.cost.eu/domains_actions/fa/Actions/FA0901 (accessed 19 September 2019).

Table 3.4 Halophytic plant species and their uses

Scientific name	Uses	References
Crithmum maritimum	Vegetable	Meot-Duros and Magne (2009)
	Edible seed oil	Zarrouk et al. (2003)
Inula crithmoides	Food and fodder	Zurayk and Baalbaki (1996)
	Ornamental	Franke (1982)
Atriplex hortensis	Vegetable	Wilson et al. (2000)
Salicornia *Sarcocornia* spp.	Biofuel/oilseed crop	Glenn et al. (1991)
	Bioremediation	Webb et al. (2013) and Shpigel et al. (2013)
	Forage	Glenn et al. (1992) and Imai et al. (2004)
	Vegetable	Ventura and Sagi (2013)
	Probiotics	Sarker et al. (2010)
	Ornamental	Ventura and Sagi (2013)
Pennisetum clandestinum	Biomass	Muscolo et al. (2013)

3.9 Salt Tolerance Mechanism in Halophytes

Halophytes are well adopted to high stress and have developed a myriad of anatomical and morphological features which assist their survival in saline environments (Grigore et al. 2014). Mechanistic approaches including osmolyte regulation, succulence development, antioxidant responses, and transport of ions via selective permeability to regulate redox reactions play pivotal roles in salinity tolerance. Moreover, signal transduction, ROS generation, and detoxification pathways are coordinately linked with salt tolerance mechanisms. Ion homeostasis (influx and efflux) includes salt tolerance in defense mechanisms. Secretion of osmoprotectants such as glycine, betaine, proline, sugars, polyphenol, and inorganic ions plays an essential role in defense mechanisms in halophytes (Patel et al. 2016; Lokhande and Suprasanna 2012). Defensive approaches against salt stress are often accompanied by salt bladders/glands, cross-talk activating genes, and antioxidant induction. The latter often results in antimicrobial activity (Himabindu et al. 2016; Shabala et al. 2014; Khan et al. 2018). Antioxidative enzymes responsible for ROS detoxification assist in deterioration of toxic radicals (Das and Strasser 2013). Furthermore, halophytes are adapted to secrete excess amounts of salt in the form of liquids, which become solid after exposure to external atmosphere and are subsequently visible to the naked eye on the leaf surface in the form of crystals. Cytosol of halophytes has great potential in managing a high ratio of Na+/K+ ions by initiating the transport of excessive salt into vacuoles (Sreeshan et al. 2014; Kronzucker and Britto 2011).

Figure 3.1 shows a breakdown of salt glands within upper layer of epidermal cells along with modified epidermal cells. With plasmodesmata, these modified epidermal cells accumulate salt from surrounding mesophyll cells and secrete salts onto the aerial surface of leaves.

Fig. 3.1 Longitudinal view of salt glands in salt-tolerant plants

3.10 Salt-Responsive Genes and Halophytes

Salicornia brachiate is reported as a hyper-accumulator halophyte native to salt marshes and also copes with various abiotic stresses. This species is well known for having stress-responsive promoters and genes which have unique properties and operate in adverse conditions such as salinity and drought stress (Singh et al. 2014; Jha et al. 2011; Chaturvedi et al. 2012; Udawat et al. 2016; Tiwari et al. 2016). Salt-responsive genes are manipulated as transgenes in plants to combat salinity, for example, in cumin, jatropha, castor, and peanuts (Pandey et al. 2016; Joshi et al. 2011). Additionally, *S. bigelovii* is a functional food due to the presence of nutritious metabolites, sugars, and seeds rich in protein and sulfur content. Moreover non-targeting metabolomics, antioxidant, and scavenging properties of *S. bigelovii* enhance its nutritional value (Mishra et al. 2013; Jha et al. 2011). *Porteresia coarctata* has a polyphyletic linkage with wild rice; the former has been reported to have high adaptability against high salt stress. This species has reserved 15,158 genes to control unraveling key metabolic pathways in the context of saline stress and submergence tolerance (Garg et al. 2014; Jha et al. 2012).

Potential uses of halophytes are summarized in Fig. 3.2.

Figure 3.2 shows potential uses of halophytes and their vital role and versatility in environmental sustainability.

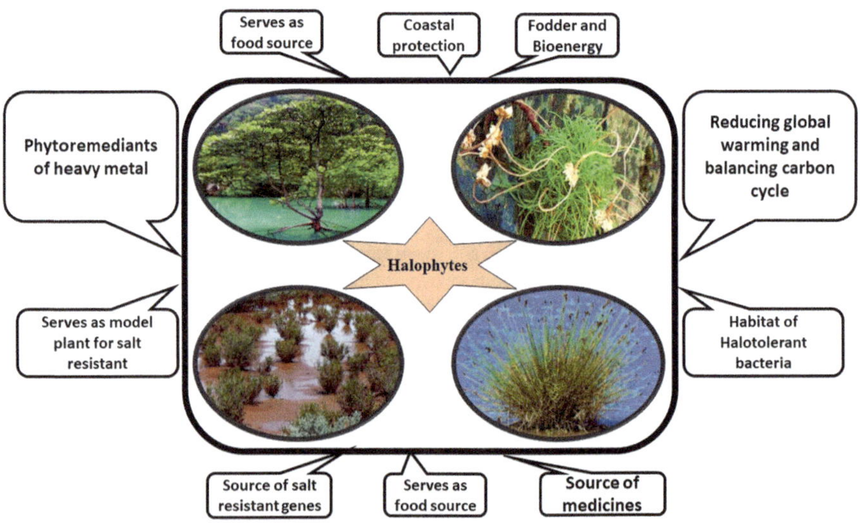

Fig. 3.2 Potential uses of halophytes

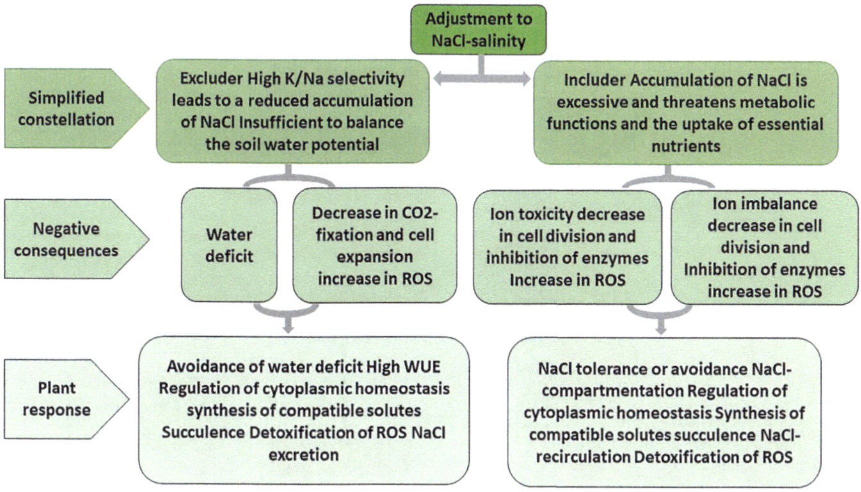

Fig. 3.3 Adaptation of resistance against NaCl salinity

3.11 Treatment of Salinity by Halophytes

In high salt conditions, halophytes are of great importance because of their yield productivity. Satisfactory yield is shown by halophytes under different degrees of salinity (Simpson et al. 2018). The most productive species yield 10–20 t ha biomass after irrigation of seawater; under high salt conditions, oil seed of *S. bigelovii* yields 2 t ha^{-1} of seed containing 31% protein and 28% oil similar to soybean seed oil quality (Glenn et al. 1995).

Figure 3.3 shows a schematic flow of how plants develop resistance to high concentrations of salt/ions and shows us how salinity is regulated going through different stages of simplified constellation, negative consequences, and plant response.

Nutritional barriers may be caused by some halophytes due to partially high antinutritional compounds and high salt contents (Khan et al. 2009). Mechanisms of salinity stress tolerance are detailed in Fig. 3.4.

Figure 3.4 represents a schematic flow of how salt tolerance is established by halophytes; osmotic and ionic imbalance is sensed by receptors which activate different genes in plants.

3.12 Halophyte Contribution to Sustaining Environmental Stability

Halophytes possess various mechanisms in order to survive in saline environments as shown in Fig. 3.4. Further, the development of a salt bladder has been reported in approximately 50% species of halophytes (Flowers and Colmer 2008). The salt

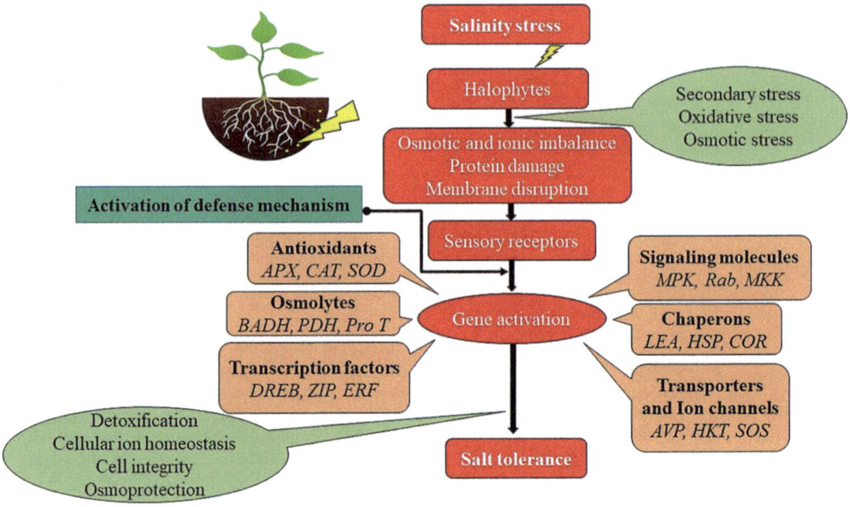

Fig. 3.4 A generalized schematic representation of salinity stress tolerance mechanism in plants

bladder is derived from epidermal glandular hairs or trichomes. These are characteristics of halophytes with the exception of those in the Pooideae family (Amarasinghe and Watson 1988; Ramadan and Flowers 2004). Findings of Shabala and Mackay (2011) depicted that epidermal bladder cells (EBCs) have a salient impact in sodium ion uptake. This process is aided by a larger size of EBCs ten times greater in size than the normal epidermal and mesophyll cells; these bladder cells assist in desalinization by sequestering 1000 times more sodium ions through each EBC channel. Bladder development is conducive to mitigate salinity issues along with water loss reduction (Adams et al. 1998).

Phytoremediation is considered as an effective amelioration strategy by several researchers for calcareous saline-sodic soils against the use of chemical amendments with comparable performance (Singh et al. 1989; Qadir et al. 1996; Ahmad et al. 1990). Besides their positive impact on salt-affected soils, some halophytes can be potentially used as oil seed crops and forages (Qadir et al. 2007). Halophytic plants can also be used to desalinate water and soil, as was firstly suggested by Boyko (1966).

3.13 Future Prospects

Phytoremediation is an environmentally sound technology and is cost-effective for remediation of saline soil, given it is properly developed. In the wake of this issue, applications of halophytes best practice are seen in not only desalinization/remediation but also in the agriculture/agroecological sector as in the forage industry (Erakhrumen and Agbontalor 2007). It is imperative to select plants which have ability to reduce salt concentration from the soil with high biomass yield depicted in Table 3.2. Such plants are selected for phytoremediation ability and tolerate high saline environments. Molecular knowledge of tolerance mechanisms and responses will pave the way for engineered plants that could be the foundation for salt-resistant crop cultivars. This would provide resilience to maintain economic (and ecological) stability. Hence, molecular and physiological studies are essential to reveal underlying mechanisms of stress and resilience processes. Signaling pathways and identification of novel genes responsible for high biomass yield require attention in halophytic species. Producing protocols toward introducing halophytic genes into transgenic crops requires further study and suitable conditions. To date, only a small number of investigations have evaluated potential genetic material in salt-tolerant plants. To combat the economic and food crises, halophyte production must be enhanced.

3.14 Encouraging a Scientific Revolution Integrating Genetic Factors of Halophytes

Genes responsible for salt-tolerant features in halophytes can be introduced into wild plant species by following cutting-edge technique like Crispr-Cas9-mediated transformation technology. Parallel complications, such as genetic misalignments, may be solved with use of computational modeling using R/MATLAB software platforms.

Halophytes not only provide aid in desalinization but can also be of use as the best indicator of natural geological salt deposits. In this regard, indication of wild halophytes provides value for the mining industry. Further mapping studies of these species are encouraged. Moreover, microbial communities associated with halophytes can be used as an alternative source of soil reclamation on one hand and for the agriculture industry on the other hand (Sáenz-Mata et al. 2016). Rhizosperic interaction studies are urgently required to assist this area. The urgency of halophytic research is underlined in that by the year 2050, population of the world is projected to stabilize at around 9.5 billion people (http://faostat3.fao.org/home/E) (accessed 17 October 2019).

3.15 Conclusion

Halophytes are "salt lovers" and are well organized in saline environments. However, there are enormous variations in the adaptability of halophytes toward salinization. Thus, a selection of halophytes adorning hyper-accumulation features, yielding high biomass and greater rates of salt amelioration from the soil, should be preferred. Application of halophytes is greatly required in the locality of arid and semi-arid regions where other measures of soil reclamation and desalination are not economically possible. Scientific revolution is still needed to introduce halophytes on a commercial scale. This will aid in removing barriers in the availability of fresh water and proper economic growth as well (Parnian et al. 2020). Parallel to this, halophytes are considered as the best source of biofuel production. Many plants are reported to have antioxidant and oil content; however more research is needed to detail quantitative and synergistic relations of salts and halophytic oil producers. Development of transgenic plants is the ultimate solution to avoid salinization, water scarcity, and global food insecurity. Research on a selection of desired genes responsible for salt-tolerant mechanism is taking place through RNA sequencing and protein expression. Despite such developments, ascertaining the correct breeding lines is hindered by lethal side effects such as sterility, which cannot be ignored in engineering transgenic cultivars.

Global agriculture must double its productivity in order to feed our population in the future; indeed this is the principal goal of sustainable development. The challenges of attaining sustainable and accelerated growth and food security have been exacerbated by global climate change and extreme weather fluctuations. The main impact of climate change is the highly variable and increasing temperature which will bring changes in precipitation patterns and ultimately increase soil salinity.

References

Adams P, Nelson DE, Yamada et al (1998) Growth and development of *Mesembryanthemum crystallinum* (Aizoaceae). New Phytol 138(2):171–190. https://doi.org/10.1046/j.1469-8137.1998.00111.x

Adolf VI, Jacobsen SE, Shabala S (2013) Salt tolerance mechanisms in quinoa (*Chenopodium quinoa* Willd.). Environ Exp Bot 92:43–54. https://doi.org/10.1016/j.envexpbot.2012.07.004

Ahmad N, Qureshi RH, Qadir M (1990) Amelioration of a calcareous saline-sodic soil by gypsum and forage plants. Land Degrad Dev 2(4):277–284. https://doi.org/10.1002/ldr.3400020404

Amarasinghe V, Watson FLSL (1988) Comparative ultrastructure of microhairs in grasses. Bot J Linn Soc 98(4):303–319. https://doi.org/10.1111/j.1095-8339.1988.tb01705.x

Aronson JA (1989) HALOPH: a data base of salt tolerant plants of the world. Arid Land Studies, University of Arizona, Tucson

Ashraf MY, Ashraf M, Mahmood K et al (2010) Phytoremediation of saline soils for sustainable agricultural productivity. In: Plant adaptation and phytoremediation. Springer, Dordrecht. https://doi.org/10.1007/978-90-481-9370-7_15

Aslam R, Bostansup N, Mariasup M et al (2011) A critical review on halophytes: salt tolerant plants. J Med Plants Res 5(33):7108–7118

Bonales-Alatorre E, Shabala S, Chen ZH et al (2013) Reduced tonoplast fast-activating and slow-activating channel activity is essential for conferring salinity tolerance in a facultative halophyte, quinoa. Plant Physiol 162(2):940–952. https://doi.org/10.1104/pp.113.216572

Bose J, Munns R, Shabala S et al (2017) Chloroplast function and ion regulation in plants growing on saline soils: lessons from halophytes. J Exp Bot 68(12):3129–3143. https://doi.org/10.1093/jxb/erx142

Boyko H (1966) Basic ecological principles of plant growing by irrigation with highly saline or seawater. In: Salinity and aridity. Springer, Dordrecht. https://doi.org/10.1007/978-94-017-6014-0_4

Brito P, Ferreira RA, Martins-Dias S et al (2021) Cerium uptake, translocation and toxicity in the salt marsh halophyte *Halimione portulacoides* (L.), Aellen. Chemosphere:128973. https://doi.org/10.1016/j.chemosphere.2020.128973

Buhmann A, Papenbrock J (2013) An economic point of view of secondary compounds in halophytes. Funct Plant Biol 40(9):952–967. https://doi.org/10.1071/FP12342

Cassaniti C, Romano D, Hop MECM et al (2013) Growing floricultural crops with brackish water. Environ Exp Bot 92:165–175. https://doi.org/10.1016/j.envexpbot.2012.08.006

Charan PD, Sharma KC (2016) Floral diversity of Thar Desert of western Rajasthan, India. J Phytol Res 29(1):55–71

Charley JL, West NE (1975) Plant-induced soil chemical patterns in some shrub-dominated semi-desert ecosystems of Utah. J Ecol 63(3):945–963. https://doi.org/10.2307/2258613

Chaturvedi AK, Mishra A, Tiwari V et al (2012) Cloning and transcript analysis of type 2 metallothionein gene (SbMT-2) from extreme halophyte *Salicornia brachiata* and its heterologous expression in *E. coli*. Gene 499(2):280–287. https://doi.org/10.1016/j.gene.2012.03.001

Cheeseman JM (2015) The evolution of halophytes, glycophytes and crops, and its implications for food security under saline conditions. New Phytol 206(2):557–570. https://doi.org/10.1111/nph.13217

Chen Y, Shen Z, Li X (2004) The use of vetiver grass (*Vetiveria zizanioides*) in the phytoremediation of soils contaminated with heavy metals. Appl Geochem 19(10):1553–1565. https://doi.org/10.1016/j.apgeochem.2004.02.003

Dajic Z (1996) Ekoloska studija halofitske zajednice *Puccinellietum limosae* (Rapcs.) Wend. (Ecological study of halophytic community *Puccinellietum limosae* (Rapcs.) Wend.). PhD thesis, Faculty of Biology, University of Belgrade FRY

Dajic Z (2006) Salt stress. In: Physiology and molecular biology of stress tolerance in plants. Springer, Dordrecht

Dansereau P (1957) Biogeography. An ecological perspective. Ronald Press, New York

Das AB, Strasser RJ (2013) Salinity-induced genes and molecular basis of salt-tolerant strategies in mangroves. In: Molecular stress physiology of plants. Springer. https://doi.org/10.1007/987-81-322-0807-5_3

Davenport R, James RA, Zakrisson-Plogander A et al (2005) Control of sodium transport in durum wheat. Plant Physiol 137(3):807–818. https://doi.org/10.1104/pp.104.057307

de Vos AC, de Almeida BR, Guerra CC et al (2013) Developing and testing new halophyte crops: a case study of salt tolerance of two species of the Brassicaceae, *Diplotaxis tenuifolia* and *Cochlearia officinalis*. Environ Exp Bot 92:154–164. https://doi.org/10.1016/j.envexpbot.2012.08.003

Devi S, Nandwal AR et al (2017) Phytoremediation potential of some halophytic species for soil salinity. Int J Phytoremed 18(7):693–696. https://doi.org/10.1080/15226514.2015.1131229

Dracup MNH, Greenway H (1985) A procedure for isolating vacuoles from leaves of the halophyte *Suaeda maritima*. Plant Cell Environ 8(2):149–154. https://doi.org/10.1111/j.1365-3040.1985.tb01222.x

El-Keblawy A, Elnaggar A, Tammam A et al (2020) Seed provenance affects salt tolerance and germination response of the habitat-indifferent *Salsola drummondii* halophyte in the arid Arabian deserts. Flora 266:151592. https://doi.org/10.1016/j.flora.2020.151592

Epstein E, Norlyn JD, Rush DW et al (1980) Saline culture of crops: a genetic approach. Science 210(4468):399–404. https://doi.org/10.1126/science.210.4468.399

Erakhrumen AA, Agbontalor A (2007) Phytoremediation: an environmentally sound technology for pollution prevention, control and remediation in developing countries. Educ Res Rev 2(7):151–156

Flowers TJ (1989) Plants under stress: biochemistry, physiology and ecology and their application to plant improvement 39. Cambridge University Press

Flowers TJ, Colmer TD (2008) Salinity tolerance in halophytes. New Phytol 179(4):945–963. https://doi.org/10.1111/j.1469-8137.2008.02531.x

Flowers TJ, Hajibagheri MA (2001) Salinity tolerance in *Hordeum vulgare*: ion concentrations in root cells of cultivars differing in salt tolerance. Plant Soil 231:1–9. https://doi.org/10.1023/A:1010372213938

Flowers TJ, Hajibagheri MA, Clipson NJW (1986) Halophytes. Q Rev Biol 61(3):313–337 https://www.jstor.org/stable/2826772. Accessed 7 Feb 2021

Flowers TJ, Galal HK, Bromham L (2010) Evolution of halophytes: multiple origins of salt tolerance in land plants. Funct Plant Biol 37(7):604–612. https://doi.org/10.1071/FP09269

Franke W (1982) Vitamin C in sea fennel (*Crithmum maritimum*) an edible wild plant. Econ Bot 36(2):163–165. https://doi.org/10.1007/BF02858711

Garg R, Verma M, Agrawal S et al (2014) Deep transcriptome sequencing of wild halophyte rice, *Porteresia coarctata*, provides novel insights into the salinity and submergence tolerance factors. DNA Res 21(1):69–84. https://doi.org/10.1093/dnares/dst042

Glenn EP, O'Leary JW, Watson MC et al (1991) *Salicornia bigelovii* Torr.: an oilseed halophyte for seawater irrigation. Science 251(4997):1065–1067. https://doi.org/10.1126/science.251.4997.1065

Glenn EP, Coates WE, Riley JJ et al (1992) *Salicornia bigelovii* Torr.: a seawater-irrigated forage for goats. Anim Feed Sci Technol 40(1):21–30

Glenn E, Hicks N, Riley J et al (1995) Seawater irrigation of halophytes for animal feed. Environmental Research Laboratory, Tucson. ISBN: 0824796640

Glenn EP, Brown JJ, Blumwald E (1999) Salt tolerance and crop potential of halophytes. Crit Rev Plant Sci 18(2):227–255. https://doi.org/10.1080/07352689991309207

Gorham J (1995) Mechanism of salt tolerance of halophytes. In: Halophytes and biosaline agriculture. Marcel Dekker, New York pp 207–223

Greenway H, Munns R (1980) Mechanisms of salt tolerance in nonhalophytes. Annu Rev Plant Physiol 31(1):149–190. https://doi.org/10.1146/annurev.pp.31.060180.001053

Grigore MN, Ivanescu L, Toma C (2014) Halophytes: an integrative anatomical study. Springer. ISBN: 978-3-319-05729-3

Gul B, Khan MA (2003) Saline agriculture: Promises and prospects for future agriculture in degraded saline lands. In: Technology & development in the new millennium. University of Karachi, Karachi

Hajibagheri MA, Hall JL, Flowers TJ (1984) Stereological analysis of leaf cells of the halophyte *Suaeda maritima* (L.) Dum. J Exp Bot 35(10):1547–1557. https://doi.org/10.1093/jxb/35.10.1547

Hassan MA, Chaura J, López-Gresa MP et al (2016) Native-invasive plants vs. halophytes in Mediterranean Salt Marshes: stress tolerance mechanisms in two related species. Front Plant Sci. https://doi.org/10.3389/fpls.2016.00473

Herbst DB (2001) Gradients of salinity stress, environmental stability and water chemistry as a templet for defining habitat types and physiological strategies in inland salt waters. Hydrobiologia 466:209–219. https://doi.org/10.1023/A:1014508026349

Himabindu Y, Chakradhar T, Reddy MC et al (2016) Salt-tolerant genes from halophytes are potential key players of salt tolerance in glycophytes. Environ Exp Bot 124:39–63. https://doi.org/10.1016/j.envexpbot.2015.11.010

Imai H, Kinoshita M, Ohnishi M (2004) Chemical characterization of glycerolipids and cerebrosides in a halophytic plant, *Salicornia europaea* L. J Oleo Sci 53(7):337–341

Jabran K, Mahajan G, Sardana V et al (2015) Allelopathy for weed control in agricultural systems. Crop Prot 72:57–65. https://doi.org/10.1016/j.cropro.2015.03.004

Jacobsen SE, Mujica A, Jensen CR (2003) The resistance of quinoa (*Chenopodium quinoa* Willd.) to adverse abiotic factors. Food Rev Int 19(1–2):99–109. https://doi.org/10.1081/FRI-120018872

James C (2011) Global status of commercialized biotech/GM crops, 2011. Brief No. 43. ISAAA, Ithaca. ISBN: 978-1-892456-52-4

Jha B, Sharma A, Mishra A (2011) Expression of SbGSTU (tau class glutathione S-transferase) gene isolated from *Salicornia brachiata* in tobacco for salt tolerance. Mol Biol Rep 38(7):4823–4832. https://doi.org/10.1007/s11033-010-0625-x

Jha B, Singh NP, Mishra A (2012) Proteome profiling of seed storage proteins reveals the nutritional potential of *Salicornia brachiata* Roxb., an extreme halophyte. J Agric Food Chem 60(17):4320–4326. https://doi.org/10.1021/jf203632v

Jolly ID, McEwan KL, Holland KL (2008) A review of groundwater–surface water interactions in arid/semi-arid wetlands and the consequences of salinity for wetland ecology. Ecohydrology 1(1):43–58. https://doi.org/10.1002/eco.6

Joshi M, Mishra A, Jha B (2011) Efficient genetic transformation of *Jatropha curcas* L. by microprojectile bombardment using embryo axes. Ind Crop Prod 33(1):67–77. https://doi.org/10.1016/j.indcrop.2010.09.002

Katschnig D, Broekman R, Rozema J (2013) Salt tolerance in the halophyte *Salicornia dolichostachya* Moss: growth, morphology and physiology. Environ Exp Bot 92:32–42. https://doi.org/10.1016/j.envexpbot.2012.04.002

Kefu Z, Hai F, Ungar IA (2002) Survey of halophyte species in China. Plant Sci 163(3):491–498

Khan MA, Ansari R, Ali H et al (2009) *Panicum turgidum*, a potentially sustainable cattle feed alternative to maize for saline areas. Agric Ecosyst Environ 129(4):542–546. https://doi.org/10.1016/j.agee.2008.10.014

Khan M, Zaman G, Jalil F et al (2018) Review on antimicrobial applications of silver nanoparticles and use of plant extracts for its synthesis. Int J Biosci 12(5):240–248. https://doi.org/10.12692/ijb/12.5.240-248

Koyro HW, Khan MA, Lieth H (2011) Halophytic crops: a resource for the future to reduce the water crisis. Emirates J Food Agric 23(1):01–16. https://doi.org/10.9755/ejfa.v23i1.5308

Kronzucker HJ, Britto DT (2011) Sodium transport in plants: a critical review. New Phytol 189(1):54–81. https://doi.org/10.1111/j.1469-8137.2010.03540.x

Ksouri R, Smaoui A, Isoda H et al (2012) Utilization of halophyte species as new sources of bioactive substances. J Arid Land Stud 22:41–44

Lee G, Carrow RN, Duncan, RR et al (2008) Synthesis of organic osmolytes and salt tolerance mechanisms in *Paspalum vaginatum*. Environ Exp Bot 63(1–3):19–27. https://doi.org/10.1016/j.envexpbot.2007.10.009

Liu XJ, Li WQ, Yang YM et al (2002) Resource agriculture in saline regions: towards sustainability. In: Halophyte utilization and regional sustainable development of agriculture. Climate Press, Beijing

Liu X, Cai S, Wang G et al (2017) Halophytic NHXs confer salt tolerance by altering cytosolic and vacuolar K+ and Na+ in Arabidopsis root cell. Plant Growth Regul 82(2):333–351. https://doi.org/10.1007/s10725-017-0262-7

Llerena VFA (1994) Massive propagation of halophytes (*Distichlis spicata* and *Tamarix* spp.) on the highly saline-alkaline soils in the ex-Lake Texcoco, Mexico. In: Halophytes as a resource for livestock and for rehabilitation of degraded lands. Tasks for Vegetation Science 32. https://doi.org/10.1007/978-94-011-0818-8_30

Lokhande VH, Suprasanna P (2012) Prospects of halophytes in understanding and managing abiotic stress tolerance. In: Environmental adaptations and stress tolerance of plants in the era of climate change. Springer, New York. https://doi.org/10.1007/978-1-4614-0815-4_2

Ma Q, Hu J, Zhou XR et al (2017) ZxAKT1 is essential for K+ uptake and K+/Na+ homeostasis in the succulent xerophyte *Zygophyllum xanthoxylum*. Plant J 90(1):48–60. https://doi.org/10.1111/tpj.13465

Maas EV, Grattan SR (1999) Crop yields as affected by salinity. In: Agricultural drainage, volume 38. https://doi.org/10.2134/agronmonogr38.c3

Mahmood K, Malik KA, Lodhi MAK et al (1994) Soil–plant relationships in saline wastelands: vegetation, soils, and successional changes, during biological amelioration. Environ Conserv 21(3):236–241. https://doi.org/10.1017/S037689290003321X

Menzel U, Lieth H (2003) Halophyte database Vers. 2.0 in alphabetical order including some updates. In: Cash crop halophytes: recent studies. Springer Science+ Business Media, Dordrecht, pp 221–250

Meot-Duros L, Magne C (2009) Antioxidant activity and phenol content of *Crithmum maritimum* L. leaves. Plant Physiol Biochem 47(1):37–41. https://doi.org/10.1016/j.plaphy.2008.09.006
Mishra A, Tanna B (2017) Halophytes: potential resources for salt stress tolerance genes and promoters. Front Plant Sci. https://doi.org/10.3389/fpls.2017.00829
Mishra A, Joshi M, Jha B (2013) Oligosaccharide mass profiling of nutritionally important *Salicornia brachiata*, an extreme halophyte. Carbohydr Polym 92(2):1942–1945. https://doi.org/10.1016/j.carbpol.2012.11.055
Munns R (2002) Comparative physiology of salt and water stress. Plant Cell Environ 25(2):239–250. https://doi.org/10.1046/j.0016-8025.2001.00808.x
Munns R, Tester M (2008) Mechanisms of salinity tolerance. Annu Rev Plant Biol 59:651–681. https://doi.org/10.1146/annurev.arplant.59.032607.092911
Muscolo A, Panuccio MR, Eshel A (2013) Ecophysiology of Pennisetum clandestinum: a valuable salt tolerant grass. Environ Exp Bot 92:55–63
Pandey S, Patel MK, Mishra A et al (2016) In planta transformed cumin (*Cuminum cyminum* L.) plants, overexpressing the SbNHX1 gene showed enhanced salt endurance. Plos One 11(7):e0159349. https://doi.org/10.1371/journal.pone.0159349
Parnian A, Furze JN (2021) Vertical phytoremediation of wastewater using *Vetiveria zizanioides* L. Environ Sci Pollut Res. https://doi.org/10.1007/s11356-020-11906-6
Parnian A, Furze JN, Parnian A (2020) Water out of waste–solar-desalination-humidification-dehumidification auxiliaries processing extractive industrial operations. Water Product J 1(1):31–38. https://doi.org/10.22034/WPJ.2020.119474
Pasternak D (1990) Fodder production with saline water. Project report, January 1982–December 1989. BGUN-ARI-35-90
Patel MK, Mishra A, Jha B (2016) Untargeted metabolomics of halophytes. In: Marine OMICS. CRC Press
Percey WJ, Shabala L, Wu Q et al (2016) Potassium retention in leaf mesophyll as an element of salinity tissue tolerance in halophytes. Plant Physiol Biochem 109:346–354. https://doi.org/10.1016/j.plaphy.2016.10.011
Pringle RM (2008) Elephants as agents of habitat creation for small vertebrates at the patch scale. Ecology 89(1):26–33. https://doi.org/10.1890/07-0776.1
Qadir M, Qureshi RH, Ahmad N (1996) Reclamation of a saline-sodic soil by gypsum and *Leptochloa fusca*. Geoderma 74(3-4):207–217. https://doi.org/10.1016/S0016-7061(96)00061-4
Qadir M, Oster JD, Schubert S et al (2007) Phytoremediation of sodic and saline-sodic soils. Adv Agron 96:197–247. https://doi.org/10.1016/S0065-2113(07)96006-X
Rajaram C, Godara SK, Ruchi W (2006) Nutritive contents of some halophytes of IGNP Phase-I Command area, Loonkaransar, Bikaner, Western Rajasthan. J Phytol Res 19(1):151–152
Ramadan T, Flowers TJ (2004) Effects of salinity and benzyl adenine on development and function of microhairs of *Zea mays* L. Planta 219(4):639–648. https://doi.org/10.1007/s00425-004-1269-7
Rausch T, Kirsch M, Löw R et al (1996) Salt stress responses of higher plants: the role of proton pumps and Na^+/H^+-antiporters. J Plant Physiol 148(3–4):425–433. https://doi.org/10.1016/S0176-1617(96)80275-6
Razzaghi F, Ahmadi SH, Adolf VI et al (2011) Water relations and transpiration of Quinoa (*Chenopodium quinoa* Willd.) under salinity and soil drying. J Agron Crop Sci 197(5):348–360. https://doi.org/10.1111/j.1439-037X.2011.00473.x
Rozema J, Schat H (2013) Salt tolerance of halophytes, research questions reviewed in the perspective of saline agriculture. Environ Exp Bot 92:83–95. https://doi.org/10.1016/j.envexpbot.2012.08.004
Sáenz-Mata J, Palacio-Rodríguez R, Sánchez-Galván H et al (2016) Plant growth promoting rhizobacteria associated to halophytes: potential applications in agriculture. In: Sabkha ecosystems. Springer, Cham
Sarker SK, Park SR, Kim GM et al (2010) Hamcho (*Salicornia herbacea*) with probiotics as alternative to antibiotic for broiler production. J Med Plants Res 4(5):415–420

Sauer JRM, Menjivar JA, Burns BA et al (1996) Snack food coating using supercritical fluid spray. US Patent 5,520,942

Serrano R (1996) Salt tolerance in plants and microorganisms: toxicity targets and defense responses. Int Rev Cytol 165:1–52. https://doi.org/10.1016/S0074-7696(08)62219-6

Seydehmet J, Lv GH, Nurmemet I et al (2018) Model prediction of secondary soil salinization in the Keriya Oasis, Northwest China. Sustainability 10(3):656. https://doi.org/10.3390/su10030656

Shabala S, Mackay A (2011) Ion transport in halophytes. Adv Bot Res 57:151–199. https://doi.org/10.1016/B978-0-12-387692-8.00005-9

Shabala S, Bose J, Hedrich R (2014) Salt bladders: do they matter? Trends Plant Sci 19(11):687–691. https://doi.org/10.1016/j.tplants.2014.09.001

Shah JP (2018) Taxonomic Status of halophytes in coastal Kachchh District, Gujarat. J Adv Plant Sci 1:1–8

Sharma V, Ramawat KG (2014) Salt stress enhanced antioxidant response in callus of three halophytes (*Salsola baryosma*, *Trianthema triquetra*, *Zygophyllum simplex*) of Thar Desert. Biologia 69(2):178–185. https://doi.org/10.2478/s11756-013-0298-8

Shpigel M, Ben-Ezra D, Shauli L et al (2013) Constructed wetland with *Salicornia* as a biofilter for mariculture effluents. Aquaculture 412:52–63. https://doi.org/10.1016/j.aquaculture.2013.06.038

Silveira JAG, Araújo SAM, Lima JPMS et al (2009) Roots and leaves display contrasting osmotic adjustment mechanisms in response to NaCl-salinity in *Atriplex nummularia*. Environ Exp Bot 66(1):1–8. https://doi.org/10.1016/j.envexpbot.2008.12.015

Simpson CR, Franco JG, King SR et al (2018) Intercropping halophytes to mitigate salinity stress in watermelon. Sustainability 10(3). https://doi.org/10.3390/su10030681

Singh K, Chauhan HS, Rajput DK et al (1989) Report of a 60 month study on litter production, changes in soil chemical properties and productivity under Poplar (*P. deltoides*) and Eucalyptus (*E. hybrid*) interplanted with aromatic grasses. Agrofor Syst 9(1):37–45. https://doi.org/10.1007/BF00120154

Singh N, Mishra A, Jha B (2014) Ectopic over-expression of peroxisomal ascorbate peroxidase (SbpAPX) gene confers salt stress tolerance in transgenic peanut (*Arachis hypogaea*). Gene 547(1):119–125. https://doi.org/10.1016/j.gene.2014.06.037

Squires VR (1994) Overview of problems and prospects for utilizing halophytes as a resource for livestock and for rehabilitation of degraded lands. In: Halophytes as a resource for livestock and for rehabilitation of degraded lands. Springer, Dordrecht, pp 1–6. https://doi.org/10.1007/978-94-011-0818-8_1

Sreeshan A, Meera SP, Augustine A (2014) A review on transporters in salt tolerant mangroves. Trees 28(4):957–960. https://doi.org/10.1007/s00468-014-1034-x

Stocker O (1928) Das halophytenproblem. In: Ergebnisse der Biologie. Springer, Berlin/Heidelberg

Tiwari V, Patel MK, Chaturvedi AK et al (2016) Functional characterization of the Tau Class Glutathione-S-Transferases Gene (SbGSTU) promoter of *Salicornia brachiata* under salinity and osmotic stress. PLoS One 11(2):e0148494. https://doi.org/10.1371/journal.pone.0148494

Tuteja N (2007) Chapter 24 mechanisms of high salinity tolerance in plants. Methods Enzymol 428:419–438. https://doi.org/10.1016/S0076-6879(07)28024-3

Udawat P, Jha RK, Sinha D et al (2016) Overexpression of a cytosolic abiotic stress responsive universal stress protein (SbUSP) mitigates salt and osmotic stress in transgenic tobacco plants. Front Plant Sci 7:518. https://doi.org/10.3389/fpls.2016.00518

Ventura Y, Sagi M (2013) Halophyte crop cultivation: the case for *Salicornia* and *Sarcocornia*. Environ Exp Bot 92:144–153

Ventura Y, Eshel A, Pasternak D et al (2015) The development of halophyte-based agriculture: past and present. Ann Bot 115(3):529–540. https://doi.org/10.1093/aob/mcu173

Vera-Estrella R, Barkla BJ, García-Ramírez L et al (2005) Salt stress in *Thellungiella halophila* activates Na+ transport mechanisms required for salinity tolerance. Plant Physiol 139(3):1507–1517. https://doi.org/10.1104/pp.105.067850

Von Sengbusch P (2003) Halophytes. Botanik Online. University of Hamburg
Wang B, Lüttge U, Ratajczak R (2001) Effects of salt treatment and osmotic stress on V-ATPase and V-PPase in leaves of the halophyte *Suaeda salsa*. J Exp Bot 52(365):2355–2365. https://doi.org/10.1093/jexbot/52.365.2355
Wang S, Wan C, Wang Y et al (2004) The characteristics of Na+, K+ and free proline distribution in several drought-resistant plants of the Alxa Desert, China. J Arid Environ 56(3):525–539. https://doi.org/10.1016/S0140-1963(03)00063-6
Wang HL, Tian CY, Jiang L et al (2014) Remediation of heavy metals contaminated saline soils: a halophyte choice? Environ Sci Technol 48(1):21–22. https://doi.org/10.1021/es405052j
Webb JM, Quintã R, Papadimitriou S et al (2013) The effect of halophyte planting density on the efficiency of constructed wetlands for the treatment of wastewater from marine aquaculture. Ecol Eng 61:145–153. https://doi.org/10.1016/j.ecoleng.2013.09.058
Weber DJ (2009) Adaptive mechanisms of halophytes in desert regions. In: Salinity and water stress. Springer, Dordrecht. https://doi.org/10.1007/978-1-4020-9065-3_18
Welch RM, Graham RD (2004) Breeding for micronutrients in staple food crops from a human nutrition perspective. J Exp Bot 55(396):353–364. https://doi.org/10.1093/jxb/erh064
Wickens GE (2013) Ecophysiology of Economic plants in arid and semi-arid lands. Springer Science & Business Media. ISBN: 978-3-662-03700-3
Wilson C, Lesch SM, Grieve CM (2000) Growth stage modulates salinity tolerance of New Zealand spinach (*Tetragonia tetragonioides*, Pall.) and red orach (*Atriplex hortensis* L.). Ann Bot 85(4):501–509. https://doi.org/10.1006/anbo.1999.1086
Woodell SR, Mooney HA, Hill AJ (1969) The behaviour of *Larrea divaricata* (creosote bush) in response to rainfall in California. J Ecol 57(1):37–44
Woodrow P, Pontecorvo G, Ciarmiello LF et al (2011) Ttd1a promoter is involved in DNA–protein binding by salt and light stresses. Mol Biol Rep 38(6):3787–3794. https://doi.org/10.1007/s11033-010-0494-3
Xia S, Song Z, Li Q et al (2020) Distribution, sources, and decomposition of soil organic matter along a salinity gradient in estuarine wetlands characterized by C: N ratio, δ13C-δ15N, and lignin biomarker. Glob Chang Biol 27(2):417–434. https://doi.org/10.1111/gcb.15403
Yamaguchi T, Hamamoto S, Uozumi N (2013) Sodium transport system in plant cells. Front Plant Sci 4:410. https://doi.org/10.3389/fpls.2013.00410
Zarrouk M, El Almi H, Youssef NB et al (2003) Lipid composition of seeds of local halophytes: *Cakile maritima*, *Zygophyllum album* and *Crithmum maritimum*. In: Cash crop halophytes: recent studies. Springer, Dordrecht, pp 121–124
Zhu JK (2001) Plant salt tolerance. Trends Plant Sci 6(2):66–71. https://doi.org/10.1016/S1360-1385(00)01838-0
Zhu JK, Hasegawa PM, Bressan RA et al (1997) Molecular aspects of osmotic stress in plants. Crit Rev Plant Sci 16(3):253–277. https://doi.org/10.1080/07352689709701950
Zurayk RA, Baalbaki R (1996) *Inula crithmoides*: a candidate plant for saline agriculture. Arid Land Res Manag 10(3):213–223. https://doi.org/10.1080/15324989609381436

Chapter 4
Drilling Waste Management Based on New Methods of Bioremediation and Solar Desalination

Amir Parnian, James N. Furze, and Amin Parnian

Abstract Drill cutting and mud waste management is the principal concern in gas and oil drilling operations. Waste management is usually based on engineering activities to reduce waste and its impact on the surrounding environment. Oil and gas drilling waste management has two operational parts which include water recovery and waste treatment. Traditional methods are costly and do not alleviate environmental concerns. A new method is proposed and organized to solve issues of pollution. The methods have been applied in Iran where there is particular demand due to high levels of pollution. The novel method is a complex of different processes that require optimization in order that they can be combined in an applied system of drilling waste management operations. Drill cuttings and mud waste are dumped into a newly designed corral/waste pit to separate water for purification and condense solids. Water is recovered by chemical treatments in cyclonic ponds and sent to the drilling rig. Solids remaining in the corral/waste pit are biologically dried by adding composted material, and the dried solid bioremediated material undergoes a process of co-composting. Waters of the treatments are sent to a solar desalination humidification-dehumidification (SDHDH) process which produces an improved purity of water and salt. The method was used to reduce the environmental impact of the oil well drilling operations in Iran and is approved by the Iranian Department

A. Parnian (✉)
National Salinity Research Center (NSRC), Agricultural Research Education and Extension Organization (AREEO), Yazd, Iran
e-mail: amir.parnian86@gmail.com

J. N. Furze
Royal Geographical Society (with the Institute of British Geographers), London, UK

Laboratory of Biotechnology and Valorization of Natural Resources,
Faculty of Sciences-Agadir, Department of Biology, Ibn Zohr University, Agadir, Morocco

Control and Systems Engineering Department, University of Technology-Iraq, Baghdad, Iraq
e-mail: james.n.furze@gmail.com; jamesfurze@hotmail.com

A. Parnian
Young Researchers Club, Masjed Soleyman Branch, Islamic Azad University,
Masjed Soleyman, Iran
e-mail: parnian.amin1988@gmail.com

J. N. Furze et al. (eds.), *Earth Systems Protection and Sustainability*,
https://doi.org/10.1007/978-3-030-85829-2_4

of Environment; functional results were verified by nationalized commercial groups of Iran. The method produces soil, salt, and fresh water and is shown to leave lower immediate adverse effects than conventional waste treatment in the local environment. Application of the methods in additional locations will offer generation of resources to enrich both human communities and ecosystem services in the face of global industrialization and pollution.

Keywords Drilling waste management · Organic absorbent · Biological drying · Composting · Bioremediation

4.1 Introduction: The Source and Impact of Petroleum Hydrocarbon on the Environment

Human industrial activity is the major source of petroleum hydrocarbons which cause disastrous contamination in the environment (Yi et al. 2016). Pollution with persistent contamination is reviled by all forms of life in all manner of ecosystems and has an enormous impact in reducing environmental diversity. Further, petroleum hydrocarbon contamination renders the ecosystem unstable and consequently dangerously vulnerable to normal changes and stresses imposed by biotic and abiotic complexes (Cheung and Kinkle 2001; Dojka et al. 1998; Kirk et al. 2005).

Petroleum hydrocarbon contamination is the result of human industrial and recreational activity. The operations related to oil and gas exploration, production, storage, refinery, and transportation lead to severe environmental pollution. Oil- and gas-related economical operations involving waste generation are associated with risks to the environment. Hazardous risks occur in the surrounding soil, air, and aquatic environments (Sharif et al. 2017). Petroleum hydrocarbon contamination has a carcinogenic effect for both humans and animals, and species diversity is being lost and prevented from regeneration (Singh and Shikha 2019; Sharif et al. 2017); further the pollution imposes some change in DNA of other forms of life (Das and Chandran 2011). The gravity of the problem is immense and justifies limiting the effects of petroleum hydrocarbon contamination. This kind of contamination puts humans in dangerous situations of environmental disorder. To avoid pollution of the environment and rectify all acts related to petroleum hydrocarbon contaminations, remediation is vital for the sustainability of the ecosystem, including that which surrounds human communities.

4.2 Drill Cuttings and Mud Waste Produced in Extractive Operations

The majority of waste generated from oil and gas well drilling operations come from exceeding or disqualified drilling fluids and drill cuttings (Onwukwe and Nwakaudu 2012). Drill cuttings are chips and particles, which become detached from the soil and geological structures during the drilling process and after drill sites are processed. Slack is removed from the drilling mud with special equipment in a physical process of solids control. Most of the mud is recirculated back to the drilling operation through the solids control process and drilling mud recycling system. Drill cuttings separated by solids control equipment are usually sent to a waste management process (Ball et al. 2012).

Drill cutting and mud waste characterization is dependent on the mud composition. There are two basic types of muds (fluids) produced in oil and gas drilling operations, water-based fluids and oil-based fluids (Caenn et al. 2011; Khodja et al. 2007); these are constituted as follows:

- Water-based systems (with freshwater and saltwater as the base for mud production) are water-based drilling fluids. These are the most widely used and are generally less expensive than other mud systems to process.
- Oil- or synthetic-based systems include the non-aqueous-based drilling fluids and have gasoline at the base. These have an oil or a synthetic base fluid in their fluid matrix phase and brine as the dispersed phase. These kinds of mud comprise 5–10% of the total composition of a common drilling operation well fluid composition.

4.3 Why Should We Manage Drill Cuttings and Mud Waste?

Drill cuttings and mud waste composition are dependent on mud composition and formation material composition. The wastes are often contaminated by harmful concentrations of trace elements and have an additional high concentration of petroleum hydrocarbons. Left unmanaged the drilling wastes flow to the surrounding environment and, as a consequence, pollute soil, surface water, and groundwater disabling the environments normal functions. As a result of public outcry, regional and national authorities and oil well drilling companies inevitably use some management techniques to reduce the drill cuttings and mud waste impacts on the environments (Caenn et al. 2011; Onwukwe and Nwakaudu 2012).

4.3.1 How Can Drilling Cutting and Mud Waste Be Managed?

Drilling operations of gas and oil wells are present in different environments. These are divided into onshore and offshore operations; produced wastes should be managed to avoid long-term environmental impacts (Bybee 2002; Veil 2002). Drilling waste management encompasses the following categories and parameterization: equipment, drilling operation and handling rates, costs, authorities, risks, and environmental impacts (Ball et al. 2012; Cripps et al. 1998).

Table 4.1 Drill cutting management operations

Number	Treatment	References
1	Leave the piles undisturbed	Gerrard et al. (1999) and Potts et al. (2019)
2	Bioremediation	Davis (2016)
3	Capping	Hess et al. (2013)
4	Gravel dumping	Cripps et al. (1998)
5	Spreading	Ball et al. (2012) and Ismail et al. (2017)
6	Retrieve with suction	Carpenter (2014)
7	Retrieve with dredging	Cripps et al. (1998)
8	Retrieve with a seafloor crawler	Cripps et al. (1998)
9	Subsea entombment in a pit	Paulsen et al. (2005)
10	Reinjection into a well	Bartko et al. (2009)
11	Bioreactor treatment	Interiano-López et al. (2019)
12	Super-critical treatment	Motamedimehr and Gitipour (2019)
13	Land-farming	Kogbara et al. (2018)
14	Mechanical treatment onshore	Mcintyre (2008)
15	Distillation	Winterbourne (2014)
16	Stabilization	Al-Ansary and Al-Tabbaa (2007)
17	Combustion and thermal treatments	Petri et al. (2015)
18	Landfill	Saeedi et al. (2020)
19	Either of treated or untreated wastes	Cripps et al. (1998), Phillips et al. (2018), and Sharif et al. (2017)

In oil extraction operations, water recovery is attempted from waste; fluids run out of the drilling rig and are recirculated to maintain the operation. Water recovery options involve physical separation by specialized equipment and may also be carried out in combination with chemicals to enhance the process. Thermal processes may also be engaged to get rid of the excess waters and fixed solids, with huge atmospheric pollution (Sharif et al. 2017).

Nineteen operation options for managing drill cutting piles are described in Table 4.1.

4.4 Solids Control: The First Step and Inevitable Part of Drilling Operations

The first/initial step in waste management of drilling operations is carried out in the solids control system. Solids control is directly attached to the drilling rig and is essential due to its reducing act in the operation's mud demand. Solid control systems include shale shakers, degassers, desanders, and desilters. These processes physically remove the outcome fluids of the coarse particles and make the mud ready for reuse. In effect drill cuttings are removed from the drilling mud at the surface, and muds are prepared for recirculation and pumping (Bybee 2002; Sharif et al. 2017). Moreover, remnants remain after the solids control step; all fluids run out of the rig operational area, which are to be managed by waste management operations. Management depends on the local situation, local law, and the decision of the drilling operations manager.

4.4.1 New Methods of Drilling Waste Management Following Solids Control

Generally, waste management steps are based on reduction, reuse, and recycling of everything disposed of by the drilling rig. The three main steps classify waste management strategies according to their desirability in terms of waste minimization. Proper management of wastes begins with pollution prevention which truly eliminates, changes, or reduces operating practices resultant discharges to the environment. Given waste elimination is not possible, minimizing the generated waste should be investigated (Arshad et al. 2018; Phillips et al. 2018).

In the majority of oil extraction methods, after solids control in consideration of waste minimization goals, water is recovered by the addition of chemicals and remaining solids are stabilized or remediated for reuse (Ball et al. 2012). In 2019, a new method was newly designed and developed for cost reduction and better remediation; the method was trialed in Naft Sefid, Shushtar County, Khuzestan province, Iran, supporting an oil well drilling operated by National Iranian Drilling Company (NIDC). The creative method reduces the environmental impact of the oil well drilling for one of Iran's oil and gas production companies.

In this method, water of the drill cuttings and mud wastes are recovered by a physical and chemical process; floated oil is removed from waters by organic oil absorbent or line-up procedure; remaining solids are bioremediated by co-composting, with super saline waters evaporated and desalinated by a new humidification-dehumidification (SDHDH); the process flow is shown in Fig. 4.1.

4.4.2 Water Recovery by Physical and Chemical Processes

In most drilling operations, after solids control, drilling, cutting, and mud wastes flow as slurry to a corral (Veil 2002). In the corral/waste pit, slurry shape materials settle to TSS (total suspended solids) and are physically reduced in time. This chapter shows a development of the corral where the corral and the waste pit are merged. Further divisions are created by cement blocks, partitioning the area into three parts, staggered with the ground slope/gradient. This new waste pit design helps to separate muddy outcomes by the water suspended solid levels, as seen in Fig. 4.2. The slurry flows to the middle part of the waste pit which is equipped with 20

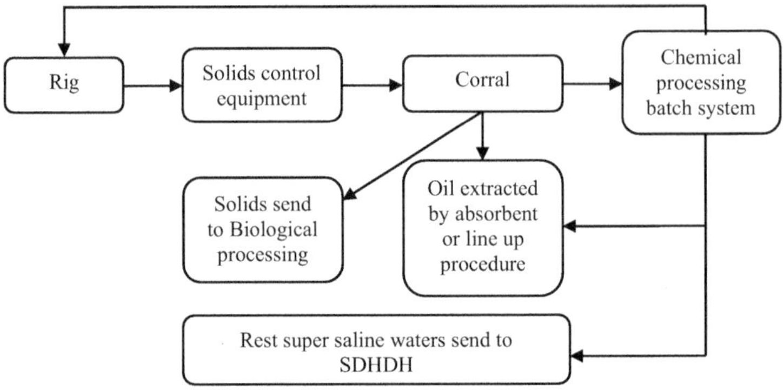

Fig. 4.1 Novel approach used to process oil extraction waste

Fig. 4.2 Combining waste pits and corrals. (Nafte Sefid, ShuShtar County, Khuzestan province, Iran, 2019)

centimeters diameter PVC agricultural drainage pipe tubes. Water flowing to the lowest elevation part of the waste pit is pumped to the batch chemical water processing ponds; solids elute from the water and are moved to the upper elevation part of the waste pit by a loader. Solids in the upper part of the pit remain and are stored for biological treatments.

In Fig. 4.2, three partitions are created by cement blocks, the large pipe transfers wastes to the middle part; the pump system and pipeline in the right side of the pit move the water to the batch chemical water processing unit. The upper left part of the image shows storage of the settled solids and a place for the waste's biological drying process.

Water treatments were established to reach three goals: avoiding problems of water source reduction, wastewater treatment, and water recycling. Chemical water treatment is a rapid method for water purification using chemicals (Gupta et al. 2012). Batch chemical water processing ponds are composed of three 80 cubic meters round ponds, shown in Fig. 4.3; the cyclonic series of ponds have the following features:

1. Production of a free vortex (for faster treatment and lower chemical use).
2. Full acid and alkali resistance as they are mainly constructed of high-density polyethylene sheets (Fig. 5.4).
3. Easy to fill and clean up due to possession of a central outlet. Hence they produce a free vortex that allows the solids to settle in the center of the pond.
4. The ponds are easy to set up and cheaper than other ponds or reservoirs used (Davarpanah et al. 2018).
5. Adding chemicals is facilitated with a pipe system due to the height and low pH of the water (around 2.5).
6. Shower systems operate over the pond and compressed air is vented into the ponds, allowing users to apply heavy aeration for odor control and biological water treatments in the system (Figs 4.3 and 4.4).

Figure 4.3 shows the pond design capable of producing a free vortex flow. Heavy aeration of waters achieves effective water treatment performance.

In the chemical water treatments, coagulation, flocculation, hardness removal, and pH setting are accomplished by adding chemicals (lime, soda, soda ash, PAC (polyaluminum chloride), PAM (polyacrylamide)) into the treatment ponds regarding a user's needs and permission, pending water properties (Davis 2010).

In Fig. 4.4 the ponds have a 7.7 m diameter and a capacity of 80 m^3.

Submersible pumps relay water to the shower, producing a free vortex flow and aeration as seen in Fig. 4.5.

The following section describes processes of oil absorption and recovery.

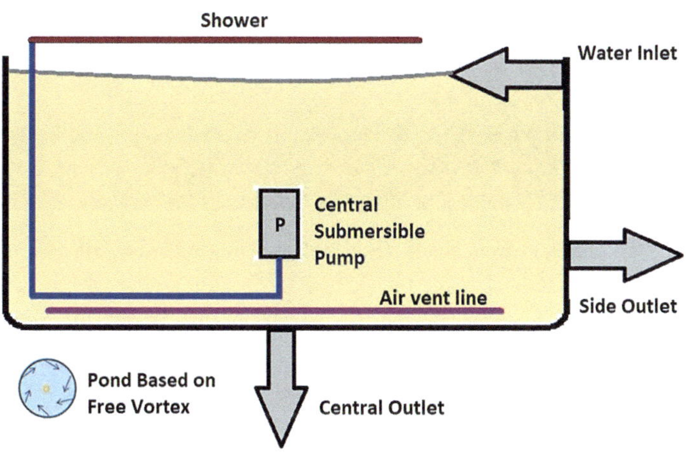

Fig. 4.3 Cyclonic pond systems

Fig. 4.4 Polyethylene cyclonic series of ponds mounted with iron structures. (Naft Sefid, Shushtar County, Khuzestan province, Iran, 2019)

Fig. 4.5 Shower operation over ponds. (Naft Sefid, Shushtar County, Khuzestan province, Iran, 2019)

4.4.3 Organic Oil Absorbent and Lineup Procedure for Oil Recovery and Removal

Oil pollution appears on water bodies through human activities (Kim et al. 2010). Drill cuttings, mud wastes, and drilling rig effluents contain different amounts of petroleum hydrocarbons. These contaminants are considered hazardous to ecosystems in close proximity or within the flow of polluted materials, which creates environmental concerns. As oil has a lower density than water, contamination stays rest on the water surface and form an oily layer mixed with water. In lineup procedures, the top oily part of the water source is pumped into a reservoir; water partitions in lower layers and drains from the tank. During this procedure, the tank fills with petroleum materials and the oil-contaminated water source be cleared. At the end of the lineup procedure, recovered petroleum materials may be reused for operational intentions, for example, as fuel for heating purposes.

Use of skimmer and lineup procedures are the most common oil removal methods; both are based on differences between water and oil density. These methods are inefficient when oil thickness on the water is low. In such scenarios, oil absorbents are made use of to enable efficient water cleanup. Commercial oil absorbents are made from different materials including cotton fibers, synthetic material, polymers, and agricultural wastes (Chai et al. 2015; Teas et al. 2001; Wang et al. 2012). In operations documented in the current chapter, to make the oil absorbent, sugarcane was used and bagasse was de-pithed/extracted by a mechanical process (Ranjbar Jafarabadi et al. 2019). Subsequently, sugarcane pith is processed in a pyrolysis reactor, during a 4-h anaerobic thermal treatment of between 170 and 210 °C (Awasthi et al. 2019). Through this process the materials obtain greater hydrophobicity properties and show oil absorption capability of seven to nine times more than their dry weight. Finally, treated sugarcane coir pith fills fluid permeable bags which are used as an oil absorbent. Figure 4.6 shows absorbents used in "low thickness" oil absorption operations.

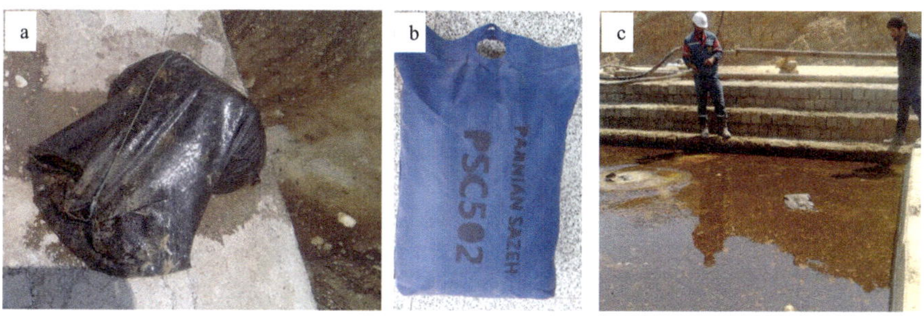

Fig. 4.6 Organic oil absorbent made by anaerobic thermal treatment of sugarcane bagasse coir (Naft Sefid, Shushtar County, Khuzestan province, Iran, 2019). (**a**) Absorbent after oil absorption is extracted out of the water; (**b**) clean absorbent before the oil absorption process; (**c**) the organic absorbent floats on the contaminated water and absorbs the oil pollution

The following section details the composting processes which may be engaged to remediate and create/restore displaced soil resources.

4.5 Co-composting of Solids and Remediated Soil Production

Oil and gas well drilling processes primarily generate two types of wastes – spent drilling fluids (muds) and a large amount of drill cuttings (solids). The fluid phase of wastes can be water, synthetic or natural oils, air, gas, or a mixture of these components (Onwukwe and Nwakaudu 2012). Due to environmental concerns, different management procedures have been detailed for the treatment of the solids remediation and fixation depending on the amount of waste and the type of contamination, authorities, and operational/financial concerns (Liden et al. 2017). The management procedures used in drilling oil and gas well waste management include thermal treatments (incineration and thermal desorption), biological treatments (composting, bioreactors, bio-slurry, and land-farming), and landfill and deep-well injection (Morillon et al. 2002; Onwukwe and Nwakaudu 2012; Paladino et al. 2016).

Biological treatment or bioremediation processes make use of living organisms to remediate or neutralize contaminant concentration to specific levels (Steliga et al. 2012). Moreover, in the matter of solid waste management, composting involves mixing drilling waste with bulking agents. Materials such as wood chips, straw, rice hulls, or husks provide increased porosity and aeration potential for biological degradation (Davis 2016; Paladino et al. 2016). Methods of treatment aim to maintain objectives toward increased time or efficiency of remediation and can be summarized in the following three steps.

1. Organic Semi-composted Material Production

To aid the process of bioremediation, agricultural waste (sugarcane wastes) and microorganisms are prepared in semi-composted heaps (Fig. 4.7). The process is similar to normal composting of agricultural waste (Zhang and Sun 2016) though materials obtained by biological drying and co-composting of solids are not fully composted; the carbon-nitrogen ratio is around 30:40 $g.g^{-1}$.

2. Biological Dehydration of Cuttings and Solids

Biological drying of cuttings and solids is the act of microorganisms. In Fig. 4.8 Organic material is mixed with wet solids; breakdown of the material leads to increased temperature and drying (Zhang et al. 2008).

3. Co-composting of Solids

After reaching the optimum material's moisture, by addition of microorganisms, nutrients (N and P, depending on the mixture composition C-N and C-P ratio) and

Fig. 4.7 Pile composting of agricultural waste (sugarcane bagasse) to produce semi-composted materials. (Amirkabir Sugarcane Cultivation and Industry Company, Ahvaz County, Khuzestan province, Iran, 2019)

Fig. 4.8 Biological drying of cuttings and solids (Naft Sefid, Shushtar County, Khuzestan province, Iran, 2019) : (**a**) Absorbing water by organic materials; (**b**) evaporation of water

Fig. 4.9 Pile composting of drilling oil and gas well solid wastes. (Naft Sefid, Shushtar County, Khuzestan province, Iran, 2019)

sugarcane molasses, the process of material mixing commences (Fig. 4.9). Davis (2016) and Paladino et al. (2016) detailed that composting of drilling oil and gas well waste has the following benefits: fixation of trace elements by $CaCO_3$ (organic Ca and CO_2) decreases metal and metalloid element bioavailability; microbial removal of petroleum hydrocarbon and other organic pollutants; and reduction of elements by dilution with organic material.

4.6 Water Evaporation and Desalination by a New Solar Desalination Humidification Dehumidification (SDHDH) System

Water desalination is a process that removes salt from the water and also demands a large amount of energy (Karagiannis and Soldatos 2008). Solar desalination methods are techniques which desalinate water using solar energy. Humidification dehumidification (HDH) systems are a type of desalination system, which are broken into an air humidifier and air dehumidifier (Giwa et al. 2016). Further a seawater greenhouse is a form of HDH system that grows plants using the sun's energy, seawater, and airflow to produce freshwater and cool air (Zarei and Behyad 2019).

To develop the seawater greenhouse in Fig. 4.10 with the aim of producing more water, a dehumidification system may be added, and purified water passes to a separate water reservoir as shown in Fig. 4.11.

In Fig. 4.11 saline water flows into the base pool that is covered by a double clear plastic curtain as a greenhouse; water is sprayed into the air by vortex nuzzles. The

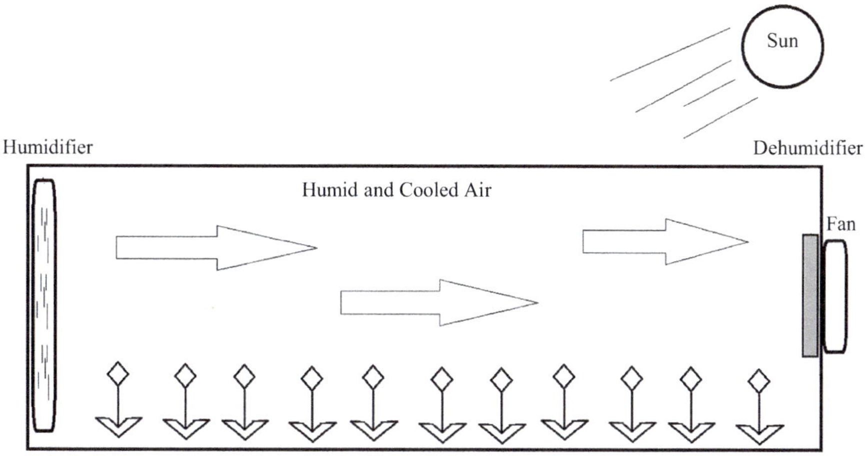

Fig. 4.10 Seawater greenhouse

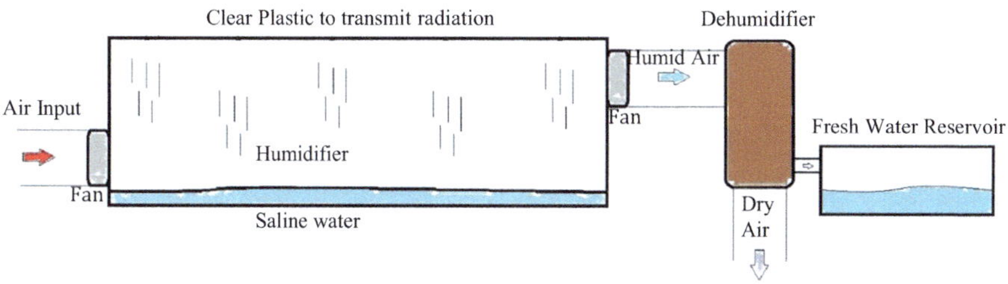

Fig. 4.11 New solar desalination humidification dehumidification (SDHDH) based on seawater greenhouse. (Parnian et al. 2020)

water and atmosphere of the semi-greenhouse (humidifier) part is warmed by solar radiation, causing the atmospheric humidity to increase. Fans push the humid air to a ground dehumidifier section and freshwater flows to the reservoir.

This technology has been used for desalination of oil and gas drilling wastewater in Iran in private oil operations; the seawater greenhouse in Fig. 4.12 of the integrated desalination plant produces 10 liters of freshwater per day.

The SDHDH is seen in Fig. 4.13(a) and pile composting is indicated in Fig 4.13(b). Bioremediation processing is seen in the heaps above the SDHDH on the right, and the waste corral is indicated (c) on the left.

This new approach developed the HDH of Fig. 4.10 into a low-cost and more efficient system; developments of the new design are mentioned below following Parnian et al. (2020).

Fig. 4.12 New humidification dehumidification (HDH) based on seawater greenhouse. (Naft Sefid, Shushtar, Khuzestan, Iran, 2019)

Fig. 4.13 Drilling solid waste bioremediation with an integrated solar desalination humidification dehumidification (SDHDH) system at Naft Sefid, Shushtar, Khuzestan, Iran, 2019

1. Merging the Solar Water Heater and the Humidifier

Inspired by greenhouses, we merged the solar water heater and the humidifier to make a low-cost SDHDH system. Saline water flows into a room covered by a clear plastic membrane, and air is pushed out to the dehumidifier by a fan. Solar radiations crossed into the heater-humidifier room, allowing the saline water temperature to rise.

2. Using Vortex Spray Nozzles to Spray Water

Spraying water into the air humidification chamber in the SDHDH system raises humidity and drives the system akin to that of a transpiration flow in plant systems. Raising humidity and extracting the air via fans maximize the growth of plants in the system and lower the super critical salt concentration in the chamber; consequently the humid air is extracted to the de-humidification chamber by fans.

3. Using a Ground Dehumidifier

The soil has a lower temperature than the air during the day in warm seasons. Inspired by house-ground heat extractors (Ali et al. 2017), a prototype of the ground dehumidifier was designed, which can be used to dehumidify air and produce fresh water. A 50 m length of a 20 cm diameter PVC tube was placed underground at a 2 m depth as the dehumidifier and water was collected in a plastic reservoir.

4. Possibility of Salt Production

The use of vortex nozzles for spraying water into the heater-humidifier room makes salt production possible as saline water may be sprayed up to the edge of salt super-saturation.

4.7 Discussion

Common methods for drilling waste management have advantages; however pollutants remain (Sharif et al. 2017; Siddique et al. 2017). After years of observation and many laboratory pilot-scale tests aiming to minimize problems at each stage, the combinatorial introduced method shown in this chapter was invented as an innovative response. Disadvantages of previous methods are as follows: requiring a place to dump/dispose solids, the cost of recovery of brines; overall high capital investment required for setup and running costs, requirement of engineering operations before running the process, and a major problem in application in high gradients of sloping topology (Ismail et al. 2017; Morillon et al. 2002; Onwukwe and Nwakaudu 2012; Saeedi et al. 2020; Veil, 2002).

Many methods have been developed to solve the problems of drilling waste (Ball et al. 2012; Jewesimi et al. 2019; Khodja et al. 2007; Napp et al. 2018; Phillips et al. 2018); demand for a new complex solution for all problems has not as yet been considered. The current chapter details a new method which has a solution for oil recovery/remediation from water (lineup procedure/oil absorbent), solid remains remediation (bioremediation through co-composting), and brine recovery/management (a new SDHDH). Additionally the current chapter created a parallel system for water recovery and treatment, making the method cheaper and more flexible than others which use high-cost equipment and extra resources.

4.8 Conclusion

This chapter introduces drilling waste management based on new methods of bioremediation and solar desalination as a new method with huge application potential. The combined method was developed with the aim to make a solution for solid waste and brine remains and meet the requirement of a flexible and low-cost waste management system for drilling oil/gas operations. The method has been successfully applied to manage oil well-drilling waste in Naft Sefid, Shushtar County, Khuzestan province, Iran in 2019. The method reduces the environmental impact of oil well drilling. Authors suggest the future application of methods documented in this chapter in alternative locations and environmental settings, although it is stated that the methods may be optimized within different biotic and abiotic parameters.

Finally, the current chapter contains innovative approaches for huge environmental problems. Out of extractive methods, new resources are generated, namely, water and soil. The authors wish to be clear that the current chapter in no way condones extractive industrial activity – "prevention is better than cure" appears to be an appropriate proverb from the Greek Goddess of Health, Sanitation and Hygiene as stated by philosopher Desiderius Erasmus in around 1566. However we do advocate soil bioremediation and recommend the use of processed materials for agricultural use, ecosystem services, and cleanup. Water reuse, recovery, and production are achieved from an unconventional source and may find uses from increasing urban water supply, to enriching droughted areas; thus increasing productivity and rehabilitation of the Earth following approximately 300 000 years of human onslaught.

References

Al-Ansary MS, Al-Tabbaa A (2007) Stabilisation/solidification of synthetic petroleum drill cuttings. J Hazard Mater 141(2):410–421. https://doi.org/10.1016/j.jhazmat.2006.05.079

Ali A, Mohamed M, Abdel-Aal M, Schellart A et al (2017) Analysis of ground-source heat pumps in north-of-England homes. Proceedings of the Institution of Civil Engineers. Energy 170(3):103–115. https://doi.org/10.1680/jener.15.00022

Arshad ABC, Yaacob MAB, Badzri SB (2018) Iraq operations' five 5 years waste management blue print to manage waste in remote environment of Garraf, Iraq operations. In: Paper presented at the SPE international conference and exhibition on health, safety, security, environment, and social responsibility, Abu Dhabi, UAE.. https://doi.org/10.2118/190588-MS

Awasthi A, Dhyani V, Biswas B et al (2019) Production of phenolic compounds using waste coir pith: estimation of kinetic and thermodynamic parameters. Bioresour Technol 274:173–179. https://doi.org/10.1016/j.biortech.2018.11.073

Ball AS, Stewart RJ, Schliephake K (2012) A review of the current options for the treatment and safe disposal of drill cuttings. Waste Manag Res 30(5):457–473

Bartko K, Al-Shobaili Y, Gagnard P et al (2009) Drill Cuttings Re-Injection (CRI) assessment for the Manifa Field: an environmentally safe and cost-effective drilling waste management strategy. In: Paper presented at the SPE Saudi Arabia Section Technical Symposium, Al-Khobar, Saudi Arabia. https://doi.org/10.2118/126077-MS

Bybee K (2002) Drilling waste management. J Pet Technol 54(11):50–52

Caenn R, Darley HC, Gray GR (2011) Composition and properties of drilling and completion fluids: Gulf Professional Publishing
Carpenter C (2014) Modeling options for drill-cuttings management. J Pet Technol 66(01):103–105. https://doi.org/10.2118/0114-0103-JPT
Chai W, Liu X, Zou J et al (2015) Pomelo peel modified with acetic anhydride and styrene as new sorbents for removal of oil pollution. Carbohydr Polym 132:245–251
Cheung PY, Kinkle BK (2001) Mycobacterium diversity and pyrene mineralization in petroleum-contaminated soils. Appl Environ Microbiol 67(5):2222–2229
Cripps S, Picken G, Aabel J et al (1998) Disposal of oil-based cuttings Report RF-98/097
Das N, Chandran P (2011) Microbial degradation of petroleum hydrocarbon contaminants: an overview. Biotechnol Res Int. https://doi.org/10.4061/2011/941810
Davarpanah A, Razmjoo A, Mirshekari B (2018) An overview of management, recycling, and wasting disposal in the drilling operation of oil and gas wells in Iran. Cogent Environ Sci 4(1). https://doi.org/10.1080/23311843.2018.1537066
Davis ML (2010) Water and wastewater engineering. Mcgraw-Hill Education. ISBN: 9780071713856
Davis DD (2016) Bioremediation of hydrocarbon contaminated soils and drill cuttings using composting with agricultural wastes. PhD thesis. Newcastle University
Dojka MA, Hugenholtz P, Haack SK et al (1998) Microbial diversity in a hydrocarbon-and chlorinated-solvent-contaminated aquifer undergoing intrinsic bioremediation. Appl Environ Microbiol 64(10):3869–3877
Gerrard S, Grant A, Marsh R et al (1999) Drill cuttings piles in the North Sea: management options during platform decommissioning, Centre for Environmental Risk Research Report No. 31. School of Environmental Sciences University of East Anglia, Norwich
Giwa A, Akther N, Al Housani A et al (2016) Recent advances in humidification dehumidification (HDH) desalination processes: improved designs and productivity. Renew Sust Energ Rev 57:929–944
Gupta VK, Ali I, Saleh TA et al (2012) Chemical treatment technologies for waste-water recycling – an overview. RSC Adv 2(16):6380–6388
Hess S, Alve E, Trannum HC, Norling K (2013) Benthic foraminiferal responses to water-based drill cuttings and natural sediment burial: results from a mesocosm experiment. Mar Micropaleontol 101:1–9. https://doi.org/10.1016/j.marmicro.2013.03.004
Interiano-López M, Ramírez-Coutiño V, Godinez-Tovar L et al (2019) Bioremediation methods assisted with humic acid for the treatment of oil-contaminated drill cuttings. Revista Mexicana de Ingeniería Química 18(3):929–937
Ismail AR, Alias A, Sulaiman W et al (2017) Drilling fluid waste management in drilling for oil and gas wells. Chem Eng Trans 56:1351–1356
Jewesimi A, Duncan C, Chokri T et al (2019) Implementation of drilling waste pneumatic transfer process for zero discharge deepwater well in Mexico. In: Conference proceedings, second EAGE workshop on deepwater exploration in Mexico: knowledge transfer and collaboration from shelf to deepwater, April 2019. https://doi.org/10.3997/2214-4609.201900366
Karagiannis IC, Soldatos PG (2008) Water desalination cost literature: review and assessment. Desalination 223(1):448–456
Khodja M, Canselier JP, Dali C et al (2007) A diagnostic of the treatment of oil well drilling waste in Algerian fields. Récens Progrèss en Génie des Procédés 94:1–11
Kim M, Yim UH, Hong SH et al (2010) Hebei Spirit oil spill monitored on site by fluorometric detection of residual oil in coastal waters off Taean, Korea. Mar Pollut Bull 60(3):383–389
Kirk JL, Klironomos JN, Lee H et al (2005) The effects of perennial ryegrass and alfalfa on microbial abundance and diversity in petroleum contaminated soil. Environ Pollut 133(3):455–465
Kogbara RB, Badom BK, Ayotamuno JM (2018) Tolerance and phytoremediation potential of four tropical grass species to land-applied drill cuttings. Int J Phytoremed 20(14):1446–1455. https://doi.org/10.1080/15226514.2018.1501337

Liden T, Clark BG, Hildenbrand ZL et al (2017) Unconventional oil and gas production: waste management and the water cycle. In: Schug KA Hildenbrand ZL (eds) Advances in chemical pollution, Environmental management and protection. Elsevier Science. ISBN: 9780128128022

Mcintyre BE (2008) Method and apparatus for cleaning drill cuttings, apparatus and process for removing liquids from drill cuttings: Google Patents. European Patent Office EP1153197B1. https://patents.google.com/patent/EP1153197B1/en. Accessed 2 Feb 2021

Morillon A, Vidalie JF, Hamzah US et al (2002) Drilling and waste management. In: SPE international conference on health, safety and environment in oil and gas exploration and production, 20–22 March 2002. https://doi.org/10.2118/73931-MS

Motamedimehr S, Gitipour S (2019) Sub and supercritical decontamination of oil-based drill cuttings: a review. Environ Energy Econ Res 3(3):225–240. https://doi.org/10.22097/eeer.2019.186444.1086

Napp AP, Pereira JES, Oliveira JS et al (2018) Comparative metagenomics reveals different hydrocarbon degradative abilities from enriched oil-drilling waste. Chemosphere 209:7–16. https://doi.org/10.1016/j.chemosphere.2018.06.068

Onwukwe S, Nwakaudu M (2012) Drilling wastes generation and management approach. Int J Environ Sci Dev 3(3):252–257. https://doi.org/10.7763/IJESD.2012.V3.226

Paladino G, Arrigoni J, Satti P et al (2016) Bioremediation of heavily hydrocarbon-contaminated drilling wastes by composting. Int J Environ Sci Technol 13(9):2227–2238

Parnian A, Furze JN, Parnian A (2020) Water out of waste – solar-desalination-humidification-dehumidification auxiliaries processing extractive industrial operations. Water Product J 1(1):31–38. https://doi.org/10.22034/WPJ.2020.119474

Paulsen JE, Dysvik A, Nilsen D et al (2005) Abandonment of seabed deposition of drill cuttings during offshore drilling. In: Paper presented at the SPE/EPA/DOE Exploration and Production Environmental Conference, Galveston, Texas. https://doi.org/10.2118/94377-MS

Petri I, Pereira MS, dos Santos JM et al (2015) Microwave remediation of oil well drill cuttings. J Pet Sci Eng 134:23–29. https://doi.org/10.1016/j.petrol.2015.07.022

Phillips L, Morris A, Innes G et al (2018) Drilling waste management – solutions that optimise drilling, reduce well cost and improve environmental performance. In: Paper presented at the Abu Dhabi International Petroleum Exhibition & Conference, Abu Dhabi, UAE. https://doi.org/10.2118/192793-MS

Potts LD, Perez Calderon LJ, Gubry-Rangin C et al (2019) Characterisation of microbial communities of drill cuttings piles from offshore oil and gas installations. Mar Pollut Bull 142:169–177. https://doi.org/10.1016/j.marpolbul.2019.03.014

Ranjbar JA, Dashtbozorg M, Mitra S et al (2019) Historical sedimentary deposition and ecotoxicological impact of aromatic biomarkers in sediment cores from ten coral reefs of the Persian Gulf, Iran. Sci Total Environ 696:133969. https://doi.org/10.1016/j.scitotenv.2019.133969

Saeedi M, Amanipoor H, Battaleb-Looie S et al M (2020) Landfill site selection for solid drilling wastes (case study: Marun oil field, southwest Iran). Int J Environ Sci Technol 17(3):1567–1590. https://doi.org/10.1007/s13762-019-02487-0

Sharif M, Nagalakshimi N, Reddy S et al (2017) Drilling waste management and control the effects. J Adv Chem Eng 7:166. https://doi.org/10.4172/2090-4568.1000166

Siddique S, Kwoffie L, Addae-Afoakwa K et al (2017) Oil based drilling fluid waste: an overview on environmentally persistent pollutants. IOP Conf Ser Mater Sci Eng 195:012008. https://doi.org/10.1088/1757-899x/195/1/012008

Singh S, Shikha (2019) Treatment and recycling of wastewater from oil refinery/petroleum industry. In: Singh R, Singh R (eds) Advances in biological treatment of industrial waste water and their recycling for a sustainable future. Applied environmental science and engineering for a sustainable future. Springer, Singapore. https://doi.org/10.1007/978-981-13-1468-1_10

Steliga T, Jakubowicz P, Kapusta P (2012) Changes in toxicity during in situ bioremediation of weathered drill wastes contaminated with petroleum hydrocarbons. Bioresour Technol 125:1–10

Teas C, Kalligeros S, Zanikos F et al (2001) Investigation of the effectiveness of absorbent materials in oil spills clean up. Desalination 140(3):259–264

Veil JA (2002) Drilling waste management: past, present, and future. In: Paper presented at the SPE annual technical conference and exhibition, San Antonio, Texas. https://doi.org/10.2118/77388-MS

Wang J, Zheng Y, Wang A (2012) Superhydrophobic kapok fiber oil-absorbent: preparation and high oil absorbency. Chem Eng J 213:1–7

Winterbourne M (2014) Superheated steam distillation for treatment of drill cuttings contaminated with oil based drilling fluids. University of Stavanger, Norway

Yi YM, Park S, Munster C et al (2016) Changes in ecological properties of petroleum oil-contaminated soil after low-temperature thermal desorption treatment. Water Air Soil Pollut 227:108. https://doi.org/10.1007/s11270-016-2804-4

Zarei T, Behyad R (2019) Predicting the water production of a solar seawater greenhouse desalination unit using multi-layer perceptron model. Sol Energy 177:595–603. https://doi.org/10.1016/j.solener.2018.11.059

Zhang L, Sun X (2016) Improving green waste composting by addition of sugarcane bagasse and exhausted grape marc. Bioresour Technol 218:335–343. https://doi.org/10.1016/j.biortech.2016.06.097

Zhang DQ, He PJ, Jin TF et al (2008) Bio-drying of municipal solid waste with high water content by aeration procedures regulation and inoculation. Bioresour Technol 99(18):8796–8802

Part II
Iran – Petroleum Pollution and Cure

Oil operations in Naft Sefid, Shushtar County, Iran, 2019 are depicted in Figs. 1–10.

Fig. 1 Salt-affected soil in a dried evaporation waste pit of approximately 20 m × 25 m, 2 m depth
Fig. 2 Petroleum contamination and salt
Fig. 3 Oil-affected soil stored in a temporary waste pit
Fig. 4 Salt- and oil-affected soil – the black marks are oil contamination
Fig. 5 Waste pits leaked to the surrounding environment
Fig. 6 After spring (February–March) rains, waste pits overflow and temporary pits are made to contain pollution
Fig. 7 Waste pit with heavy contamination of oil on water covering 1km^2, 2m depth
Fig. 8 Oil infiltrates the soil
Fig. 9 Oily soil and sludge in the dried pit
Fig. 10 The area is cleaned by collecting and gathering the contaminated water and soil in autumn before the rainy season. The excavated area is filled with clean soil of the surrounding area

Oil spill of April 2020 of the main transfer pipe in Jofeir, Hoveizeh county, Khuzestan province, Iran; soil remediation is shown in Figs. 11–21.

Fig. 11 Polluted area
Fig. 12 Polluted area
Fig. 13 Gathering of contaminated soil
Fig. 14 Piles of contaminated soil
Fig. 15 Agricultural waste and compost
Fig. 16 Contaminated soil piles
Fig. 17 Contaminated soil piles are mixed with agricultural waste
Fig. 18 Mixed material is treated with microorganisms, sugar and urea
Fig. 19 Turning and aeration of compost material
Fig. 20 Adding water for composting
Fig. 21 Growth of fungi and other microbes

Over a period of three months, 1200 m^3 of contaminated soil was co-composted and 'ready to use' in local green spaces and landscape restoration. These efforts have been endorsed by the Environmental Department of Khuzestan, Iran. Research is ongoing to ensure optimal soil conditions to enable diversity; further details are given in Chap. 4.

Fig. 1

Fig. 2

Fig. 3

Fig. 4

Fig. 5

Fig. 6

Fig. 7

Fig. 8

Fig. 9

Fig. 10

Fig. 11

Fig. 12

Fig. 13

Fig. 14

Fig. 15

Fig. 16

Fig. 17

Fig. 18

Fig. 19

Fig. 20

Fig. 21

Chapter 5
Orchid Diversity, Conservation, and Sustainability in Northeastern India

Sanjeet Kumar, Rajkumari Supriya Devi, Rikina Choudhury, Manisha Mahapatra, Susanta Kumar Biswal, Navneet Kaur, Jamuna Tudu, and Sakti Kanta Rath

Abstract The northeastern region (NER) of India is bestowed by the great Himalayan landscapes and their unique bioresources. Among the bioresources, orchid species are revered and well-known indicators of climatic change. Orchids are monocotyledonous plants known for their sui generis and alluring flowers. About 1484 orchid species are reported to be in India including 856 in the NER. Environmental sustainability requires an understanding of climatic change; hence conservation of orchid bioresources has heightened importance. Consequently, an attempt has been made to gather information of orchid species present in the NER of India from extensive field survey and available secondary sources. New additions, bar coding, and mass propagation protocols of orchid species available in the region are discussed.

The survey results revealed that 249 common species are enumerated by primary and secondary sources of the NER of the country. Among these, a number of species were found to be sensitive to microclimatic change. In the study period, many species (including *Bulbophyllum hookeri*, *Vanda coerulea*, *Renanthera imschootiana*) were observed in different threatened categories of the International Union for Conservation Nature (IUCN). Through vegetative propagation, selected species were reintroduced.

S. Kumar (✉) · R. S. Devi
Biodiversity and Conservation Laboratory, Ambika Prasad Research Foundation, Regional Centre, Imphal, Manipur, India
e-mail: sanjeet.biotech@gmail.com; supriyark91@gmail.com

R. Choudhury · M. Mahapatra · N. Kaur · J. Tudu · S. K. Rath
Department of Life Sciences, Rama Devi Women's University, Bhubaneswar, Odisha, India
e-mail: rikinachoudhury3012@gmail.com; mahapatramanisha36@gmail.com; navneetkaur@rdwu.ac.in; jamunatudu205@gmail.com; saktirath@gmail.com

S. K. Biswal
School of Applied Sciences, Centurion University of Technology and Management, Bhubaneswar, Odisha, India
e-mail: dr.skbiswal@cutm.ac.in

J. N. Furze et al. (eds.), *Earth Systems Protection and Sustainability*,
https://doi.org/10.1007/978-3-030-85829-2_5

This chapter highlights the importance of orchids as key indicators of climate change and details their uses for medicinal, food, and cultural significance. The study recommends that there is a need to develop a strategic protocol for conservation and a digital documentation of the species available in study areas to ensure sustainable usage and conservation.

Keywords Orchid diversity · Climate change · Bar-coding · Mass propagation

5.1 Introduction

Orchids are monocot flowering plants that belong to the world's largest families of angiosperms, the Orchidaceae (Zhang et al. 2018; Misra 2007, 2014, 2019). They are unique in many aspects as they have diverse shapes, colors, and forms, representing highly evolved flagship species across 750–800 genera and about 30,000 to 35,000 species throughout the world (Paul and Kumaria 2017). Orchids have trimerous, pentacyclic, and zygomorphic flowers with two lateral sepals, one dorsal sepal, two petals, and a labellum. They are terrestrials, lithophytes, epiphytes, and rarely saprophytes including *Didymoplexis seidenfadenii* and *Gastrodia silentvalleyana* (Kumar et al. 2008; Jalal and Jayanthi 2012). They are categorized as monopodial and sympodial types and are grouped as tropical and temperate species (Hedge 1997). With the exceptions of Arctic, Antarctica, and a few isolated island regions, orchids are distributed worldwide. It was observed that Columbia and Indo-Malaysian regions show the highest distribution of orchids. The genus *Bulbophyllum* contains about 2000 species followed by *Epidendrum* (1500 species), *Dendrobium* (1400 species), and *Pleurothallis* (1000 species) throughout the world having 10,000 hybrid species. As per genera, orchids are concentrated in Tropical Asia and South America followed by Tropical Africa, Oceania, Europe, Temperate Asia, and North America. India is a major habitat of orchid species having about 190 genera including about 400 endemic species. They have medicinal, food, and sociocultural values as shown in Table 5.1 and Fig. 5.2. The most common genera in India are *Habenaria*, *Zeuxine*, *Spiranthes*, *Calanthe*, *Phaius*, *Eria*, *Liparis*, *Oberonia*, *Dendrobium*, *Luisia*, *Vanda*, *Acampe*, *Aerides*, *Bulbophyllum*, *Coelogyne*, and *Cymbidium* (De and Medhi 2014a). In India, the northeastern region (NER) shows the highest number of orchid species due to its favorable climatic and microclimatic environment. The most common species are *Acampe ochracea*, *Aerides multiflora*, *Calanthe biloba*, *Coelogyne ochracea*, *Cymbidium aloifolium*, *Cymbidium elegans* (Fig. 5.1f), *Cymbidium sinense* (Fig. 5.1g), *Dendrobium infundibulum*, *Dendrobium fimbriatum* var. oculatum (Fig. 5.1i), *Bulbophyllum careyanum* (Fig. 5.1e), *D. densiflorum*, *Epidendrum radicans*, *Eria pannea*, *Renanthera imschootiana*, *Pleione praecox*, *Spiranthes spiralis*, *Zeuxine strateumatica*, *Vanda coerulea*, *Paphiopedilum insigne*, *Phaius mishmensis* (Fig. 5.1h), *P. hirsutissimum*, *B. hirtum*, *C. fuscescens*, *C. giganteum*, *Dendrobium chrysanthum* (Fig. 5.1a), *V. parviflora*, *E. bambusifolia*, *D. nobile*, and *Bletilla striata* (Fig. 5.1d).

Among the states of the NER of the country, there are 600 species reported from Arunachal Pradesh followed by 580 species in Sikkim, 450 species in Meghalaya,

Table 5.1 Common orchid species of the Northeastern region of India and their habits and uses

Plant name	Habit	Uses
Acampe carinata	Epiphytic	Medicinal (snake bite, stomach disorder, uterine disease)
Acampe ochracea	Terrestrial/epiphytic/ lithophyte/mycotrophic	Anti-malaria
Acampe papillosa	Epiphytic	Medicinal (rheumatism)
Acampe rigida	Epiphytic	Medicinal (muscle pain, joint pain, promotes blood circulation)
Acampe praemorsa	Epiphyte	Medicinal (rheumatism, arthritis)
Acanthephippium sylhetense	Epiphytic	Aesthetic
Aerides fieldingii	Epiphytic	Ornamental
Aerides multiflora	Epiphytic	Medicinal (earache)
Aerides odoratum	Epiphytic	Medicinal (antibacterial, tuberculosis)
Aerides rosea	Epiphytic	Trade
Aerides vandarum	Epiphytic	Ornamental
Aerides odoratum	Epiphyte	Ornamental
Agrostophyllum callosum	Epiphytic	Medicinal (skin disorder and diabetes)
Agrostophyllum planicaule	Epiphytic	Medicinal
Aldrovanda vesiculosa	Epiphytic	Trade
Anoectochilus roxburghii	Terrestrial	Medicinal
Anoectochilus tetraplerus	Terrestrial	Trade
Anthogonium gracile	Terrestrial	Medicinal (bone fracture and cracked heels)
Apostasia wallichii	Terrestrial	Aesthetic
Arachnis labrosa	Epiphytic	Medicinal
Arundina graminifolia	Terrestrial	Medicinal (body ache and cracked heels)
Ascocentrum ampullaceum	Epiphyte	Trade, ornamental
Ascocentrum ampullaceum var. auruanticum	Terrestrial	Perfume, ornamental
Ascocentrum himalaicum	Terrestrial/epiphytic / lithophyte/mycotrophic	Trade, ornamental
Bulbophyllum cariniflorum	Moss-covered tree trunk	Medicinal, ornamental
Brachycorythis galeandra	Terrestrial	Used as charm
Brachycorythis obcordata	Terrestrial	Expectorant
Bryobium pudicum	Epiphytic	Ornamental
Bulbophyllum affine	Moss-covered tree trunk	Perfume, trade
Bulbophyllum careyanum	Moss-covered tree trunk	Traditional medicine (abortion)
Bulbophyllum elatum	Moss-covered tree trunk	Ornamental
Bulbophyllum forrestii	Pseudobulbs	Folk medicine
Bulbophyllum guttulatum	Moss-covered tree trunk	Ornamental
Bulbophyllum helenae	Pseudobulbs	Trade
Bulbophyllum hirtum	Moss-covered tree trunk	Used in wildlife research, ornamental

(continued)

Table 5.1 (continued)

Plant name	Habit	Uses
Bulbophyllum khasyanum	Moss-covered tree trunk	Trade
Bulbophyllum leopardinum	Moss-covered tree trunk	Medicinal and ornamental
Bulbophyllum lobbii	Moss-covered tree trunk	Trade
Bulbophyllum manipurense	Moss-covered tree trunk	Horticulture
Bulbophyllum moniliforme	Moss-covered tree trunk	Trade
Bulbophyllum odoratissimum	Moss-covered tree trunk	Medicinal (tuberculosis)
Bulbophyllum penicillium	Pseudobulbs	Medicinal
Bulbophyllum picturatum	Terrestrial	Medicinal and trade
Bulbophyllum polyrhizum	Moss-covered tree trunk	Medicinal
Bulbophyllum propinquum	Moss-covered tree trunk	Research in wildlife and ornamental
Bulbophyllum reptans	Moss-covered tree trunk	Medicinal
Bulbophyllum scabratum	Moss-covered tree trunk/ lithophyte	Medicinal and ornamental
Bulbophyllum trichocephalum	Moss-covered tree trunk	Ornamental
Bulbophyllum triste	Moss-covered tree trunk	Ornamental
Bulbophyllum umbellatum	Moss-covered tree trunk	Medicinal, ornamental
Bulbophyllum xylophyllum	Moss-covered tree trunk	Medicinal, ornamental
Calanthe alpina	Terrestrial	Medicinal
Calanthe angusta	Terrestrial	Ornamental
Calanthe anjanii	Terrestrial	Ornamental
Calanthe biloba	Terrestrial	Ornamental
Calanthe brevicornu	Terrestrial	Ornamental
Calanthe clavata	Terrestrial	House plant
Calanthe densiflora	Terrestrial	Medicinal, ornamental
Calanthe mannii	Terrestrial	Medicinal, ornamental
Calanthe sylvatica	Terrestrial	Medicinal (nose bleeding, cold, and cough)
Calanthe tricarinata	Terrestrial	Medicinal (eczema)
Cephalanthera longifolia	Terrestrial	Food and medicinal (wound healer)
Cleisostoma brevipes	Epiphyte	Trade
Cleisocentron pallens	Epiphyte	Trade, ethnobotany
Cleisostoma paniculatum	Epiphyte	Trade
Coelogyne barbata	Epiphyte/lithophyte	Trade
Coelogyne calcicola	Epiphyte/lithophyte	Trade
Coelogyne corymbosa	Epiphyte	Medicinal (analgesic and headache)
Coelogyne cristata	Epiphyte	Medicinal (constipation), aphrodisiac, and ornamental
Coelogyne elata	Epiphyte	Medicinal
Coelogyne fimbriata	Epiphyte	Medicinal (headache, fever, and indigestion) and ornamental

(continued)

Table 5.1 (continued)

Plant name	Habit	Uses
Coelogyne flaccid	Epiphyte	Medicinal (headache and fever) and Ornamental
Coelogyne fuscescens	Epiphyte	Medicinal (abdominal pain) and ornamental
Coelogyne ghatakii	Epiphyte	Trade
Coelogyne graminifolia	Epiphyte	Medicinal
Coelogyne griffithii	Epiphyte	Trade
Coelogyne hitendrae	Epiphyte	Trade
Coelogyne ochracea	Epiphyte	Ornamental
Coelogyne ovalis	Epiphyte	Medicinal (cough, urinary infection, and eye disorder) and ornamental
Coelogyne stricta	Epiphyte/lithophyte	Medicinal (headache and fever)
Coelogyne suaveolens	Epiphyte	Perfume
Coelogyne flexuosa	Epiphyte	Ornamental
Coelogyne viscosa	Epiphyte	Ornamental
Cymbidium aloifolium	Epiphyte	Medicinal (bone fracture and nervous system disorder)
Cymbidium gammieanum	Epiphyte	Medicinal and ornamental
Cymbidium giganteum	Epiphyte	Food and ornamental
Cymbidium longifolium	Epiphyte/terrestrial/ lithophyte	Ornamental
Cymbidium maladimum	Epiphyte	Medicinal
Cymbidium tracyanum	Epiphyte/terrestrial/ lithophyte	Ornamental
Cymbidium whiteae	Terrestrial	Ornamental, horticulture
Cypripedium cordigerum	Terrestrial	Medicine, ornamental
Cypripedium elegans	Terrestrial	Medicinal (nervous system disorder), ornamental
Cypripedium himalaicum	Terrestrial	Medicinal, food, and ornamental
Cypripedium tibeticum	Terrestrial	Trade
Dactylorhiza hatagirea	Terrestrial	Food (leaves), medicinal (bone fracture), and ornamental
Dendrobium angulatum	Epiphyte	Medicinal
Dendrobium infundibulum	Epiphyte	Ornamental
Dendrobium ochreatum	Epiphyte	Trade
Dendrobium porphyrochilum	Epiphyte/lithophyte	Stimulant
Dendrobium primulinum	Epiphyte/terrestrial	Ornamental, trade
Dendrobium acinaciforme	Epiphyte	Ayurvedic medicine
Dendrobium aduncum	Epiphyte	Medicinal
Dendrobium aggregatum	Epiphyte	Medicinal
Dendrobium amoenum	Epiphyte	Medicinal (burned skin and dislocated bone)
Dendrobium anceps	Epiphyte	Trade

(continued)

Table 5.1 (continued)

Plant name	Habit	Uses
Dendrobium aphyllum	Epiphyte	Ornamental, trade
Dendrobium cathcartii	Epiphyte	Food flavoring
Dendrobium chrysanthum	Epiphyte	Medicinal (antipyretic), ornamental
Dendrobium denudans	Epiphyte	Medicinal (cough and tonsil)
Dendrobium fimbriatum	Epiphyte	Medicinal (earache)
Dendrobium lituiflorum	Epiphyte	Trade
Dendrobium moschatum	Epiphyte	Medicinal (earache and bone fracture)
Dendrobium ochreatum	Epiphyte	Ornamental
Dendrobium pendulum	Epiphyte/terrestrial/ lithophyte	Trade
Dendrobium podagraria	Epiphyte	Aesthetic and stimulant
Dendrobium primulinum	Epiphyte	Trade and ornamental
Dendrobium pulchellum	Pseudobulb	Trade
Dendrobium ruckeri	Epiphyte	Ornamental
Dendrobium salaccense	Pseudobulb	Food and fragrance
Dendrobium thyrsiflorum	Epiphyte	Trade
Dendrobium transparens	Epiphyte	Medicinal (bone fracture), ornamental
Dendrobium aphyllum	Epiphyte	Medicinal (anal fistula)
Dendrobium chrysotoxum	Epiphyte	Medicinal (antipyretic), trade
Dendrobium densiflorum	Epiphyte	Medicinal (bone fracture, pimple)
Dendrobium fimbriatum	Epiphyte	Medicinal (bone fracture, nervous debility)
Dendrobium formosum	Epiphyte	Ornamental, Trade
Dendrobium lindleyi	Epiphyte	Ideal for hanging basket and pot plant
Eria ferruginea	Epiphyte	Ornamental
Eria flava	Epiphyte	Perfume
Esmeralda cathcartii	Terrestrial	Ornamental, trade
Eulophia bicarinata	Terrestrial	Ayurvedic medicine
Gastrochilus calceolaris	Epiphyte	Ornamental
Gastrochilus inconspicuus	Epiphyte	Ethnobotany and pharmacology
Gastrodia dyeriana	Epiphyte	Ornamental, Trade
Gastrodia elata	Epiphyte	Medicinal (boils and pimples)
Geodorum densiflorum	Terrestrial	Medicinal (skin disorder and improves fertility)
Geodorum recurvum	Terrestrial	Ornamental
Goodyra procera	Terrestrial	Herbal medicine
Hetaeria affinis	Terrestrial	Ornamental
Holcoglossum amesianum	Epiphyte/lithophyte	Ornamental
Hygrochilus parishii	Epiphyte	Medicinal
Liparis formosana	Epiphyte	Medicinal
Liparis gamblei	Medicinal/terrestrial	Medicinal, trade
Liparis glossula	Terrestrial	Trade

(continued)

Table 5.1 (continued)

Plant name	Habit	Uses
Liparis longipes	Epiphyte	Aesthetic, ornamental
Liparis lydiaii	Epiphyte	Ornamental
Liparis mannii	Epiphyte/terrestrial	Medicinal
Liparis nervosa	Terrestrial	Medicinal (stomachache and malignant ulcer)
Liparis odorata	Terrestrial	Medicinal (cancerous ulcer and throat cancer)
Luisia filiformis	Epiphyte/terrestrial/ lithophyte	Trade
Luisia inconspicuous	Epiphyte	Trade
Microtis unifolia	Terrestrial	Food
Nervilia aragoana	Terrestrial	Medicinal (enhance sperm formation)
Oberonia pachyrachis	Epiphyte	Medicinal (boils)
Oncidium sphacelatum	Epiphyte	Ornamental
Ornithochilus fuscus	Epiphyte	Medicinal, ornamental
Panisea uniflora	Epiphyte/pseudobulb	Ornamental
Paphiopedilum hirsutissimum	Lithophyte	Trade
Papilionanthe teres	Epiphyte	Medicinal (fever, cough, and cold)
Peristylus constrictus	Terrestrial	Medicinal (boils)
Phaius wallichii	Terrestrial	Ornamental
Phaius flavus	Terrestrial	Trade
Phaius tankervilleae	Terrestrial	Medicinal (ethnobotany), horticulture
Phalaenopsis braceana	Epiphyte	Perfume, Trade
Pholidota articulate	Epiphyte	Medicinal (cancer, skin ulcer, bone fracture)
Pholidota imbricata	Epiphyte	Medicinal (nasal pain and arthritis)
Phreatia elegans	Epiphyte	Ornamental
Pinalia stricta	Epiphyte	Herbal medicine
Podochilus cultratus	Epiphyte	Medicinal, ornamental
Pomatocalpa mannii	Epiphyte	Trade, Ornamental
Pomatocalpa spicatum	Epiphyte	Ornamental
Pomatocalpa undulatum	Epiphyte	Medicinal
Porpax elwesii	Epiphyte	Ornamental
Porpax fibuliformis	Epiphyte	Trade
Porpax gigantean	Lithophyte	Medicinal
Pteroceras suaveolens	Epiphyte	Medicinal
Pteroceras teres	Epiphyte	Ornamental
Rhomboda lanceolata	Terrestrial	Trade
Rhomboda longifolia	Terrestrial	Trade
Rhynchostylis gigantean	Epiphyte	Ornamental
Rhynchostylis retusa	Epiphyte	Medicinal (constipation and rheumatism), ornamental

(continued)

Table 5.1 (continued)

Plant name	Habit	Uses
Rhytionanthos cornutum	Epiphyte	Trade
Risleya atropurpurea	Terrestrial	Trade
Robiquetia spatulata	Epiphyte	Ornamental
Robiquetia succisa	Epiphyte	Ornamental
Saccolabiopsis pusilla	Epiphyte	Trade
Sarcoglyphis manipurensis	Epiphyte	No report
Satyrium nepalense	Terrestrial	Medicinal (diarrhea, dysentery, and malaria)
Schizostachyum pergracile	Epiphyte	Medicinal
Schoenorchis fragrans	Epiphyte	Perfume, trade, and ornamental
Schoenorchis gemmata	Epiphyte	Trade
Schoenorchis manipurensis	Epiphyte	Perfume
Smitmandia micrantha	Epiphyte	Medicinal (antibacterial), ornamental
Spathoglottis ixioides	Terrestrial	Ornamental
Spathoglottis plicata	Terrestrial	Poultice and horticulture
Spathoglottis pubescens	Terrestrial	Food
Spiranthes sinensis	Terrestrial	Medicinal (kidney disease, fever, and fatigue)
Staurochilus ramosus	Epiphyte	Horticulture, trade
Stereochilus hirtus	Epiphyte	Ethnobotany
Sunipia bicolor	Epiphyte	Trade
Sunipia candida	Epiphyte	Medicinal (antibiotic)
Sunipia intermedia	Epiphyte	Trade
Sunipia scariosa	Epiphyte	Medicinal
Taeniophyllum glandulosum	Epiphyte	Medicinal
Tainia angustifolia	Epiphyte	Medicinal
Tainia latifolia	Terrestrial	House plant
Tainia minor	Terrestrial	Medicinal
Thelasis pygmaea	Epiphyte	Medicinal, ornamental
Thelasis longifolia	Epiphyte	Medicinal, ornamental
Thelasis pygmaea	Epiphyte	Medicinal, ornamental
Thrixspermum musciflorum	Epiphyte	Medicinal (poultice to ulcers of nose)
Thunia alba	Epiphyte/lithophyte/terrestrial	Medicinal (dislocated bone), ornamental
Thunia marshalliana	Epiphyte/terrestrial	Horticulture
Trichotosia pulvinata	Epiphyte	Trade
Tropidia angulosa	Terrestrial	Trade
Tropidia curculigoides	Terrestrial	Medicinal (diarrhea and malaria)
Tylostylis discolor	Epiphyte/lithophyte	Medicinal
Uncifera acuminate	Epiphyte	Ornamental
Uncifera lancifolia	Epiphyte/lithophyte	Folk medicine
Uncifera obtusifolia	Epiphyte	Ornamental

(continued)

Table 5.1 (continued)

Plant name	Habit	Uses
Vanda alpine	Epiphyte	Medicinal
Vanda amesiana	Epiphyte	Medicinal
Vanda bicolor	Epiphyte/lithophyte/ terrestrial	Indoor plant and horticulture
Vanda coerulescens	Epiphyte	Anti-aging skin treatment, ornamental
Vanda cristata	Epiphyte	Medicinal (dry cough and dislocated bone), Ornamental
Vanda griffithii	Epiphyte/lithophyte/ terrestrial	Medicinal
Vanda liouvillei	Epiphyte	Trade
Vanda motesiana	Epiphyte	Trade
Vanda parviflora	Epiphyte	Medicinal (ear infection)
Vanda roxburghii	Epiphyte	Medicinal (pimple and boils)
Vanda stangeana	Epiphyte	Trade
Vanda teres	Epiphytic	Horticulture, trade
Vanda tessellata	Epiphyte	Medicinal (back joints and nervous system disorder), ornamental
Vanda testaceae	Epiphyte	Medicinal, ornamental
Vandopsis parishii	Epiphyte	Medicinal (immunity booster)
Vandopsis undulate	Epiphyte	Ornamental
Yoania japonica	Terrestrial	Ornamental
Yoania prainii	Terrestrial	Trade
Zeuxine affinis	Terrestrial	Ornamental
Zeuxine flava	Epiphyte/lithophyte/ terrestrial	Medicinal, ornamental
Zeuxine goodyeroides	Terrestrial	Ornamental
Zeuxine longilabris	Terrestrial	Medicinal (whooping cough)
Zeuxine nervosa	Terrestrial	Medicinal
Zeuxine pulchra	Terrestrial	Trade
Zeuxine reflexa	Terrestrial	Trade
Zeuxine seidenfadenii	Terrestrial	Trade
Zeuxine strateumatica	Terrestrial	Medicinal
Zygopetalum intermedium	Terrestrial	Perfume, trade, and ornamental

389 species in Manipur, 300 species in Nagaland, 250 species from Assam, 70 species from Tripura, and 251 species from Mizoram (De and Medhi 2014a; Rao and Kumar 2018). The common orchid species available in Arunachal Pradesh are *Rhynchostylis retusa*, *A. odorata*, *C. aloifolium*, *D. aphyllum*, *D. fimbriatum*, *D. densiflorum*, *V. coerulea*, *R. imschootiana*, and *P. fairrieanum*. The common orchids in Assam are *S. sinensis*, *Goodyera procera*, *Vanilla pilifera*, *B. obrienianum*, *C. odora*, *E. pumila*, and *Luisia macrotis*. The common reported orchids from Manipur are *V. coerulea*, *Renanthera imschootiana*, *Z. strateumatica*, *D. fimbriatum*, *P. tankervilleae*, *Dendrobium nobile*, *B. umbellatum*, *C. aloifolium*, *V. testacea*, *A. carinata*,

Fig. 5.1 Common orchids of study areas in Sikkim and Manipur, India. (**a**) *Dendrobium chrysanthum*, (**b**) *Arundina graminifolia*, (**c**) *Bulbophyllum hirtum*, (**d**) *Bletilla striata*, (**e**) *Bulbophyllum careyanum*, (**f**) *Cymbidium elegans*, (**g**) *Cymbidium sinense*, (**h**) *Phaius mishmensis*, (**i**) *Dendrobium fimbriatum* var. oculatum

and *A. odorata*. The common orchids available in Meghalaya are *Anoectochilus brevilabris*, *Paphiopedilum insigne*, *C. elegance*, and *C. corymbosa* (De and Medhi 2014a). The state of Mizoram is also known as a home to a diverse group of orchid species including *E. spicata*, *Arundina graminifolia*, *V. coerulea*, *R. imschootiana*, *Eulophia nuda*, *Malaxis acuminate*, *P. tankervilleae*, *Pholidota imbricata*, and *V. testacea*. The common orchids available in Nagaland are *Paphiopedilum hirsutissimum*, *A. carinata*, *A. odorata*, *C. biloba*, *C. mannii*, *C. cristata*, *C. elegans*, and *D. aphyllum*. Sikkim is the second largest home for orchids in the NER. The most common orchid species are *C. alpine*, *Uncifera lancifolia*, *C. cristata*, *C. devonianum*, *D. falconeri*, and *V. pumila*. The orchid species reported from Tripura are *A. papillosa*, *A. praemorsa*, *A. multiflorum*, *C. aloifolium*, *D. densiflorum*, *E. ferruginea*, *Hetaeria affinis*, *Liparis odorata*, *S. sinensis*, and *V. teres* (De and Medhi 2014a) (Fig. 5.2).

The objectives of this chapter are to (i) bring attention toward the diversity of orchid in the NER of India, (ii) establish the importance of orchid services to ecology and human beings, and (iii) bring attention to mitigation of threats and conservation strategy of orchid diversity in the NER of India.

Fig. 5.2 Sociocultural and economic importance of orchid species and ex situ conservation along with reintroduction of species back into study areas

5.2 Threats to Orchid Diversity in the Northeastern Part of India

Threats are major factors causing extinction of different plant species. Threats can be of different types from overexploitation of natural resources, anthropogenic activities, plant trade, climate change to biotic and abiotic stresses that push orchids to extinction. Orchids include many rare and endangered species, which have been identified as under threat of extinction (Swarts and Dixon 2009; Sarvalingam and Rajendran 2016). It is reported that 50% of vascular plant species are endemic to 25 global biodiversity hotspots, each with at least 1500 native plant species. Orchidaceae have complex interactions with pollinators, mycorrhizal fungi, and host trees; hence their risk of extinction is compounded and greater due to dependence on other organisms, such as insect pollinators for growth and reproduction. Orchids are highly sensitive to climate change and habitat disturbance and loss.

Additional threats to orchids involve unsustainable harvesting and indiscriminate collection for horticulture, which has an extended impact on genera such as *Cattleya*, *Laelia*, and *Renanthera* and slipper orchids (*Cypripedium*, *Paphiopedilum*, and *Phragmipedium*). Orchids will be eventually stripped from their wild habitats. Illegal smuggling is a major threat of orchid extermination and has resulted in them being placed on the appendices of the Convention on International Trade in endangered species (CITES-IUCN status). Orchid species are transported not only to different states but also across international borders (Fay 2018). With liberalization of export policy and the boom in the floriculture industry, considerable importance has been given to the development of the orchid trade. Trading and business strategies constitute alarming situations in conservation of orchids.

Basically there are two types of trade:

- Plant trades affecting wild and cultivated orchids
- Cut flowers affecting hybrid orchid species production, playing a significant role in the floriculture trade

The following genera of orchids are heavily affected by trading: *Acampe*, *Aerides*, *Armodorum*, *Ascocentrum*, *Bulbophyllum*, *Chrysoglossum*, *Cleisocentron*, *Cottonia*, *Cymbidium*, *Dendrobium*, *Diplomeris*, *Gastrochilus*, *Habenaria*, *Micropera*, *Pomatocalpa*, *Paphiopedilum*, *Pholidota*, *Pleione*, *Thunia*, and *Vanda* (Hedge 1997). Encroachment of land is an additional common reason for extinction of orchids; this includes habitat degradation, fragmentation, and illegal collection for trading and consumption. Species such as *P. olivacea* have been forced into extinction in the wild.

There is a routine activity of collection and selling of medicinal orchids which involves uprooting of the whole plant resulting in their extinction. Due to this extreme disturbance, destruction of a number of economically important plants in alpine meadows continues, largely untapped. The activity has resulted in the reduction of *Dactylorhiza hatagirea*, a highly valued medicinal orchid, the population of which has been decreasing in the Himalayas according to the Convention on International Trade in endangered species (Pant 2013).

Orchids include epiphytes which share nutrition, light, temperature, and moisture in the complex web of the plant canopy. Hence, they are greatly affected by climatic change, global warming, and biotic and abiotic stresses. Orchids have natural hardiness and can tolerate slight variations of temperature and light, but climatic variations over the decades has led to migration of orchids to alternative locations. These factors also disrupt patterns of flowering time due to the fluctuation in spring temperatures; in winter flowering is hastened (Barman and Devadas 2013).

5.3 Threatened Orchids of Northeastern India

Northeastern India is a center of mega diversity and is comprised of eight states (Arunachal Pradesh, Assam, Manipur, Meghalaya, Mizoram, Nagaland, Sikkim, and Tripura). Overall, 50 % of the area supports flora of which 31.58 % of species are endemic. Megadiverse regions are rich in orchids, ferns, bamboos, magnolias, and rhododendrons.

Thirty-four species of orchids have been identified among the threatened plants of India. Habitat loss, deterioration, fragmentation, introduction of exotic species, overexploitation, environmental pollution, global warming, commercialization of agriculture and forestry, and jhum cultivation are major causes of loss of orchid diversity in study areas (De and Medhi 2014a; Tripathy et al. 2016). A large number of ornamental, rare, endangered, and threatened orchid species are gradually vanishing from northeastern regions due to human exploitation. Common threatened orchid species (Fig. 5.3; Table 5.2) include *A. grandiflorus*, *Arachnanthe cathcartii*, *Arundina graminifolia*, *B. candidum*, *C. mannii*, *Cleisostoma paniculatum*, *C. cochleare*, *C. giganteum*, *C. grandiflorum*, *D. densiflorum*, *D. devonianum*, *D. falconeri*, *D. wardianum*, *Galeola falconeri*, *L. delicatula*, *V. coerulea*, and *Tainia minor*. The NER is the home of the threatened lady's skipper orchids, *Paphiopedilum* species, *P. fairrieanum*, *P. venustum*, and *P. spicerianum*. Saprophytic genera like *Epipogium*, *Gastrodia*, *Eulophia*, and *Galeola* grow on decaying organic matter on the forest floor. Assam has as many as 26 species of threatened orchids, including *Bulbophyllum obrienianum*, *C. odora*, *D. pauciflorum*, *D. spatella*, *E. pumila*, *E. candida*, *L. plantaginea*, *L. macrotis*, and *Phalaenopsis mastersii*.

Due to overexploitation orchids have been considerably depleted and many are only present in orchidaria. Taxa severely threatened are *A. tetraplerus*, *Ascocentrum miniatum*, *D. sonia*, *D. draconis*, *D. heterocarpum*, and *D. wardianum*. Threatened species of orchid in Mizoram are *A. ochracea*, *Arundina chinensis*, *B. parryae*, *C. nitida*, *C. ovalis*, *C. eburneum*, *C. macrorrhizum*, *D. falconeri*, *P. charlesworthii*, *P. villosum*, and *R. imschootiana*. Threatened species in Nagaland are *C. eburneum* or ivory-colored Cymbidium, red Vanda or *R. imschootiana*, and *C. tigrinum*. The most threatened species of Sikkim include *Satyrium nepalense*, *A. sikkimensis*, *C. cristata*, *C. eburneum*, *C. devonianum*, *D. chrysotoxum*, *D. densiflorum*, *D.*

Fig. 5.3 Most threatened orchid species of study areas of Sikkim and Manipur in Northeastern India, (**a**) *Renanthera imschootiana*, (**b**) *Dendrobium falconeri*, (**c**) *Vanda coerulea*

wardianum, and *V. pumila*. Threatened orchids of Tripura are endangered blue Vanda or *V. coerulea* and *R. imschootiana*.

Further species under extreme threat are listed in Table 5.2.

5.4 Impacts of Climate Change on Bio-wealth and Orchid Diversity in Northeastern Region of India

Extinction of species is the most dangerous consequence of climate change to global biodiversity. Change in climate leads to loss, degradation, and fragmentation of habitat (Whiteside and Ward 2011). Variation in climate induces flooding and hinders nutrient retention and instigates soil erosion, pest attack, and loss of purity of

Table 5.2 Threatened orchid species in NER of India

Name	Status	Distribution(s)
Aerides vandara	Threatened	Assam, Nagaland, and Meghalaya
Aerides fieldingii	Threatened	Assam, Meghalaya, and Sikkim
Anoectochilus rotundifolius	Threatened	Sikkim
Arachnis clarkei	Threatened	Sikkim
Bulbophyllum hookeri	Threatened	Sikkim
B. albidum	Threatened	Sikkim
B. yunnanensis	Threatened	Sikkim
B. moniliforme	Endangered	Manipur
B. obrienianum	Threatened	Assam
B. rothschildianum	Threatened	Nagaland
Calanthe alpina	Threatened	Sikkim
C. herbacea	Threatened	Sikkim
Coelogyne arunachalensis	Threatened	Arunachal Pradesh
C. nitida	Threatened	Assam, Meghalaya, Sikkim
Cymbidium devonianum	Threatened	Sikkim
C. hookerianum	Vulnerable	Sikkim
C. whiteae	Endangered	Sikkim
Dendrobium heterocarpum	Threatened	Sikkim
D. bensoniae	Threatened	Manipur
D. draconis	Threatened	Manipur
D. wardianum	Threatened	Manipur
D. falconeri	Threatened	Manipur, Sikkim
D. nobile	Threatened	Arunachal Pradesh, Sikkim
H. richardiana	Threatened	Sikkim
H. khasiana	Threatened	Meghalaya
Liparis plantaginea	Threatened	Meghalaya, Nagaland
L. pulchella	Threatened	Assam
Paphiopedilum fairrieanum	Endangered	Arunachal Pradesh, Sikkim
P. hirsutissimum	Threatened	Sikkim
P. wardii	Threatened	Arunachal Pradesh
Phaius tankervilleae	Threatened	Manipur, Mizoram
Pleione lagenaria	Threatened	Sikkim and Meghalaya
Renanthera imschootiana	Critically endangered	Manipur, Nagaland, Tripura
Tainia khasiania	Threatened	Meghalaya
Vanda coerulea	Critically endangered	Manipur, Arunachal Pradesh
V. pumila	Threatened	Sikkim, Manipur
Zeuxine pulchra	Endangered	Sikkim

both air and water which have potential repercussions of biodiversity loss. Severe flooding causes loss of coral reef and mangrove forest which buffers the strong ocean storm against coastal erosion. The mass destruction of mangrove and coral reefs directly affects the animals which use them for productive breeding grounds, leading to loss of biodiversity (Chen et al. 2017). Loss of vegetation induces soil erosion directing destructive landslides which lead to mass killing and mass extinction. Climate change largely affects the aquatic life because aquatic forms of life are more susceptible to changing environmental conditions (Mitchell et al. 2006).

After the Arctic, the Himalayan region is the most sensitive climate change hotspot, though is less observed and very little data is available of this youngest mountain range. In the last century, average global temperature elevated by 0.70 °C, but in the Himalayas an escalation of 1°C greatly affected diversity. The fragile Himalayan region is adversely affected by the global warming hiatus period, which is the dirty result of anthropogenic climate change. Rapid conversion of solid state water to melting water causes flash floods, disaster, and livestock disease. The outcome of climate variation in the water tower of Asia is extreme precipitation during the monsoon season in the eastern Himalaya and wetter cooler conditions in western Himalaya (Panday et al. 2015).

Precipitation across the Himalayas is not uniform. Climate change in the Himalayan region results in a decrease of wind speed because of an asymmetric decline in latitudinal temperature and pressure gradients (You et al. 2010a). Solar radiation and sunshine duration decreases change the climate as the deposition of dust particles on glacial areas or snow alters the surface, leading to a variation in surface radiation balance which increases the temperature of the region (You et al. 2010b; Ji et al. 2016). Hydrological responses to climate change worsen day by day. Climate change directly or indirectly influences temperature, average annual rainfall, modification of weeds or microbes, changes in carbon dioxide level, and variations in sea level. Extreme temperature decreases plant productivity by inhibiting photosystem II functionality and induces sterility. Water deficit conditions caused by changes in global climate affect the inflorescence pattern of flower and growth of plants. Fertilization is affected by drought stress. Elevated carbon dioxide and ozone increase plant disease susceptibility. Climate change affects pollination and seed dispersal of both domestic and wild plants which ultimately results in decreasing the reproductive capability of plants (Chen et al. 2017).

Variation in climate across prolonged periods induce abiotic stress factors in plants leading to a decline in plant productivity due to generation of reactive oxygen species (ROS), which upset cellular mechanisms (Raza et al. 2019). Climate change drives orchid species to move pole-ward in a specific manner. The particular response of orchids depends on geographical area and species biology. Generally species in flat areas migrate toward hilly regions during warmer seasons (Jackson and Overpeck 2000), and hilly areas' migration occurs upward along the elevation gradient (Thomson 1990). In mixed topographical areas, populations of rare species reduce and face local extinction due to unknown complex causes. Warming of the climate causes a decline in the populations of wild orchid species (Barman and Devadas 2013).

Orchids present in upper forest canopies are sensitive to desiccation and heat as the availability of light is higher than that of other regions of the forest. Epiphytes share light, temperature, nutrition, and moisture in a complicated manner; global warming affects the availability of all factors directly or indirectly. Mild change in temperature does not affect the survival capacity of orchids, but increases in temperature may affect flowering period adversely; further increased levels of carbon dioxide inhibit flowering. Climate change, loss of habitat, and fragmentation may decline the population of orchids. Evaporation and unpredictable rainfall may change the moisture content of soil which ultimately leads to changes in microclimatic vegetation affecting the terrestrial orchid population (De and Medhi 2014b; Kaye et al. 2019; Wraith et al. 2020). Elevated precipitation causes soil erosion which negatively affects orchid plants in mountain areas, including *Calanthe*, *Pholidota*, and *Eria*. Drought conditions also affect survival capacity of orchids such as *Melaleuca*. Orchids have specific pollination patterns requiring specific pollinators. Environmental changes upset the longevity and evolutionary potential of specific pollinators. Furthermore, flowering is affected by enhanced temperature due to global warming. High temperature and low rainfall lead to forest fire which is one of the main causes of loss of orchid species and extinction. Extreme dry spells in the spring season lead to drying of epiphytic orchids growing on host plants. Mycorrhizal fungi assist in orchid germination and growth and are regulated by specific temperature, rainfall, and reduced moisture.

5.5 Climate Change, Species Loss, and Desired Species Conservation

Climate change occurs due to different natural factors over a period of centuries. Changes force many plant species to migrate pole-ward, toward areas more suited for growth. The positive and negative responses of different plant species in relation to climate change depend on the biology of the species and the geographical location of the population. Orchids are suffering and some are at the verge of extinction due to overexploitation, habitat loss due to anthropogenic activities, illegal smuggling or plant trade, and climate change.

Orchid species are at risk of extinction as their rate of evolution does not keep in pace with local or wider climate change. Global climatic change leads to habitat loss, degradation, fragmentation, phonological change, and alterations in population levels which have profound impacts in ecosystems and evolutionary processes. New species may be introduced with existing species, which in turn disturbs ecosystem development and therefore threatens orchids. Global warming mainly affects orchid populations. Factors affecting rarity of orchids involve both biotic and abiotic factors. Biotic factors involve symbiotic relationships between orchids and mycorrhizal fungi, which are vital for natural seed germination, seedling, and post-seedling growth of all orchids. Mycorrhizal fungi in orchid germination and growth are easily disturbed with rising temperature, erratic rainfall, and reduced moisture.

Changes in environment affect the evolutionary potential of pollination. Changes in breeding system involve self-pollination mechanisms. Species of orchids may be pollinated by specific species of pollinators; hence long-term flowering behaviors are rare for orchids. Flowering fluctuations occur due to fluctuations in spring temperatures resulting in delayed pollination. Rise in temperature with increased CO_2 levels not only results in vegetative plant growth but also has an adverse affect on flowering periods. Early or late flowering of any particular orchid species indirectly affects its pollination. Imbalances in pollination affect orchid growth, which is governed by light, nutrition, and moisture. Heavy rainfall and evaporation rates imbalance soil moisture, vegetation, and the microclimate of both forest areas and grassland. Orchid populations are severely affected by microclimate (Barman and Devadas 2013).

Strategies and conservation methods opted for to counter pressures on orchid populations serve to restore and maintain native ecosystem; manage habitats for rare, threatened, and endangered species; rank vulnerability of species; carry out long-term phonological monitoring of plants and pollinators; assist migration of orchids; establish symbiotic seed germination and seedling growth for restoration of orchids; enable intraspecific hybridization; and facilitate seed storage and banking. The following section details conservation methods which may be practiced to counter losses in India.

5.6 Methods of Conservation for Orchids

Conservation of orchid species in the NER of India is imperative due to rapid degradation through climate changes and anthropogenic activities. There is a need to address all possible solutions using both traditional methods and advanced techniques in contemporary strategies of conservation.

5.6.1 Propagation of Orchids

Conservation through propagation is through sexual or by asexual methods. Propagation by sexual methods is difficult as seeds lack an endosperm and need a fungal stimulant for natural germination (Pant 2013). Common trading orchids are heterozygote and cannot be grown from seeds; vegetative methods are the only option to propagate plants. Hence the convention of using shoot cuttings to propagate orchids is adopted. Vegetative propagation through bulb clumps of rhizomes or by the rooting of the shoot is time-consuming and less efficient in getting adequate numbers of orchids. Problems lead to some medicinally important orchids being threatened, and some are on the verge of extinction. Mass propagation can be carried out to reintroduce species in their natural habitat (Sharma and Mohan 2016).

5.6.2 Mass Propagation

There are various orchidaria and orchid sanctuaries available worldwide. In India, mainly in the Northeast, orchids are treated reverently as the most beautiful, ornamental, and medicinal flowers. As a result of their outstanding beauty, many people seek to trade and exploit these vulnerable species. Many species are endangered and a few are already extinct. Protection and conservation of orchids is urgently required. Some states have special initiatives for protection of orchids in their natural habitats, i.e., in situ methods of conservation (e.g., Appangala in Karnataka, Lolegaon, and Darjeeling areas in West Bengal have been designated as orchid reserves by state government). Orchidaria have been set up in Deorali and Singtam in Sikkim. Sessa in Arunachal Pradesh and Hararo at Imphal and Manipur have taken initiative to protect orchids by applying tissue culture techniques. Preservation and maintenance are important, and appropriate management of native orchid habitats is a prime motive of orchid conservation programs (Paul et al. 2017). Aside from preservation and maintenance, in situ and ex situ conservation approaches can be opted for.

In in situ conservation methods, plants are allowed to interact and co-evolve, hence protection of genetic resources of the natural environment by protection of the environment itself. The government of India supports in situ conservation and has already practiced it in many regions of study areas in the NER. Additionally, ex situ methods involve in vivo technologies, cryopreservation, molecular marker technology, and molecular diagnostics. Ex situ conservation plays an important role in rehabilitation, multiplication, and judicious exploitation of the plant resource. An array of culture techniques have evolved using explants of orchids (Paul and Kumaria 2017).

In situ and ex situ methods are further discussed in the following section.

5.6.3 In Situ and Ex Situ Conservation of Orchid Species

In situ conservation is considered as the most appropriate way for conserving biodiversity. This involves protection of habitats for orchids. Intensive care of habitat management is required due to small population sizes and restricted distribution. The disadvantage of in situ conservation is such that it is not always a viable option due to existing or ongoing modification of habitats and migration or absence of pollinators due to unfavorable conditions (Behera et al. 2013). Interactions occurring in defined orchid habitats in situ are termed as ecologically specific and more accurately reflect the specificity of mycorrhizal associations.

In ex situ method of preservation of components, collections of biological diversity are made outside their natural habitat. Ex situ conservation is an important aspect of orchid conservation which may include both seed banks and in vitro culture plant tissue collections. Such collections are a very good method for long-term conservation for threatened species (Behera et al. 2013).

5.6.4 Cryo-seed Bank Methods

Cryopreservation entails a unique conservation strategy for storage of orchid seeds under liquid nitrogen at −196°C. Seeds are stored at ultra-low temperature to pause metabolism so that cells can be preserved without deterioration. Cryopreservation of tropical plants is rare as they are more vulnerable to desiccation and chilling temperatures than temperate species (Yam and Thame 2005). Cryopreservation is justified as there is reduction of accidental loss of seedlings and incidental confirmation of seed germination; storage is easier for a large number of seeds in small containers for a long time in place of plants in nursery; hybrid seeds of orchids can be stored and exchanged with other institutions; long-term preservation of germplasm is a very basic strategy of conservation; and storage of germplasm as seeds are efficient and cost-effective.

5.6.5 Propagation by Symbiotic and Asymbiotic Seed Germination

The mechanisms of orchid seed germination are adaptive and unique. The small seeds of orchids are highly fragile and are produced in large numbers (Arditti 1967). The unique character of orchid seed is it lacks storage tissue which is essential for germination and seedling growth. For natural reproduction, orchids require association with specific fungal partners and non-sporic groups of mycorrhizal associations such as those of *Rhizoctonia* (including *Rhizoctonia mucoroides*, *Rhizoctonia repens*, *Rhizoctonia lanuginosa*). Seeds of orchids are unable to utilize their own reserve material and cannot hydrolyze starch or cellulose.

Asymbiotic germination without the availability of sugar sources proceeds only up to early protocorm stages, after which there is a requirement of external sugars from mycorrhizal partners. Fungal association primarily augments the transportation of vitamins, carbohydrate, and auxin in orchids and is known as symbiotic germination (Pant 2013). In the symbiotic association, fungi reside inside the cellular cytoplasm of orchids where nutrition and growth of the fungus occurs, though can be inhibited by an antifungal compound secreted by the plant. Orchid seed germination is a long process (Rao 1977), and any slight change in the environment can destroy the entire population. Asymbiotic germination patterns of orchid seeds are the primary stage of in vitro propagation of pure cultures (Yam and Arditti 2009). Classically, Knudson (1922) confirmed that in seed germination in in vitro culture, the fungal association can be bypassed. However in vitro cultures provide a widely accepted method of orchid propagation.

5.6.6 *Plant Tissue Culture Conservation of Orchid Species*

Biotechnological plant tissue culture technique can be applied for large-scale propagation and conservation of threatened, rare, and endangered species of orchid. Exploitation of orchids is very common as they have commercial, medicinal, and aesthetic values. Due to in vitro propagation, numbers of many wild orchid species can be salvaged; using cloning techniques a large number of identical clones may be produced from protocorm explants (Deb and Pongener 2012a, b).

5.6.7 *Mass Propagation and Micropropagation*

Mass propagation and micropropagation are basic tissue culture techniques. These are in vitro methods used in ex situ methods of conservation. These extensively used techniques are not only for the rapid and large-scale production of threatened plants but also for conservation of each and every species of orchid. Orchids follow sexual reproduction processes but also reproduce through conventional means such as back bulbs and shoot division. Root and shoot tips show very slow growth with low yield. Nodes, meristems, and axillary buds can be used in propagation. Mass propagation using conventional and tissue culture techniques is thus an important strategy to save natural populations from the pressures of commercial collection (Mohanty et al. 2012). In vitro techniques involve embryo culture and meristem culture.

Formation of embryos is a post-pollination phenomenon; fertilization is important for obtaining seedlings. Young ovules do not form suitable explants from plants. However, ovules can be used for raising new plants through tissue culture. Success of ovule tissue culture varies from species to species. The asymbiotic potential of seeds representing different developmental stages has been observed in species such as *C. aloifolium*, *P. tankervilleae*, *Spathoglottis plicata*, and *A. odorata* (Mohanty et al. 2012). In meristem culture, the pure line of a desired species is used for culturing the meristem. Orchids produce a large number of heterozygotes due to their out-breeding nature. Propagation through the embryo is disadvantageous for the cut flower industry, and meristem culture is a preferred method. Axillary tips, shoot buds, and shoot tip meristems are used for the development of whole plants in the cases of *D. sonia*, *A. formosanus*, *Hayata*, and *Phaius tankervilleae* (Das et al. 2008). Steps involved for in vitro techniques are collection of seeds, root, shoot, or any explants of plant; sterilization of explants; preparation of culture media and culture condition; addition of growth regulators to culture media (such as indoleacetic acid, 6-benzylaminopurine); inoculation of explants; rooting or hardening of plants; and acclimatization and data analysis.

5.6.8 Population Inventory and Reintroduction

Degradation of orchid habits is occurring at an alarming rate. Habitat restoration for indigenous species requires a sound strategy accompanied with monitored population inventories. Research activities involve population inventories, though are insufficient in the NER of India. In 1995, an orchid conservation program commenced along with restoration of germplasm into natural habitats (Pant 2013; Lesar et al. 2012; Aggarwal and Zettler 2010; Stewart and Kane 2006).

In future efforts, line transect and quadrant methods could be used with "Google Earth" mapping to produce location-specific population inventories. Orchid seeds may be produced naturally and transported to the exact location where their mycorrhizal association is present. This will lead to successful germination of seeds with seedlings transferred to their original regions for reintroduction. The best time for orchid reintroduction is from mid-October to December with a plentiful supply of water for seedling settlement (Yam et al. 2011). Reintroduction of medicinally important orchids to natural habitats and cultivation by in vitro propagation are very rare.

5.6.9 Monitoring of Introduced Plants

Monitoring of introduced orchid species in their natural habitat is essential following their reintroduction. After the relocation of orchid samplings, periodic observations should be made and logged and database maintenance and accurate, consistent growth observation are routine recommendations. Reintroduced orchids grow with a lush habit without superficial application of chemical fertilizers, as dead leaves gravitate in trunk crevices and act as nutritious decayed humus. Faster growth and settlement may be achieved using a foliar spray of a limited amount of fertilizer after 1 month of introduction to 6 months of settlement. Proper exposure to light should also be ensured (Yam et al. 2011).

5.6.10 Seed Culture Technique and Artificial Seeds

Orchid species have numerous seeds in a single pod which favors propagation by seed culture. Conservation of orchids through artificial seeds and their propagation are established long-term strategies (Pant 2013). Noticeable aspects are that without the help of fungus, orchid seeds can grow in vitro under seed culture techniques.

5.6.11 Conservation Through Value Addition

Value addition of indigenous species plays a vital role in the conservation of plant taxa. Orchids provide a sound platform for this in different aspects. They could be food additives, medicinal agents, or ornamental plants. Floriculture is an ultra-modern technique to conserve and commercialize orchids in controlled environmental conditions. In India, floriculture can be considered as a dynamic business for growing industries. Orchids hold a significant place because of their life span, easy packing, handsome look, and high productivity and transportability (De et al. 2014a; Zettler et al. 2017). Heightened efforts can be made to conserve orchids by developing methods of orchid breeding and ultimately lead to growth of the industry (Hossain et al. 2013). Value addition of orchid species can be achieved by the transfer of technology in growing and post-harvest handling of mercantile orchids to farmers and entrepreneurs to settle a small-scale industry with respective government organizations (Pant 2013; De et al. 2014a, b). The value addition through orchidarium and cultivation of endemic species is useful in reducing the over-extraction of wild species from nature.

5.6.12 Conservation Through Database Development

Database development is an important key for the conservation of taxa. Survey works, collected from secondary sources, and their gathering may be completed with use of software, applications, or webpages to provide details of species which is an ongoing conservation strategy with future potential (Gerz et al. 2018). The need for a central database arises to compile all required available information in a single knowledgebase, accessible for researchers and other stakeholders. Local institutes and international organizations may be added to a web-managed platform for development of a database, including that of the NER of India. For conservation purposes, an interactive user interface to use readily available data for omitting biological risk and easy decision making is required. A BIOPOP (Poschlod et al. 2003)-style database of plant traits and Internet application for nature conservation can be developed for orchid species conservation to establish better knowledge of biology and ecology. Such efforts will assist detailed analysis of factors causing species reduction and help to identify the required factors responsible for their conservation (Poschlod et al. 2003; Fay 2018).

5.6.13 Apps and Other Software

Mobile applications and software can create a comprehensive web-accessible database of at-risk plant species which may be continuously updated for professionals and trained volunteers for use in conservation. This facilitates early detection and rapid action to save vulnerable orchid species. The use of smart phone, web technology, and digital camera will enhance the scope of vulnerable orchid species management and their conservation. Like "win PTS" (Portable Technology Solutions), software tools can be developed to freely access information and suggestion from exports for conservation of the orchid plant species.

5.6.14 Awareness Programs

Public awareness is a base step to sensitize and mobilize the community residing near the habitats of vulnerable orchid regions for their conservation. Such programs need periodic general observation. Continuous interaction with local communities, group discussion with the community-based organization (CBO), and establishment of a community-based local committee for creating awareness through periodic meetings on local resources and sustainable uses are vigilant approaches that will ensure orchid protection. Public awareness approaches including personal contact (Fig. 5.4), mass media dissemination and group discussion, rural poetry, folk song, diversity fairs, rural dramatization, exchange visits, and village workshops are fundamental steps toward conservation. Public awareness mostly conserves species by strengthening the CBO, enhancing in situ conservation, and spreading sensation among locals (Sthapit et al. 2003).

5.7 Model of Conservation and Recommendations

Conservation models of wild orchids focus on two major activities: (i) securing the plan and management of action to save orchid which naturally needs specific strategy and (ii) establishment of mycorrhizal bank and ex situ seed banks to conserve orchids in a state of immediate threat. Formulation of different horticulture techniques to conserve orchids and their high-class propagation as well as translocation are also essential. In certain taxa, species with complex mycoheterotrophic association can be managed well in greenhouse condition but cannot be relocated into field. Much research is required to fill knowledge gaps. In conservation there are a lot of aspirations but less implementation.

The scientific community should work on gray areas and set some clear and clarified criteria to meet global needs. Strict policy should be made and rectification of poorly formulated rules is a must. Plant conservation biologists and conservation practitioners should collaborate to recommend comprehensive and realistic strategies and plans of action to address these deficiencies. Due to gradual climate change and overexploitation of flagship species, orchid species are disappearing from their natural habitat and some are on the verge of extinction. Different kinds of research-based conservation strategies are being employed to resolve this problem (Fig. 5.5).

Fig. 5.4 Discussion and awareness on ex situ propagation and sustainable harvesting of wild orchids at Imphal, Manipur, India. (**a**), (**b**), (**c**) Species is *Coelogyne nitida*; (**d**), (**e**) species is *Pholidota* sp.; (**f**), (**g**) species is *Pholidota* sp.

Figure 5.5 shows that the orchid population of the NER of India is subjected to different categories of threats which are seriously impacting the populations, leading to a requirement for restorative programs. Conservation methods may be applied according to the impact on the species. All species should be subject to economic programs, resulting in production of orchidaria; educational awareness programs must detail the risk of over-extraction of species and propagation which ensures vegetative propagation of plant matter. Species identified as those with threatened

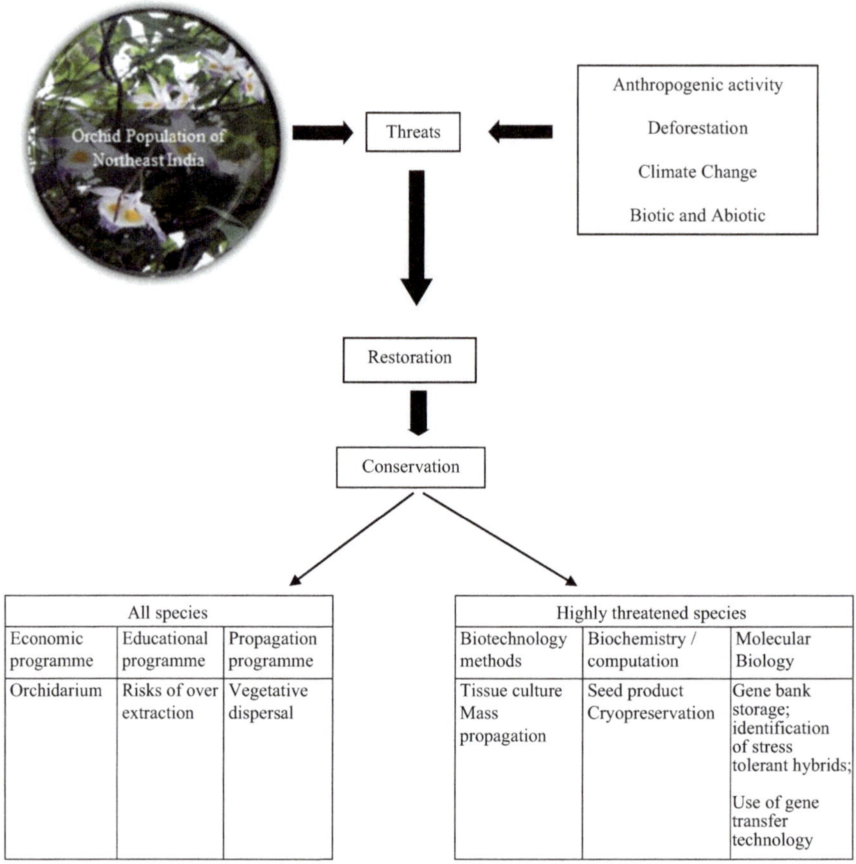

Fig. 5.5 Conservation model of orchid species

classification are subject to more refined, sensitive methods of biotechnology, biochemistry, and molecular biology. Example techniques applied are also shown for each of these three in Fig. 5.5.

Biotechnological approaches, tissue culture, and mass propagation require prime focus. Embryo culture and meristem culture are important themes of tissue culture and propagation-based conservation as through these techniques mass production and reintroduction of a maximum number of plantlets can be realized. By applying molecular biology to conserve orchid population, methods such as gene deposition in gene banks for lifelong preservation can be achieved. By utilizing (molecular) genetic breeding, stress-tolerant hybrids of orchid can be introduced to avoid extinction. Bacteria-mediated gene transfer and recombination methods can be applied to conserve orchid populations (Broothaerts et al. 2005; Hwang et al. 2017).

Artificial seed/seed production is an ideal means for orchid introduction into the wild, and cryopreservation of seeds under ultra-low temperatures provides further conservation opportunities. By applying various artificial intelligent techniques such as the provision of easily available open source software, database tools, and skill-based work through artificial intelligent "robots," orchid extinction can be robustly prevented. Boosting the economy and creating job opportunities will establish orchid-based industries. Additionally new orchidarium established to create awareness and care for nature among people should ensure protection of valuable bioresources for future generations to come.

Acknowledgments Authors wish to thank the local people of the Northeastern Region (NER) of India; authorities of the Forest Department of NER; the Director of Institute of Bioresources and Sustainable Development, Imphal, India; and team members of Ambika Prasad Research Foundation, Regional Centre, Imphal, India.

References

Aggarwal S, Zettler LW (2010) Reintroduction of an endangered terrestrial orchid, *Dactylorhiza hatagirea* (D. Don) Soo, assisted by symbiotic seed germination-First report from the Indian subcontinent. Nat Sci 8(10):139–145

Barman D, Devadas R (2013) Climate change on orchid population and conservation strategies: a review. J Crop Weed 9(2):1–12

Behera D, Rath CC, Mohapatra U (2013) Medicinal orchids in India and their Conservation: a review. Floriculture Ornamental Biotech 7(1):53–59

Broothaerts W, Mitchell HJ, Weir B et al (2005) Gene transfer to plants by diverse species of bacteria. Nature 433:629–633. https://doi.org/10.1038/nature03309

Chen WY, Suzuki T, Lackner M (eds) (2017) Handbook of climate change mitigation and adaptation. Springer, Cham

Das MC, Kumaria S, Tandon P. (2008) In vitro propagation and conservation of Dendrobium lituiflorum Lindl. through protocorm-like bodies. J Plant Biochem & Biotechnol 17:177–180

De LC, Medhi RP (2014a) Diversity and Conservation of rare and endemic orchids of North East India-A Review. Indian J Hill Farming 27(1):81–89

De LC, Medhi RP (2014b) Climate change and its impact on orchid productivity. Int J Sci Res 3(10):500–504

De LC, Pathak P, Rao AN et al (2014a) Global orchid industry. In: Commercial orchids. De Gruyter, Berlin

De LC, Vij SP, Medhi RP (2014b) Post-harvest physiology and technology in orchids. J Hortic. https://doi.org/10.4172/2376-0354.1000102

Deb CR, Pongener A (2012a) Development of a cost effective in vitro regenerative protocol of *Cymbidium aloifolium* (L.) Sw. using nodal segments as an explants source. Int J Chem Biol Sci 1:77–84

Deb CR, Pongener A (2012b) Studies on the in vitro regenerative competence of aerial roots of two horticultural important *Cymbidium* species. J Plant Biochem Biotechnol 21(2):235–241. https://doi.org/10.1007/s13562-011-0099-5

Fay MF (2018) Orchid Conservation: how can we meet the challenges in the twenty first century? Bot Stud. https://doi.org/10.1186/s40529-018-0232-z

Gerz M, Bueno CG, Ozinga WA et al (2018) Niche differentiation and expansion of plant species are associated with mycorrhizal symbiosis. J Ecol 106:254–264. https://doi.org/10.1111/1365-2745.12873

Hedge SN (1997) Orchid wealth of India. Proc Indian Natl Sci Acad 63:229–244

Hossain MM, Kant R, Van PT et al (2013) The application of biotechnology to orchids. Crit Rev Plant Sci 32(2):69–139. https://doi.org/10.1080/07352689.2012.715984

Hwang HH, Yu M, Lai EM (2017) *Agrobacterium*-mediated plant transformation: biology and applications. Arabidopsis Book. https://doi.org/10.1199/tab.0186

Jackson ST, Overpeck JT (2000) Responses of plant populations and communities to environmental changes of the late Quaternary. Paleobiology 26(S4):194–220. https://doi.org/10.1017/S0094837300026932

Jalal JS, Jayanthi J (2012) Endemic orchids of Peninsular India: a review. J Threat Taxa 4(15):3415–3425. https://doi.org/10.11609/JoTT.o3091.3415-25

Ji Z, Kang S, Zhang Q et al (2016) Investigation of mineral aerosols radiative effects over High Mountain Asia in 1990–2009 using a regional climate model. Atmos Res 178:484–496. https://doi.org/10.1016/j.atmosres.2016.05.003

Kaye TN, Bahm MA, Thorpe AS et al (2019) Population extinctions driven by climate change, population size, and time since observation may make rare species databases inaccurate. Plos One 14(10):e0210378. https://doi.org/10.1371/journal.pone.0210378

Knudson L (1922) Nonsymbiotic germination of orchid seeds. Bot Gaz 73(1):1–25

Kumar CS, Kumar PCS, Sibi M et al (2008) A new species of *Gastrodia* R. Br. (Orchidaceae) from Silent Valley, Kerala, India. Rheedea 18(2):107–110

Lesar H, Hlebec B, Ceranic N et al (2012) Acclimatization of terrestrial orchid *Bletilla striata* Rchb. f. (Orchidaceae) propagated under *in vitro* conditions. Acta Agric Slov 99(1):69–75

Misra S (2007) Orchids of India A glimpse. Bishen Singh Mahendra Pal Singh, Dehra Dun

Misra S (2014) Orchids of Odisha A Handbook. Bishen Singh Mahendra Pal Singh, Dehra Dun

Misra S (2019) Orchids of India A Handbook. Bishen Singh Mahendra Pal Singh, Dehra Dun

Mitchell T, Tanner T, Wilkinson E et al (2006) Overcoming the barriers. Mainstreaming climate change adaptation in developing countries. Institute of Development Studies, Brighton. https://doi.org/10.13140/RG.2.2.17956.37765

Mohanty P, Paul S, Das M et al (2012) A simple and efficient protocol for the mass propagation of *Cymbidium mastersii*: an ornamental orchid of Northeast India. AoB Plants. https://doi.org/10.1093/aobpla/pls023

Panday PK, Thibeault J, Frey KE (2015) Changing temperature and precipitation extremes in the Hindu Kush-Himalayan region: an analysis of CMIP3 and CMIP5 simulations and projections. Int J Climatol 35(10):3058–3077. https://doi.org/10.1002/joc.4192

Pant B (2013) Medicinal orchids and their uses: tissue culture a potential alternative for conservation. Afr J Plant Sci. https://doi.org/10.5897/AJPS2013.1031

Paul P, Kumaria S (2017) Orchids the marvelous plants. NEHU J 15(1):31–40

Paul P, Joshi M, Gurjar D et al (2017) *In vitro* organogenesis and estimation of β-sitosterol in *Dendrobium fimbriatum* Hook.: an orchid of biopharmaceutical importance. S Afr J Bot 113:248–252. https://doi.org/10.1016/j.sajb.2017.08.019

Poschlod P, Kleyer M, Jackel AK et al (2003) BIOPOP—a database of plant traits and internet application for nature conservation. Folia Geobot 38:263–271. https://doi.org/10.1007/BF02803198

Rao VR (1977) Effect of Temperature on the Nitrogenase Activity of Intact and Detached Nodules in *Lotus* and *Stylosanthes*. J Exp Bot 28(2):261–267. https://doi.org/10.1093/jxb/28.2.261

Rao AN, Kumar V (2018) Updated checklist of orchid flora of Manipur. Turczaninowia 21(4):109–134. https://doi.org/10.14258/turczaninowia.21.4.12

Raza A, Razzaq A, Mehmood SS et al (2019) Impact of climate change on crops adaptation and strategies to tackle its outcome: a review. Plants 8(2):34–34. https://doi.org/10.3390/plants8020034

Sarvalingam A, Rajendran A (2016) Rare, Endangered and Threatened (RET) climbers of Southern Western Ghats, India. Rev Chil de Hist Nat 89(9). https://doi.org/10.1186/s4693-016-0058-6

Sharma P, Mohan TM (2016) Role of DBT in promoting biotechnology-based development in North East India. Curr Sci 110(4):562–572. https://doi.org/10.18520/cs/v110/i4/562-572

Stewart SL, Kane ME (2006) Asymbiotic seed germination and in vitro seedling development of *Habenaria macroceratitis* (Orchidaceae), a rare Florida terrestrial orchid. Plant Cell Tiss Org Cult 86(2):147–158. https://doi.org/10.1007/s11240-006-9098-y

Sthapit BR, Upadhyay MP, Baniya BK et al (eds) (2003) On-farm management of agricultural biodiversity in Nepal. Proceedings of a National Workshop, 24–26 April 2001, Lumle, Nepal. In: On-farm management of agricultural biodiversity in Nepal. National Workshop. LIBIRD and IPGRI

Swarts ND, Dixon KW (2009) Terrestrial orchids conservation in the age of extinction. Ann Bot 104:543–556. https://doi.org/10.1093/aob/mcp025

Thomson DJ (1990) Time series analysis of Holocene climate data. Philos Trans R Soc Lond Ser A Math Phys Sci 330:601–616. https://doi.org/10.1098/rsta.1990.0041

Tripathy SK, Roy A, Kushwaha D et al (2016) Perspectives of forest biodiversity conservation in Northeast India. J Biodivers Biopros Dev. https://doi.org/10.4172/2376-0214.1000157

Whiteside JH, Ward PD (2011) Ammonoid diversity and disparity track episodes of chaotic carbon cycling during the early Mesozoic. Geology 39(2):99–102. https://doi.org/10.1130/G31401.1

Wraith J, Norman P, Pickering C (2020) Orchid conservation and research: an analysis of gaps and priorities for globally Red Listed species. Ambio 49:1601–1611. https://doi.org/10.1007/s13280-019-01306-7

Yam TW, Arditti J (2009) History of orchid propagation: a mirror of the history of biotechnology. Plant Biotechnol Rep. https://doi.org/10.1007/s11816-008-0066-3

Yam TW, Thame A (2005) Conservation and reintroduction of the native orchids of Singapore. Selbyana 26:75–80

Yam TW, Tay F, Ang P et al (2011) Conservation and reintroduction of native orchids of Singapore–the next phase. European. J Environ Sci 1(2):38–47. https://doi.org/10.14712/23361964.2015.45

You Q, Kang S, Flugel WA et al (2010a) Decreasing wind speed and weakening latitudinal surface pressure gradients in the Tibetan Plateau. Climate Res 42(1):57–64. https://doi.org/10.3354/cr00864

You Q, Kang S, Flugel WA et al (2010b) From brightening to dimming in sunshine duration over the eastern and central Tibetan Plateau (1961–2005). Theor Appl Climatol 101(3-4):445–457. https://doi.org/10.1007/s00704-009-0231-9

Zettler LW, Rajavelona L, Yokoya K et al (2017) Techniques for the collection, transportation and isolation of orchid endophytes from afar: a case study from Madagascar. Bot Stud 58. https://doi.org/10.1186/s40529-017-0209-3

Zhang S, Yang Y, Li J et al (2018) Physiological diversity of orchids. Plant Divers 40:196–208. https://doi.org/10.1016/j.pld.2018.06.003

Chapter 6
Effectiveness of the Role Technology Plays in Tackling Climate Change

Kafayat O. Shobowale, Abdullatif A. Olanrewaju, and A. A. Muftau

Abstract The fact that industrial sectors contribute significantly toward climate change and its adverse effect on the environment, properties, and sustainability is no longer news; the effect is felt all over the world. There have been studies on the role of different technologies in energy including carbon capture and storage (CCS), biofuel, biomass power generation, wind power generation and solar power generation, and information and communications technology (ICT) including satellite technology, remote sensing (RS), wireless sensor networks, and web-based applications, all of which play a role in monitoring climate change and minimizing its environmental impact. These technologies have made data collection, processing, and analysis in real time possible. This has enabled more informed decision-making. The impact of the role technologies play in mitigating climate change needs to be analyzed based on successes and failures to identify those that require improvement. Technologies have been reported to be under-utilized, or their applications in tackling climate change have not been sufficiently reported. This chapter reviews the impact of energy and ICT-based technological applications in climate change, how technology innovations have helped in monitoring the climate, efficiently managing energy, and adapting to the effect and mitigation of climate change toward sustainable development. Through technology, developments reduce greenhouse gas emissions responsible for global warming. A healthy environment, a sustainable economy, and technological innovations are paramount to all countries' well-being.

K. O. Shobowale (✉)
Department of Mechatronics Engineering, Air Force Institute of Technology, Kaduna, Nigeria
e-mail: kshobowale@gmail.com

A. A. Olanrewaju
Department of Construction and Green Engineering, Universiti Tunku Abdul Rahman, Kampar, Malaysia
e-mail: abdullateef.olanrewaju@ymail.com; abdullateef.olanrewaju@gmail.com

A. A. Muftau
Baniyas East, Abu Dhabi, United Arab Emirates
e-mail: larabiose@yahoo.com

J. N. Furze et al. (eds.), *Earth Systems Protection and Sustainability*,
https://doi.org/10.1007/978-3-030-85829-2_6

Keywords Energy · Sustainable Development Goal · Climate change · Technology · Renewable Energy · ICT

6.1 Introduction

The sustainable development goal (SDG) remit of the United Nations General Assembly has 17 goals; the 13th goal focuses on climate action whose primary objective is to expediently tackle climate change and its impacts on the environment and human lives (UN 2019). The industrial era that revolutionized mass production of products and services to cater for the increasing world population also introduced and left large-scale anthropogenic atmospheric emissions in worrying high quantities. Energy used by infrastructures, industry sectors, and commuting is the main cause of greenhouse gas emissions in the cities (Maric et al. 2015). A number of authors (Huenteler et al. 2015; Kriegler et al. 2014; De Coninck and Sagar 2014) have argued that technology is the key to mitigation and adaptation of the GHG effect in climate change. Climate change detection was only possible through technological enhancement and use of enhanced technological tools to detect the change (Bhandari 2018); renewable energy, for example, is driven by technology.

Renewable energy global power capacity has steadily increased; in 2015, its range covered 1064 gigawatt (GW) (hydro-power), 433 GW (wind power), 227 GW (solar PV), 106 GW (bio-power), 13.2 GW (geothermal), and 4.8GW (solar thermal) (Kabir et al. 2018). The global carbon emission in 2012 was 35.6 billion metric tons (Hosenuzzaman et al. 2015). Increases in temperature (giving rise to heat waves, bushfires), flooding, landslides, rise in seawater level, drought, earthquake, and hurricane are some of the consequences of climate change linked with emissions.

According to the 5th assessment of the Intergovernmental Panel on Climate Change (IPCC, 2014), a body that reports regular assessment of the scientific basis of climate change, global temperature could rise up to more than 4-degree Celsius by 2100 which makes further control of emissions more difficult. As energy was primarily generated from fossil fuels (including coal, oil, and natural gas), the emissions of anthropogenic gases such as CO_2 from fossil fuels led to the search for alternative clean energy (renewable energy). Figure 6.1 depicts global electricity production in 2017, which shows that fossil fuel and nuclear accounted for almost 80% of electricity generation.

Technologies have been proposed to tackle emissions from nonrenewable forms of energy, i.e., by carbon capture and storage (CCC). Renewable energies that have been used in combating climate change include wind, solar PV/solar thermal, geothermal, bioenergy, and hydropower. According to Pan et al. (2018), CO_2 reduction mitigation should be tailored toward improving energy efficiency by 32%, with renewable energy at 32%, carbon capture and storage at 15%, nuclear at 11%, and fuel switching at 10%. Toward a green economy and clean energy supply, environmental friendliness, pollution-free, and relative lower operation and maintenance cost characterize renewable energies (Pan et al. 2018).

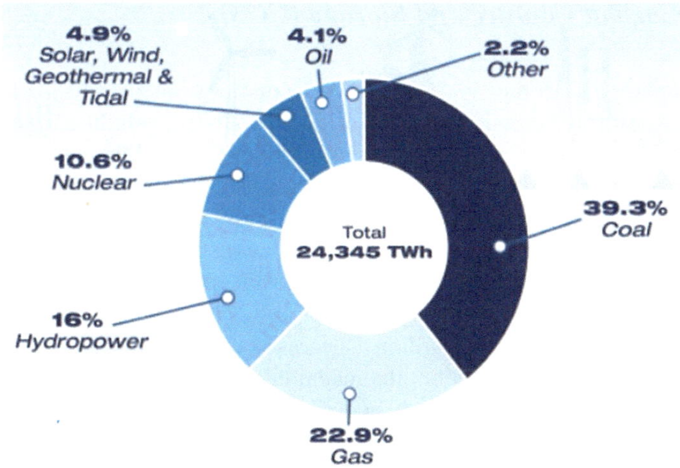

Fig. 6.1 Global electricity production in 2017. (World Nuclear News 2019)

Information and communications technology (ICT) have mainly been used in monitoring and data collection through sensors and other devices in order to minimize environmental impacts. In Rotterdam smart power and water systems are being put in place to use data and information for infrastructure supervision and systems which can increase the release of greenhouse gases (Maric et al. 2015). Climate Pathways, for example, is an application developed to make people understand the need to reduce greenhouse gas emissions to stabilize the Earth's temperature and what effects not reducing emissions has on the environment at large (Climate Pathways 2019). It simulates global temperature change which can be viewed on a mobile device. The app gives answers to how much greenhouse gas emissions have to fall in order to stabilize the Earth's temperature and the implications of limiting reductions of emissions to less than 2 degree Celsius.

6.2 Energy Technologies Combating Climate Change

The prominent energy we use in our everyday lives is fossil fuel supplied (coal, oil, and natural gas) and renewable energy technologies that use natural sources such as in photovoltaic cells to generate solar energy, the use of wind to generate wind energy, the use of water to generate hydropower, the use of steam of depleted oil and gas fields to generate geothermal energy, and the use of biological sources to generate bioenergy. Renewable energies are coming of age and emerge in both household and industrial usage. In the US Energy Information Administration (2012) report, 22% of the total world energy generated was from renewable energy; this, they stated, would have been impossible in the previous 10 years.

6.2.1 Carbon Capture and Storage (CCS)

Carbon capture and storage is a low-carbon technology entailing technologies that are used to absorb CO_2 emissions from different sources (including fossil fuels, as more than 90% of anthropogenic gas emissions released are from the combustion of fossil fuel power plants) and pass them into storage (geological storage, ocean storage, storing by mineral carbonation, storing in terrestrial ecosystems) or keeping them for use (e.g., producing methanol or producing urea as a fertilizer for feedstock). CCS technologies can be integrated gasification combined cycle (IGCC) power plants, oxy-coal fired plants. Braun et al. (2017) reported the negative perception of the public on CSS regarding concerns as to the safety of the storage of the stored CO_2. Capital cost, duration, the magnitude of the CO_2, electricity, and efficiency penalties are highly variable and impact the cost of deployment of CCS (Thronicker et al. 2016; Reiner 2016). Although CSS technology is established in enhanced oil recovery (EOR), its large-scale implementation is yet to be established in CO_2 mitigation in fossil fuel burning. Major global CSS projects are depicted in Fig. 6.2.

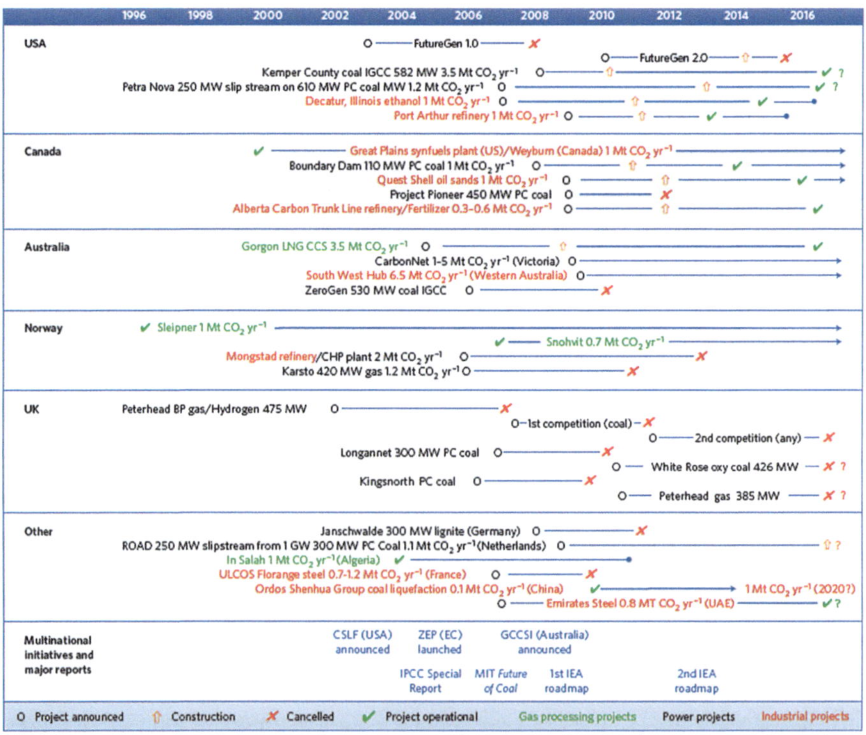

Fig. 6.2 Major CCS projects. (Reiner 2016)

CO_2 has been reported (Rackley 2017) to be the most severe of anthropogenic emissions that cause greenhouse gas (GHG). Human activities, such as energy usage through the combustion of fossil fuel, acidify the ocean; the way land is used/managed and emissions from industries distort the absorption of CO_2 in the ocean and on the land. The generation and use of energy and electricity from fossil fuels are estimated at 85% globally (Wilberforce et al. 2019; Rackley 2017), increasing the rate at which CO_2 is emitted into the atmosphere, which consequently induces climate change. With the estimated 85% use and reliance on energy from fossil fuels, mitigation of CO_2 emissions is imperative and highlights the role of CCS to reduce CO_2 absorption in the atmosphere (Wilberforce et al. 2019; Muratori et al. 2017; Thronicker et al. 2016; Bui et al. 2018; Dowell et al. 2017; Wennersten et al. 2014).

The estimated global number of commercial carbon capture facilities in operation and daily carbon capture capacity is depicted in Fig. 6.3.

Factors that determine the severity of CO_2 emissions are (Rackley 2017) industrial energy usage, quantity of fossil fuel within the total energy supply, population, economy growth/trade globalization, energy efficiency/primary energy production technology growth, policy initiatives, and incentives from environmental pressures.

Carbon capture and storage technologies are still not fully commercialized; the technologies are still undergoing pilot and demonstration testing. The early demonstration project costs have an adverse effect on deployment. CCS demonstration project plant costs could amount to US$1 billion (Reiner 2016). CSS technology has been deployed in Norwegian waste-to-energy (WtE) by reducing the emissions of climate-induced gases. Waste is diverted from landfills to energy and material recovery. With WtE technologies, energy is produced without the need for land occupation such as that of woody biomass (Lausseleta et al. 2017).

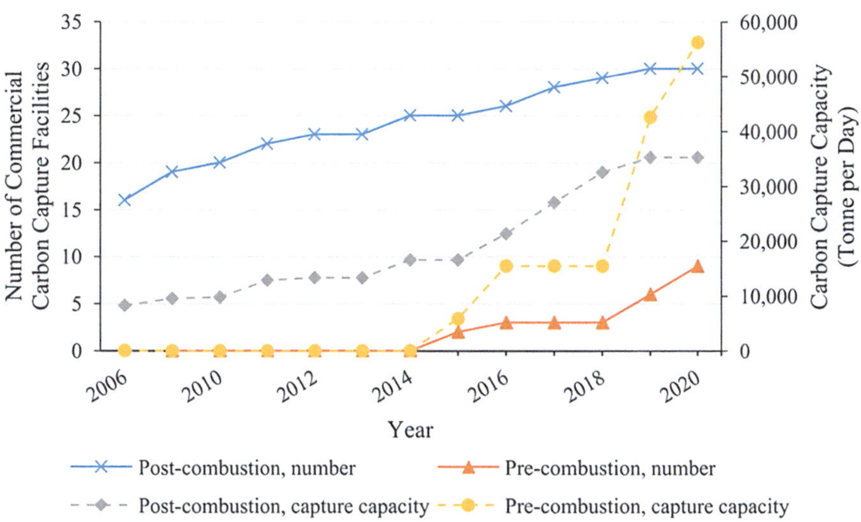

Fig. 6.3 World commercialized CSS technology. (Wilberforce et al. 2019)

Combination of CCS with sources of energy other than fossil fuels has given rise in complementary advantages; bio-energy with CCS (BECCS) has been reported to provide reliable low-carbon electricity (Bui et al. 2018; Muratori et al. 2017), which has been likened to seasonal dependent wind or solar PV renewable energy sources. Fridahl and Lehtveer (2018) affirmed that BECCS is in the development phase with uncertainties on aspects such as storage capacity quantities, biomass availability, biodiversity fights and concerns over food security, deployment expenses and funding opportunities, and rivalry for land and its associated components. Five Bio-CCS projects are in operation (Kemper 2015); each has a capturing capacity between 0.1 and 0.3MtCO2/year of biogenic origin, the largest of all being the Illinois Basin-Decatur Project (IBDP). This is being expanded to Illinois industrial CCS (IL-ICCS) project with a capturing capacity of up to 1MtCO2/year of CO_2. The operational Bio-CCS projects have been reported to have an ethanol plant as the CO_2 source as CO_2 capture from ethanol production is a commercially proven technology (Kemper 2015). CO_2 is used for enhanced oil recovery (CO_2-EOR) in three of the five projects. Kemper (2015) has stressed the importance of EOR as a driver for commercial Fossil-CCS and Bio-CCS.

6.2.2 Renewable Energy Technologies

Renewable energy technologies produce energy from natural sources such as wind, sunlight, and water. They do not face depletion, produce CO_2 emission, and generate liquid or solid waste products (Hosenuzzaman et al. 2015).

6.2.2.1 Solar Energy Technology

The sun is said to produce 1.4 x 10^5 TWh of power with 3.6 x10^4 of the power useable for conversion into energy and other uses (Hosenuzzaman et al. 2015); in 2017, global power consumption was around 25TWh (World Nuclear News 2019) and CO_2 emissions are 33 gigatons (Naturgy 2019). Table 6.1 ranks solar energy acceptance by developed countries as a viable means to reduce carbon emission. Solar energy technology is comprised of either photovoltaic or solar thermal technology (Kabir et al. 2018; Kannan and Vakeesan 2016). Solar photovoltaic technology uses semiconductors to convert sunlight directly into electrical energy and is the most widely used. Photovoltaic cells are the elements used to convert solar energy into electricity. Solar and wind power reduces CO_2 emissions; it is dependent on the weather conditions (MacDonald et al. 2016). Solar photovoltaic technologies in demand are those based on "wafer-thin" cells, commercial thin-film cells, and new thin-film technologies (Kabir et al. 2018). Energy powered by solar is coupled with thermal energy for domestic and commercial applications such as drying, heating, and cooling. Industry-wise, concentrated solar thermal (CST) technologies are used

Table 6.1 Solar power generation capacity, global ranking in 2015

Ranking	Country	Total capacity (MW)	Installed (MW)
1	China	43180	15130
2	Germany	39553	1418
3	Japan	33300	10000
4	USA	27400	7260
5	Italy	19160	700
6	UK	8437	3109
7	Spain	6967	6946
8	France	6680	1020
9	Australia	5049	913
10	India	4680	2048

in heating, while concentrated solar power (CSP) technologies are used for electricity generation (Kabir et al. 2018).

A scenario was used to model the US solar and wind generation by MacDonald et al. (2016); they reported that with the use of future anticipated costs for wind and solar, up to 80% of CO_2 emissions from the electricity sector in the USA could be reduced relative to their 1990 levels. They stated that reductions are made possible with current technologies without electrical storage. The share of electricity production of solar and wind power increases as the system grows to include large-scale weather patterns (MacDonald et al. 2016). From a report by Kabir et al. (2018), American Solar Energy Industries Association projected that the total solar PV capacity of the USA could reach 45 GW by 2017 due to the rate of its deployment. Solar power has become the foremost source of new power in Australia producing 913 MW against 774 MW derived from wind power in 2015. 1300 MW of coal power was decommissioned in Australia to pave way for renewable energy in the same year. Kabir et al. (2018) report that in India, the capacity of the solar power grid installed increased from 3743 (March 2015) to 6762 (March 2016) and 8062 MW (July 2016). Plans are in development in France to construct a 1,000 kilometer roadway powered by solar; each kilometer could provide renewable energy to power around 5000 residential homes (Kabir et al. 2018).

Developed countries that have produced PV power-based electricity with approximate capacities are Germany (7.6GW), China (5GW), Italy (3.4GW), the USA (3.3GW), Japan (2GW), Spain (2.3GW), and the UK (12.8GW at the end of 2017). The PV market is 55% in Europe and is primarily used for electricity production (Hosenuzzaman et al. 2015). Countries are ranked in their generation of solar power in 2015 (Kabir et al. 2018) in Table 6.1.

PV does not add to global warming as its operation gives zero releases of CO_2, NOx, and SO_2 gases. For every kWh of electricity produced, PV saves 0.53 kg of CO_2 emissions (Hosenuzzaman et al. 2015). Reduction in CO_2, NOx, and of SO_2 is estimated at 69–100 million tons, 68,000–99,000 tons, and 126,000 to 184,000 tons, respectively, with projected use of PV by 2030, thus reducing diseases such as heart attacks from 720 to 490 and asthma by 470–320 annually. Annual fuel costs and power cost reductions can be reduced by US$ 220,875.00 and US $26,800.00, respectively. Hosenuzzaman et al. (2015) stressed that investments now are tilted toward solar PV as its manufacturing is the fastest-growing industry coupled with the fact that it has increased by 44% amounting to revenue of US$128 billion. Efficiencies have been coupled; application of solar energy can be simply complemented by rechargeable batteries (Lib et al. 2017).

6.2.2.2 Wind Energy

Wind energy has been used with other types of renewable energies as a complement. François et al. (2017) demonstrated how wind energy generation can augment Norway's hydrothermal energy generation. Wind and solar energies are estimated to reach 420GW and generate around 25% of electricity in Europe by 2030 (Grams et al. 2017); local co-deployment of wind and solar PV can balance seasonal variability. Solar PV and wind generator system could produce reliable energy supply generating approximately 30% more electrical energy (around 100 TWh/year.), if supplemented with 30 days of storage as compared to the current electricity supply system (Fragaki et al. 2019). Wind and solar energy power has been soaring in the USA (Millstein et al. 2017), estimated total wind and solar air quality benefits as of 2015 were between US$29.7 and 112.8 billion mostly from 3,000 to 12,700 avoided premature mortalities, and total climate benefits as of 2015 amount to between US$5.3 and 106.8 billion, with ranges based on air quality, health impact models, and the social cost of carbon estimates (Millstein et al. 2017).

6.2.2.3 Geothermal Energy

Geothermal energy is produced from underground hot water reservoirs, and heat or steam is used to drive an electrical generator. Irrespective of meteorological conditions, geothermal resources can be produced (Moya 2018; Tomasini-Montenegro et al. 2016). The geothermal power plant is classically defined as (Moya 2018) binary (14%), backpressure (1%), single flash (41%), double flash (19%), triple flash (2%), and dry steam plant (23%). The steps in developing geothermal energy are shown in Fig. 6.4

According to the International Energy Agency roadmap for technology on geothermal heat and power documented by Pan et al. (2018), by 2050, geothermal electricity generation could grow up to 1,400 TWh per year, about 3.5% of global

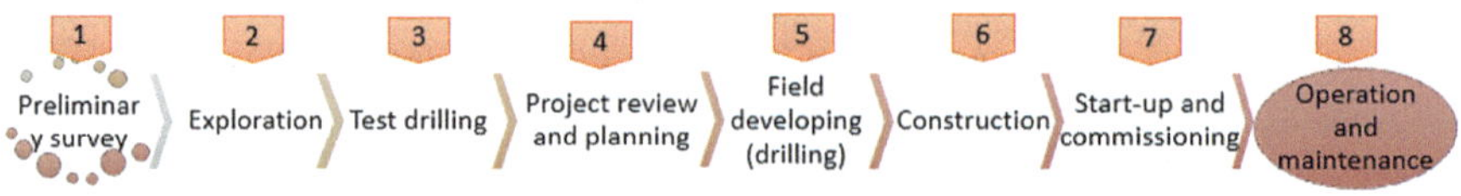

Fig. 6.4 Stages of developing geothermal energy. (Moya et al. 2018)

electricity production, thus avoiding approximately 800 Mt of CO_2 emissions per year. Geothermal energy could generate electricity continuously for the whole day increasing energy security with operational flexibility, compared to that of wind and solar energy; it is predictable and its resource is flexible and provides sustainable and clean energy, CO_2 emissions, and air and water pollutant reductions (Pan et al. 2018; Moya et al. 2018).

6.2.2.4 Hydropower Energy

Hydropower energy is derived from the movement of water that moves from high- to low-pressure gradients. Hydropower is reported to be a mature source enough to generate electricity and turn turbines. It is reported that the current global installed capacity of hydropower is very little compared to its potential (Owusu and Asumadu-Sarkodie 2016). The World Energy Council report of 2013 stated that approximately 50% of installed hydropower capacity dwells among four countries which are China, Brazil, Canada, and the USA (World Energy Council 2013). 95.3% of Norwegian electricity generation is derived from hydropower (François et al. 2017). Calderón et al. (2016) reported that Columbia generates 76% of its electricity from hydropower; the sector that consumes the most fossil fuel is transportation.

6.2.2.5 Bioenergy

In biomass for energy production, we recall that carbon taken from the atmosphere during photosynthesis is the carbon that is released to the atmosphere in the course of energy conversion, as such, the system is seen as carbon neutral. Bioenergy is usually used in conjunction with carbon capture and storage. Bioenergy with carbon capture and storage (BECCS) is achieved when energy is generated from the conversion of biomass to produce CO_2 that is captured, transported, and permanently kept in geological formations (Pour et al. 2017). Using BECCS to generate energy and remove CO_2 from the atmosphere makes it one of the most promising global warming mitigation options as a low-carbon energy system (Pour et al. 2017; Quader and Ahmed 2017). Variables include (Pour et al. 2017) the type of biomass resource, the conversion technology, the CO_2 capture process used, and storage options, together with their environmental, economic, and social impacts. According to Pour et al. (2018), there are about 20 BECCS projects globally, the majority of which are in Europe, North America, and Scandinavia. However, with only five operating, the range of the capacity is between 0.1 and 0.3MtCO2/year negative CO_2 emission, capturing CO_2 from production plants of ethanol.

In 2010, bioelectricity provided only 1.5% of global electricity production (Pour et al. 2017), 3.1% between 2011 and 2012 (68% of renewable energy production). It is estimated that by 2035, world electricity generation by bioenergy should be around 4.1%, 20% of bioenergy is used for biofuel, while the remaining 80% is used for heat and energy production (Pour et al. 2018; Ayoub and Abdullah 2012).

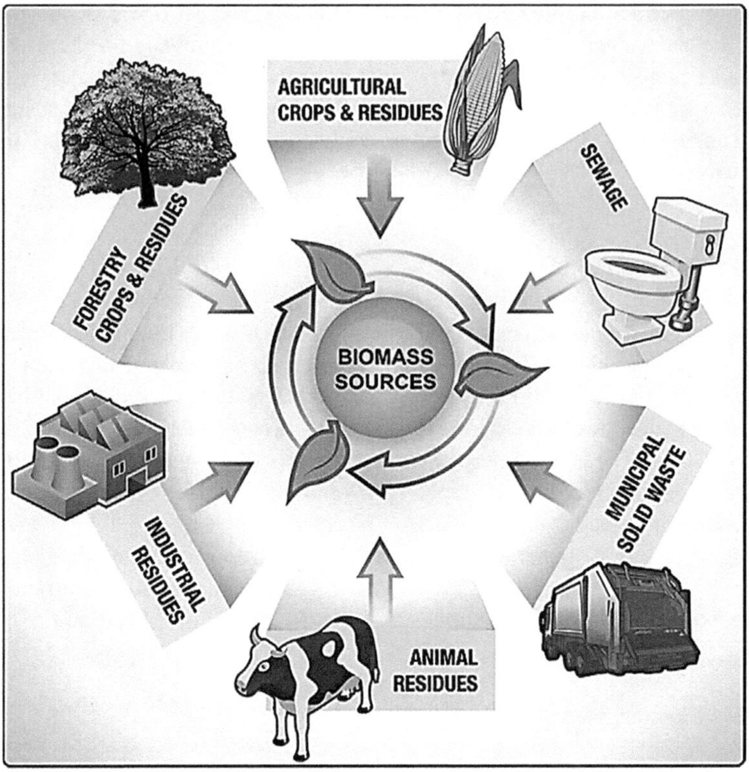

Fig. 6.5 Sources of biomass. (Quader and Ahmed 2017)

Bioelectricity is generated in 62 countries around the world (Pour et al. 2018); the USA (69.1 TWh), Germany (49.1 TWh), and Brazil (32.9 TWh) have the highest share of bioelectricity in their power sector. The biomass sources that produce bioenergy are depicted in Fig. 6.5.

6.3 Information and Communications Technology Combating Climate Change

The impact of information and communications technology (ICT) in our everyday lives cannot be over-emphasized. From suggestions on what to eat, what to wear, navigating through cities, shopping for items, and so on, ICT is ever present. In the same vein, developers are leveraging the versatility of ICT to solve problems arising from the challenges of climate change and environmental degradation facing our dear planet. Advances in development in ICT have been used to mitigate the effects of climate changes in different facets of human life.

The different types of models for generating scenarios for climate change are products of ICT. These models give insight used to build further actions for mitigating climate change and its effects. Technological computer hardware and software handle complex tasks that help to manage optimization difficulties using computational resources applicable to the sustainable energy field (Baños et al. 2011). A study conducted by Fernando and Okuda (2009) reported that it is possible to remove around 7.8 gigatons of CO_2 emission through ICT deployment by 2020 considering aspects such as environmental impacts in designing, implementing, and evaluating ICT policies and initiatives. Preuschmann et al. (2017) discussed the web-based climate service product IMPACT2C; different climate projects and their outputs are bundled into visualizing the effects of CO_2-induced global warming climate from energy, health, agriculture, forestry, ecosystem, water, and tourism. Bachelet et al. (2017) designed web-based applications that can be used as a knowledge base for understanding the impact of climate change and charting appropriate actions to mitigate its effect.

Green ICT can be used to raise climate change awareness, to gather climate data/information (from sensors and other technologies), to change people's perception on climate change through education/training, and to develop techniques and tools (smart/intelligent cities, web applications, communication tools) to yield better productivity toward mitigating climate change (Pattinson 2017).

Climate information through ICT can be used in a lot of sectors such as (Kundzewicz et al. 2017) health, construction, agriculture and forestry, flood risk reduction, and water and sewage sectors. Forecasting is one aspect of ICT; projecting incidence of landslides, earthquake, heat wave, or cold wave will help the public and authorities be prepared. It is reported that cold waves in winter are still a major killer in Poland (Kundzewicz et al. 2017); updated information through ICT could mitigate this. Priority is given to the transformation of climate data by the climate system services in Europe to produce customized products and services (Swart et al. 2017), projections, trends, forecasts, information, economic analysis, technology assessments, best practices counseling, development and evaluation of solutions, and other services related to climate that could be used for the benefit of all. Broadly, the services include data, information, and knowledge that support adaptation, mitigation, and disaster risk management (DRM). Some information portals developed in Europe and other countries are (Swart et al. 2017) the Climate Information Portal for Copernicus (CLIPC) project, DRIAS, UKCP09 (UK Climate Projections), Climate Change Knowledge Portal, Climate4impact (IS-ENES), Climate Explorer, Climate.gov, Climate Adapt, GCM Data Portal, and ECA&D. Data sources include global Coupled Model Intercomparison Project, Phase 6 (CMIP6), Inter-Sectoral Impact Model Intercomparison Project (ISI-MIP), and the Agricultural Model Intercomparison and Improvement Project (AgMIP – a modeling comparison program of impacts of climate change on agricultural systems).

Several authors have reported that the reduction of emissions that can cause GHG can be achieved with the use of ICT (Swart et al. 2017; Stewart 2015; Pattinson 2017; Monzon et al. 2017, Maric et al. 2015) and green ICT (Suryawanshia and Narkhedeb 2015; Radu 2014). Stewart (2015) states that in the transportation case,

ICT tools have further helped in sustainable transport system mobile applications and online systems that promote a reduction in private vehicles through public transports and car-pooling. Further, Stewart (2015) identified carbon emission reduction through use of ICT; in simulation results, the transportation sector contributes approximately 25% of CO_2 emissions in the UK. Their work based on the European Union COMPASS project was used in one of the objectives determining the role of ICT in sources of transportation data and carbon reduction quantifications that could be achieved with the use of ICT. They used a LATIS (land use and transport integration in Scotland) model, an integrated land-use and transport model for strategic assessment across Scotland. Using the model, the effects of the relative impact of possible ICT scenario could be generated spanning sub-regions such as in urban, inter-urban, and rural regions (Stewart 2015).

Investigating the role ICT plays toward the sustainable development of energy efficiency in urban planning, authors (Maric et al. 2015) state that ICT is mainly used in this sector for infrastructure, renewable energy, and transportation and that ICT is a necessity as an important tool which should be used to develop model, simulations, and measurement for an energy-efficient ecosystem for managing the surroundings to reduce greenhouse gas emissions. ICT is incorporated into urban planning systems with computer tools and software, modeling, numerical modeling of the energy performance of buildings, and advanced computer electronic hardware.

Singapore, Stockholm, and Brisbane are using intelligent systems to reduce traffic jams and air pollution (Maric et al. 2015). The creation of databases for the spatial and urban plans is achieved using mapping-based technology of geographic information systems (GIS) (Maric et al. 2015), for spatial and statistical analyses, remote detections, 3D analyses and modeling, and plan visualization. For adaption to climate change, control and monitoring of climate change indicators, development of prediction model, development of scenarios, economic analyses, monitoring of the impact of extreme temperature change, modification of current adaptation strategies and measures, and stakeholder inclusion are all powered by ICT (Maric et al. 2015). Aleke and Nhamo (2016) reported through literature review how ICT is used in climate change adaptation in the mining industry to reduce its risk of exposure of the industry to climate change. They reported that the risk associated with climate change is threatening the survivability of the mining industry.

6.4 State of the Perceived Adoption of Technology in Combating Climate Change

The level of technology adoption for climate change depends on the availability of the natural resources that enables renewable energy. Each country is blessed with an abundance of one or more of these natural resources. Harnessing these natural resources for renewable energy depends on each country's seriousness in tackling climate change.

All the technologies have their pros and cons, one technology cannot provide the energy the world needs (Wilberforce et al. 2019; Huisingh et al. 2015; Kalkuhl et al. 2014), and the use of fossil fuel will continue to grow. It is recommended that a mixture of all technologies should be used with the ones that have the most adverse effects both to the environment and human survival such as through the use of fossil fuel and bioenergy, which should be further explored through research and development to find a balance. Kundzewicz et al. (2017) state that even though Poland and Norway have abundance of fossil fuel (oil and gas in Norway and coal in Poland), Norway is more proactive in climate change than Poland. Also François et al. (2016) reported that solar and hydropower dual combination performed better in case study scenarios. There is a 30% share target for renewable energy in Turkey's national energy sources by the Turkish government by 2030 (Pan et al. 2018).

6.4.1 Renewable Energy Effectiveness

Wind and solar energy deployment has been reported to increase over time. Millstein et al. (2017) reported that in 2015 in the USA, due to the associated air quality that is of benefit of the deployment of wind and solar power, 3000 to 12,700 premature mortalities were avoided which resulted in cost benefits of between $29.7 and $112.8 billion. Also, in the same period, savings in climate issue were between $5.3 and $108.8 billion.

The results of a statistical model by Millstein et al. (2017) to determine the percentage of CO_2, SO_2, NOx, and PM2.5 emissions reduced due to solar and wind generation in the period between 2007 to 2015 are 20%, 72%, 50%, and 46%, respectively. Utility of wind increased; use of utility solar and distributed photovoltaic power was from approximately 10GW in 2007 to approximately 100GW in 2015. Energy generation grew from 35,000GWh/y in 2007 to 227,000GWh/year in 2015. Seventeen percent of total wind and solar generated in 2015 was from solar, up from less than 5% in 2007. SO_2 emissions at 9.0 million tons in 2007 became 2.5 million tons in 2015 due to optimized control technologies on coal power plants (Millstein et al. 2017). With the installation of 113,533 household solar systems in California, USA, 696,544 metric tons of CO_2 emissions have been reduced. CO_2 emission for each KW/hr generated from coal, natural gas, and solar is estimated (Kabir et al. 2018) to be approximately in the range 0.64 to 1.63, 0.27 to 0.91, and 0.03 to 0.09 kg (emission ratio of 18:9.5:1), respectively. The top seven countries with renewable power capacities in 2015 are shown in Fig. 6.6, wind power is the most abundant of all the renewable energy, and China tops the list.

In 2017, the country with the largest installed capacity of geothermal power is the USA (Pan et al. 2018), with approximately 3.72 GW, the Philippines have approximately 1.93 GW, Indonesia approximately 1.86 GW, Turkey approximately 1.06 GW, New Zealand approximately 0.98 GW, Mexico approximately 0.92 GW, Italy approximately 0.92 GW, and Iceland approximately 0.71 GW all in 2017.

It was reported that geothermal energy in New Zealand in 2017 produces approximately 13% of the country's electricity supply (Pan et al. 2018). In 2010, the

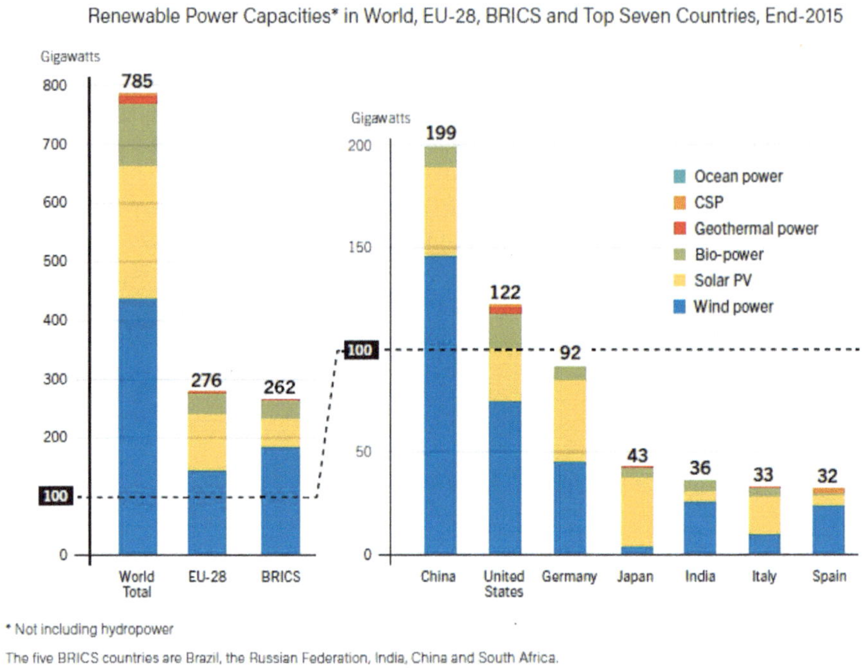

Fig. 6.6 2015 top seven countries with renewable energy capacities. (Kabir et al. 2018)

installed geothermal capacity in the Philippines is approximately 1.87 GW (12% of the national power mix). Indonesia is estimated to have 40% (29GW) of the world's geothermal resources, with about 4% of the total capacity in use. Turkey in 2015 installed 0.9% (approximately 620 MW). About 13% of New Zealand's electricity supply is produced from geothermal energy, with a target of 90% renewable electricity by 2025 (Pan et al. 2018). From these analyses, Kabir et al. 2018 reiterated that renewable energies are yet to produce optimum capacity as a lot still needs to be done to increase the share of renewable energies by all parties.

Only a few countries (such as China, Germany, Greece, Japan, the USA, Italy) have been reported (Creutzig et al. 2017) to be aggressive in their drive to increase power generation using renewable energy, and some meet more than 5% of their electricity production from photovoltaic solar energy and other renewable energies.

6.4.2 *Carbon Capture and Storage Effectiveness*

According to Thronicker et al. (2016), more than a quarter of CCS projects have been put on hold, postponed, or canceled globally due to factors such as high risk, high cost, and economic viability. Their empirical finding suggests that CCS projects are likely to succeed if CO_2 storage sites are confirmed and technology to use in capturing carbon is carefully looked at as the first step in its deployment. 152–240 Gt of CO_2 can be sequestered globally with CSS (Muratori et al. 2017).

A case of successful CSS projects in enhanced oil recovery is Statoil's Sleipner project (stores 1 million tons/year CO_2 since 1996, shipped hundreds of kilometers for use in enhanced oil recovery operations for over 30 years). BP initiated the first large-scale CCS power project in 2002. The first fully integrated CCS power project that captures, transports, and stores was by the Boundary Dam in Saskatchewan in late 2014. CCS is currently at the demonstration stage with issues such as unproven business models, skeptical investors, novel technology integration challenges, and the need to deliver on a commercial scale (Reiner 2016). Stewart (2015) reported that in Glasgow public transport, in-vehicle time could decrease by 10% with a reduction of 108 tons/year of CO_2. Reduction in CO_2 in urban areas equates to around 38 tons/year, in inter-urban areas 18 tons, and in rural areas 4 tons/year. Using information technology tools in a case study, Maric et al. (2015) reported that the location of spatial characteristics detailed the study was enhanced with the use of ICT for the selection of the final location spatial characteristics. With this, energy consumption of buildings was reduced taking into consideration the urban measures of the building and building natural features into the design phase. Global geothermal power plant installation is 12.729MW and estimated to be 21.443MW by 2020, while the global thermal capacity installation is 70.885 MWT distributed in nine direct-use geothermal energy applications (Moya et al. 2018; Pan et al. 2018).

The lack of CCS projects in commercialization for the energy sector may suggest some form of unseriousness on the part of different countries' goals on mitigating climate change and CCS technologies (Reiner 2016), while Wennersten et al. (2015) claimed that since CSS has been successfully used in enhanced oil recovery, the experience can be bought to place in reducing emissions by fossil fuels which can yield commercialization. Cooperation between parties can bring a lot of benefits. The CCS process is energy-intensive and has its own fair share of CO_2 emission and other environmental impacts (Rackley 2017), which as Bennett et al. (2017) reported may be tackled with the use of solar energy to power the CCS and (Kemper 2015; Bui et al. 2018; Fridahl and Lehtveer 2018), using bio-energy with CCS, provides electricity that is reliable and has low carbon, which is not weather dependent as wind or photovoltaic. This is the same for solar energy as seen in Table 6.1, where the installed amount is small for some countries as compared to their potential.

6.4.3 ICT Effectiveness

ICT in climate change mitigation can be awareness-raising on climate-related risks, providing data to make decision for adaptation, giving early warning systems for flooding in preparation for timely evacuations, knowledge sharing on adaptation, coordinating disaster recovery information, developing adaptation policies, providing training in flood and risks management, and gathering and analyzing information for vulnerability assessments (Aleke and Nhamo 2016). Ghana mining sector uses AFIS (advanced fire information systems) to check the negative impacts of wildfires on the environment (Aleke and Nhamo 2016). Stewart (2015) reported in their study that in Glasgow public transport in-vehicle time could get a reduction of 1.06 million kms/year using ICT, which will, in turn, reduce CO_2 emissions. Swart et al. (2017) recommended that for the climate information portal to be effective, different proposed users should be taken into consideration, developing extension of existing portals rather than reinventing the wheel of developing a new portal that already exists, and data and information needs to be correct, maintaining user evaluation of the portal open for improvement.

Radu (2014) reported that anthropogenic emissions can be reduced with ICT-based solutions such as "smart grid" (renewable energy and reduction on fossil fuel technologies), to save an estimated release of approximately 2 billion tons of CO_2 with cost savings; smart logistics (optimizing logistics network), to save the release of around 1.68Gt CO_2 and $341 billion; smart building (optimization of energy, using technologies for building management systems), to save the release of around 1.52Gt CO_2 and $442 billion; and smart motor systems (automation through ICT of motor systems manufacturing), to save the release of around 1.0Gt CO_2 and $107 billion. Further dematerialization (using technologies such as ICT for teleconferencing, encouraging online media) is estimated to save the release of around 1.0Gt CO_2.

6.5 Conclusion

Technology is one of the keys to sustainable development. Carbon capture and storage, renewable energy, and information and communications technology all work together for the benefit of the environment and humanity. Data can be collected from CCS and renewable energy systems through sensors and ICT processes giving insights and informed decisions toward the goal of tackling climate change. Since there is an abundance of climate data, ICT is very important in climate change forecast of future trends; software simulation tools help analyze climate change scenarios. Mobile technology helps in communication such that a potential disaster is communicated beforehand and adequate preparation is made for any eventuality. Perception by those vested with the authority to make decisions on climate change mitigation is very important, for instance, in setting mitigation goals by the International Energy Agency (IEA), Intergovernmental Panel on Climate Change (IPCC), United Nations (UN), and other agencies. Real-time challenges are escalating costs and technical challenges; further authorities' perception impedes project progress, while these same challenges are felt by others who forge ahead with the development. All technologies have their pros and cons, one technology cannot provide the energy the world needs, and the use of fossil fuel will continue to grow. It is recommended that a mixture of all technologies discussed in this chapter should be used with the ones that have the most adverse effects both to the environment and human survival such as through the use of fossil fuel and bioenergy, which should be further explored through research and development to find a balance. Sustainable Development Goals recommend measures are needed to reduce emissions such as the carbon footprint of fossil fuels together with the use of renewable energy sources. Different integrated assessment models have been used for scenario analysis and have proven the efficacy of these technologies.

References

Aleke BI, Nhamo G (2016) Information and communication technology and climate change adaptation: evidence from selected mining companies in South Africa, Jàmbá. J Disaster Risk Stud 8(3). https://doi.org/10.4102/jamba.v8i3.250

Ayoub M., A.Z. Abdullah (2012) Critical review on the current scenario and significance of crude glycerol resulting from biodiesel industry towards more sustainable energy industry. Renewable and Sustainable Energy Reviews 16(5), https://doi.org/10.1016/j.rser.2012.01.054

Bachelet D, Gough M, Sheehan T et al (2017) Climate consoles: pieces in the puzzle of climate change adaptation. Clim Serv 8:36–43

Baños R, Manzano-Agugliaro F, Montoya FG, Gil C, AlcaydevA and Gómez J (2011) Optimization methods applied to renewable and sustainable energy: A review. Renewable and Sustainable Energy Reviews, 2011, vol. 15, issue 4, 1753–1766

Bennett R, Clifford S, Anderson K et al (2017) Carbon capture powered by solar energy. Energy Procedia 114:1–6

Bhandari MP (2018) Climate change science: a historical outline. Adv Agric Environ Sci 1(1):5–12

Braun C, Merk C, Pönitzsch G et al (2017) Public perception of climate engineering and carbon capture and storage in Germany: survey evidence. Clim Pol. https://doi.org/10.1080/14693062.2017.1304888

Bui M, Fajardy M, Dowell NM (2018) Bio-energy with carbon capture and storage (BECCS): opportunities for performance improvement. Fuel 213:164–175

Calderón S, Alvarez AC, Loboguerrero AM et al (2016) Achieving CO2 reductions in Colombia: effects of carbon taxes and abatement targets. Energy Econ 56:575–586

Climate Pathways (2019) Mobile App. https://www.climateinteractive.org. Accessed on 27 July 2019

Creutzig F, Agoston P, Goldschmidt JC et al (2017) The underestimated potential of solar energy to mitigate climate change. Nat Energy 2:17140

De Coninck H, Sagar A (2014) Making sense of policy for climate technology development and transfer. Clim Pol 15(1):1–11

Dowell NM, Fennell PS, Shah N et al (2017) The role of CO_2 capture and utilization in mitigating climate change. Nat Clim Chang 7:243–249

Fernando P, Okuda A (2009) Green ICT: a cool factor in the wake of multiple meltdowns, Xuan Zengpei, Place, ESCAP technical paper, December 2009. Available via https://www.unescap.org/sites/default/files/Green%20ICT.pdf. Accessed 21 May 2021

Fragaki A, Markvart T, Laskos G (2019) All UK electricity supplied by wind and photovoltaics – the 30–30 rule. Energy 169:228–237

François B, Borga M, Creutin JD et al (2016) Complementarity between solar and hydro-power: sensitivity study to climate characteristics in Northern-Italy. Renew Energy 86:543–553

François B, Martino S, Tøfte LS et al (2017) Effects of increased wind power generation on mid-Norway's energy balance under climate change: a market based approach. Energies 10(2):227. https://doi.org/10.3390/en10020227

Fridahl M, Lehtveer M (2018) Bioenergy with carbon capture and storage (BECCS): global potential, investment preferences, and deployment barriers. Energy Res Soc Sci 42:155–165

Grams CM, Beerli R, Pfenninger S et al (2017) Balancing Europe's wind-power output through spatial deployment informed by weather regimes. Nat Clim Chang 7:557–562. https://doi.org/10.1038/nclimate3338

Hosenuzzaman M, Rahim NA, Selvaraj J et al (2015) Global prospects, progress, policies, and environmental impact of solar photovoltaic power generation. Renew Sust Energ Rev 41:284–297

Huenteler J, Schmidt TS, Ossenbrink J et al (2015) Technology life-cycles in the energy sector – technological characteristics and the role of deployment for innovation. Technol Forecast Soc Change. https://doi.org/10.1016/j.techfore.2015.09.022

Huisingh D, Zhang Z, Moore JC et al (2015) Recent advances in carbon emissions reduction: policies, technologies, monitoring, assessment, and modeling. J Clean Prod 103:1–12

IPCC (2014) Climate change 2014: synthesis report. Contribution of working groups I, II and III to the fifth assessment report of the intergovernmental panel on climate change [Core Writing Team, RK Pachauri and LA Meyer (eds)]. IPCC, Geneva, Switzerland, 151 pp. Available via https://www.ipcc.ch/site/assets/uploads/2018/02/SYR_AR5_FINAL_full.pdf. Accessed 21 May 2021

Kabir E, Kumar P, Kumar S, Adelodun AA, Kim KH (2018) Solar energy: potential and future prospects. Renew Sust Energ Rev 82:894–900

Kalkuhl M, Edenhofer O, Lessmann K (2014) The role of carbon capture and sequestration policies for climate change mitigation. Environ Resour Econ

Kannan N, Vakeesan D (2016) Solar energy for the future world: a review. Renew Sust Energ Rev 62:1092–1105

Kemper J (2015) Biomass and carbon dioxide capture and storage: a review. Int J Greenhouse Gas Control 40:401–430

Kriegler E, Weyant JP, Blanford GJ et al (2014) The role of technology for achieving climate policy objectives: overview of the EMF 27 study on global technology and climate policy strategies. Clim Chang 123:353–367

Kundzewicz ZW, Førland EJ, Piniewski M (2017) Challenges for developing national climate services – Poland and Norway. Clim Serv 8:17–25

Lausseleta C, Cherubinia F, Oreggionia GD et al (2017) Norwegian waste-to-energy: climate change, circular economy and carbon capture and storage. Resour Conserv Recycl 126:50–61

Lib Q, Liu Y, Guo S et al (2017) Solar energy storage in the rechargeable batteries. Nano Today. https://doi.org/10.1016/j.nantod.2017.08.007

MacDonald AE, Clack CTM, Alexander A et al (2016) Future cost-competitive electricity systems and their impact on US CO_2 emissions. Nat Clim Chang 6:526–531

Maric I, Pucar M, Kovacevic B (2015) Reducing the impact of climate change by applying information technologies and measures for improving energy efficiency in urban planning. Energy Build. https://doi.org/10.1016/j.enbuild.2015.04.044

Millstein D, Wiser R, Bolinger M et al (2017) The climate and air-quality benefits of wind and solar power in the United States. Nat Energy 2:17134. https://doi.org/10.1038/nenergy.2017.134

Monzon A, Garcia-Castro A, Valdes C (2017) Methodology to assess the effects of ICT-measures on emissions. The Case Study of Madrid. Proc Eng 178:13–23

Moya, Diego & Aldás, Clay & Kaparaju, Prasad, (2018) "Geothermal energy: Power plant technology and direct heat applications," Renewable and Sustainable Energy Reviews, Elsevier, vol. 94(C), pages 889–901

Muratori M, Kheshgib H, Mignoneb B et al (2017) Carbon capture and storage across fuels and sectors in energy system transformation pathways. Int J Greenhouse Gas Control 56:34–41

Naturgy (2019). https://naturgy.ie/solar-pv-ireland-finally-takes-significant-step-right-direction/. Accessed 21 May 2021

Owusu PA, Asumadu-Sarkodie S (2016) A review of renewable energy sources, sustainability issues and climate change mitigation. Cogent Eng. https://doi.org/10.1080/23311916.2016.1167990

Pan SY, Gao M, Shah KJ et al (2018) Establishment of enhanced geothermal energy utilization plans: barriers and strategies. Renew Energy. https://doi.org/10.1016/j.renene.2018.07.126

Pattinson C (2017) ICT and green sustainability research and teaching. IFAC Papers On Line 50-1:12938–12943

Pour N, Webley PA, Cook PJ (2017) A sustainability framework for bioenergy with carbon capture and storage (BECCS) technologies. Energy Procedia 114:6044–6056

Pour N, Webley PA, Cook PJ (2018) Potential for using municipal solid waste as a resource for bioenergy with carbon capture and storage (BECCS). Int J Greenhouse Gas Control 68:1–15

Preuschmann S, Hänsler A, Kotova L et al (2017) The IMPACT2C web-atlas – conception, organization and aim of a web-based climate service product. Clim Serv 7:115–125

Quader MA, Ahmed S (2017) Bioenergy with Carbon Capture and Storage (BECCS): future prospects of carbon-negative technologies. Clean Energy Sustain Dev. https://doi.org/10.1016/B978-0-12-805423-9.00004-1

Rackley SA (2017) Carbon capture and storage, 2nd edn. Elsevier, Butterworth Heinemann

Radu LD (2014) Green ICTs potentials in emerging economies. Proc Econ Financ 15:430–436

Reiner DM (2016) Learning through a portfolio of carbon capture and storage demonstration projects. Nat Energy 1:15011

Stewart K (2015) Assessing the carbon impact of ICT measures: a case study investigation using latis1 model. Int J Trans Sci Technol 4(2):277–134

Stigson P, Hansson A, Lind M (2012) Obstacles for CCS deployment: an analysis of discrepancies of perceptions. Mitig Adapt Strat Glob Change 17(6):601–619. https://doi.org/10.1007/s11027-011-9353-3

Suryawanshia K, Narkhedeb S (2015) Green ICT for sustainable development: a higher education perspective. Proc Comput Sci 70:701–707

Swart RJ, de Bruin K, Dhenain S et al (2017) Developing climate information portals with users: promises and pitfalls. Clim Serv 6:12–22

Thronicker D, Lange I, Pless J (2016) Determining the success of carbon capture and storage projects. The Electricity Journal, 29(7):1–4

Tomasini-Montenegro C, Santoyo-Castelazo E, Gujba H et.al (2016) Life Cycle Assessment of geothermal power generation technologies: an updated review. Appl Therm Eng. https://doi.org/10.1016/j.applthermaleng.2016.10.074

UN (2019) Sustainable Development Goals Goal 13. Available via United Nations https://www.un.org/sustainabledevelopment/climate-change/ Accessed 27 July 2019

Wennersten R, Sun Q, Li H (2014) The future potential for Carbon Capture and Storage in climate change mitigation – an overview from perspectives of technology, economy and risk. J Clean Prod. https://doi.org/10.1016/j.jclepro.2014.09.023

Wennersten R, Sun Q, Li H (2015) The future potential for Carbon Capture and Storage in climate change mitigation – an overview from perspectives of technology, economy and risk. J Clean Prod. https://doi.org/10.1016/j.jclepro.2014.09.023

Wilberforce T, Baroutaji A, Soudan B et al (2019) Outlook of carbon capture technology and challenges. Sci Total Environ 657:56–72

World Energy Council (2013). (https://www.worldenergy.org/) Assessed on 6th April, 2020

World Nuclear News (2019). http://world-nuclear-news.org/Articles/Nuclear-Innovation-Clean-Energy-Future-(NICE-Futur. Accessed 4 Aug 2019

Chapter 7
Social and Environmental Imperatives for Risk Management: Lessons from the Rohingya Refugee Crisis

Sudhir Kumar

Abstract The world is witnessing the highest levels of displacement on record. An unprecedented 70.8 million people around the world were forced from their homes by conflict and persecution in 2018. Among them were nearly 30 million refugees, over half of whom were under the age of 18. Since late 2017, over a million Rohingya refugees have been present in the Cox's Bazar district of Bangladesh. Cox's Bazar is one of the poorest districts of the country and is prone to natural hazards including cyclone, floods, and landslides. Bangladesh is one of the world's most densely populated countries in the world, and influx of over one million refuges added pressure on natural resources. The influx due to displacement occurred over a short duration. Limited space and time provisions existed for the planned settlement of refugees. Herein the Kutupalong camp is highly congested and located on unstable slopes, though is considered the world's biggest refugee camp. The humanitarian response to this crisis took several measures, which ranged from understanding risk to setting up risk governance mechanisms, thus reducing risk and managing residual risk. The response involved the government, UN agencies, academia, NGOs, and community volunteers. This chapter draws key lessons from risk governance and coordination among partners effecting management of risk. Addressing risk includes structural and non-structural measures pertaining to policy formation and coordination. Key challenges faced in addressing the risks imposed during flood and cyclone periods in 2018 are detailed. The measures and challenges are context specific and comprehensive background is given. The management of risk in Rohingya refugee crisis is a continuous exercise as the overall policy to govern this crisis evolves. The chapter is closed with a set of broad recommendations for managing disaster risk in a conflict/refugee context, which can be customized as per the ground realities.

Keywords Refugee crisis · Disaster · Climate change · Social and environmental impact-risk management

S. Kumar (✉)
Asian Disaster Preparedness Centre (ADPC), Bangkok, Thailand
e-mail: sudhir.ise@gmail.com

J. N. Furze et al. (eds.), *Earth Systems Protection and Sustainability*,
https://doi.org/10.1007/978-3-030-85829-2_7

7.1 Introduction

The world is witnessing the highest levels of displacement on record; 70.8 million people around the world were forced from their homes at the end of 2018 (UNHCR 2019). Among the displaced, nearly 25.9 million were refugees and more than half of them were children. Displaced populations, including refugees, are often concentrated in disaster-prone areas. The displacement of an individual or a community is a consequence of its vulnerability to withstand shocks. Thus, displacement is both a consequence of the vulnerability and a cause of future vulnerability (Pinto et al. 2014).

The relationship between displacement, including refugee crises, and vulnerability is complex and nuanced as they are multifaceted. Climate, enviornmental degradation and natural hazards induced disasters are increasingly interacting with the drivers of refugee movements. A survey by the Office of the United Nations Refugee Agency suggests that refugees living in camp settings are much more likely to be displaced by disasters than the general population in the world. There are many reasons for the high vulnerability of refugee camps to disasters associated with natural hazards. Refugee camps are often located in remote areas with limited access by road and to infrastructure. For example, in July 2014, 47,000 refugees in Leitchuor camp in Ethiopia were cut off from the world for more than a month due to a flood that affected the main access road to the camp. Camps may be located on land that is not traditionally considered suitable for human settlements, being too arid or too hilly. The camps in Rwanda, for example, are located on steep land prone to the risk of landslides, which in 2014 destroyed 250 shelters (UNHCR 2015). Camps may be densely populated. A single fire can spread to dozens of houses in a flash. In a refugee camp in Thailand, one kitchen fire claimed the lives of 37 refugees and made 2300 refugees homeless in March of 2013 (UNHCR 2013). Areas surrounding refugee camps, especially forests, often experience acute environmental degradation, mainly because refugees need wood as a source of cooking fuel and building materials. Deforestation increases the risk of disasters such as flood and drought. Finally, shelters for refugees and internally displaced persons are generally not resistant to natural hazards. In many emergency situations, refugees are only given simple shelter materials of plastic sheets, wood, or tents, which can have a lifespan as short as only a few months (UNHCR 2015). According to the Norwegian Refugee Council, in Nigeria, over 6800 people living in displacement camps in Maiduguri were impacted by floods, which entailed displacement within a displacement camp (NRC 2019).

Issues of conflict, disaster, and displacement are gaining importance. The New York Declaration for Refugees and Migrants highlighted the issue of disaster risk in fragile contexts (UNGA 2016). The Global Compact on Refugees, endorsed by United Nations in 2018, has acknowledged the importance of risk management in refugee situations. Its "Comprehensive Refugee Response Framework" component calls for preparedness, including contingency planning, to strengthen comprehensive responses to large refugee situations (UN 2018).

The management of natural hazard risk in conflict-affected settings including a refugee crisis is challenging. There remains a dearth of practical and policy advice on how to devise and implement disaster risk management strategies for complex risk context, including where violent conflict forms part of the broader environment in which disaster risk management takes place. This theme is still to be learned and warrants further attention (UNDRR 2019). The management of disaster risk in the Rohingya refugee crisis in Cox's Bazar, Bangladesh, offers some valuable insights for future risk management in fragile (refugee) contexts.

7.1.1 Rohingya Refugee Context in Bangladesh

Located in South Asia, Bangladesh has a total area of 147,570 km^2 and a population of 161.4 million which ranks as the eighth most populous country in the world. The flat topography, low-lying land, and climatic features, combined with its population density and socioeconomic environment, make the country highly susceptible to many natural hazards, including floods, droughts, cyclones, and earthquakes. More than 80% of the population is potentially exposed to floods, earthquakes, and droughts and more than 70% to cyclones. On average, the country experiences severe tropical cyclones every 3 years, and about 25% of the land mass is inundated with flood waters every year. Severe flooding occurs every 4–5 years and covers 60% of the land mass (World Bank 2018).

The country is administratively divided into 64 districts; Cox's Bazar is located at the southern tip of the country. This coastal district is administratively subdivided into eight subdistricts (upzilas) including Ukhiya and Teknaf, where the Rohingya refugee camps are located.

As of January 2019, over 900,000 Rohingya refugees reside in the Cox's Bazar district of Bangladesh. While most arrived between August and December 2017, arrivals have continued. Most refugees live in 34 camps, in two sub-districts of Cox's Bazar, namely, Ukhiya and Teknaf. Out of 34 camps, 23 camps form the Kutupalong-Balukhali expansion site in Ukhiya subdistrict, which hosts more than 600,000 refugees. This influx has almost tripled the total population in Ukhiya and Teknaf subdistricts, with sizable consequences for the environment and livelihoods of Bangladeshi residents (SEG et al. 2019). The refugee population remains 100% reliant on food assistance to sustain the minimum daily required kilocalories per individual. The location of camps and their populations are shown in Fig. 7.1.

The sudden large-scale influx of Rohingya refugee over such a short duration did not offer a window for shelter planning. Moreover, planning for a million refugees in a densely populated developing country is extremely challenging. The camps lack planning and are extremely congested with an average area of 18.76 m^2 per person, while the standard allowance is 45 m^2 (Sphere Association 2018). It is important to note that Sphere Standards are a set of principles and minimum humanitarian standards in four technical areas of humanitarian response: water supply, sanitation and hygiene promotion, food security and nutrition, and shelter and

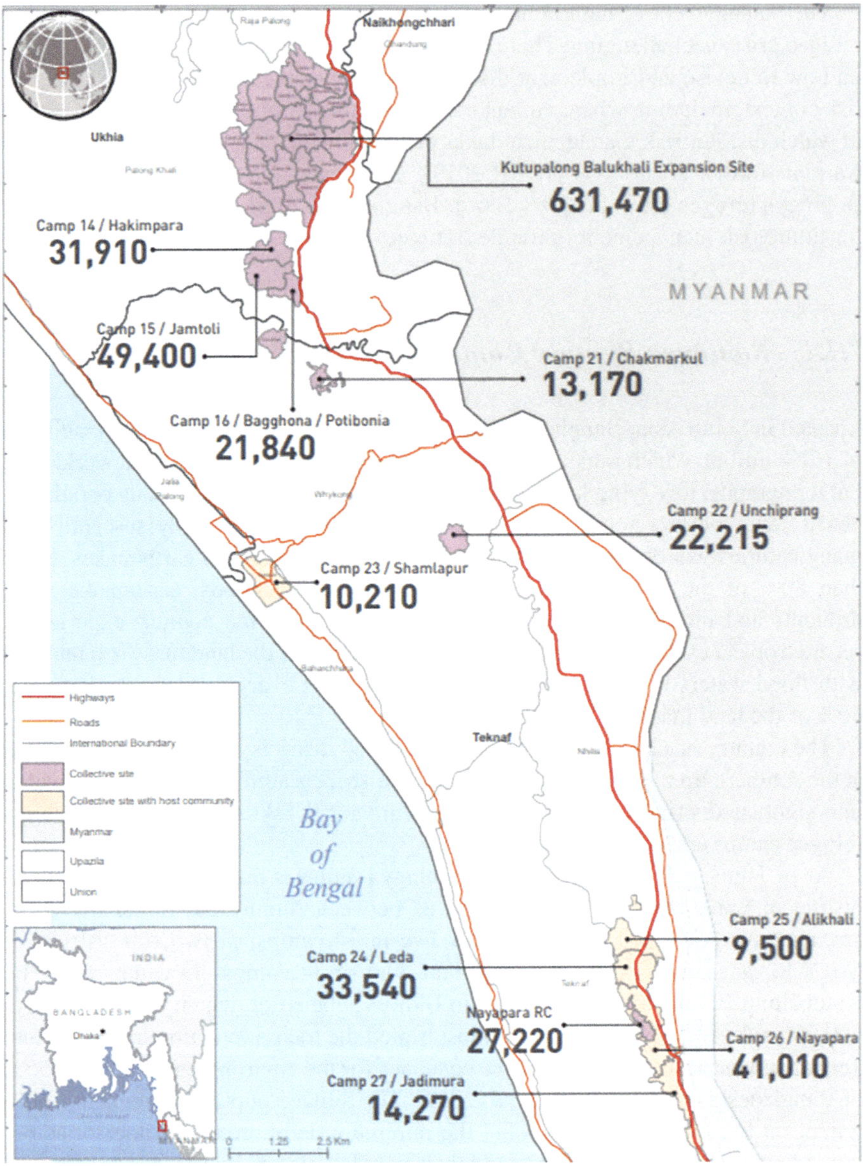

Fig. 7.1 Refugee population and camps in Cox's Bazar, Bangladesh. (ISCG et al. 2019)

settlement. Most of the shelters do not adhere to the covered living space of 3.5 m^2 per person (Oxfam 2020). Shelters and blocks of shelters also lack fire break standards, which help in reducing fire risk.

The large-scale construction of shelters has severe environmental impact. According to the Cox's Bazar Forest Department, the influx has destroyed about

4818 acres of forest reserves worth 55 million US dollars (UNDP 2018a). The excessive dependence on groundwater has drastically reduced the water level in the region (Akhter et al. 2020). Outside the refugee camps, areas of Teknaf have been affected by the extra use of groundwater by the refugee community. Agricultural prospects in the area, which have always suffered due to a lack of fresh water, are increasingly threatened (Rahman 2019a, b). The damage to hills caused by construction of refugee shelters and related facilities has led to a flow of sandy soil down the hills, which has impacted grazing lands.

The influx of large numbers of refugees has social and economic impacts. The drop in groundwater levels impacts the life and livelihoods of the host community. Wages have depressed due to the availability of an additional labor force at a lower price. The Bangladesh government has put a ban on fishing in the Naf River bordering Bangladesh and Myanmar, for security and border control reasons. The fishing community of Cox's Bazar has been impacted. The shrinking livelihoods, opportunity, and wages have increased poverty per head count in Teknaf and Ukhiya subdistricts of Cox's Bazar (UNDP 2018a). Overall, the Rohingya refugee influx has led to profound impacts on the social, environmental, economic, and physical spheres of Cox's Bazar.

7.2 Vulnerability and Hazard Analysis of Rohingya Refugee Site

The vulnerability of an individual or a group requires analysis of two factors: threats or hazards and factors which make them vulnerable to the threat or hazard. Key vulnerabilities of Rohingya refugee sites include social, economic, environmental, and physical factors and are detailed in Sect. 7.2.1. The potential hazards are detailed in Sect. 7.2.2 and shown in Fig. 7.3.

7.2.1 Vulnerability Analysis of Rohingya Refugee Camps

7.2.1.1 Environmental Degradation

The 4800 acres of undeveloped forest land, allocated for a new camp by the government of Bangladesh in September 2017, is now densely populated. Firewood is the main source of cooking fuel across all refugee settlements, which further exacerbates deforestation on a substantial scale. The collection of fuel wood from the natural forest within a 5 km perimeter around the camps will sustain fuel supply for approximately 4 months, but the forested area of 14,000 hectares will be degraded and converted into shrub-dominated areas with low biomass and productivity. The deforestation has exacerbated risks of hazards (SEG et al. 2018).

There has been excessive dependence on groundwater in refugee camps as several thousand tube wells have been installed. The water levels around the camp areas are reported to have fallen between 5 m and 9 m (UNDP 2018a). Irrigation wells are gradually drying up as the water table is falling as a result of watershed destruction, coupled with a significant reduction in the recharge of groundwater reserves. The continued pressure on the aquifer may result in saltwater intrusion, rendering it unusable for the district. Pressure increases to the point such that it impacts the social cohesion between the host community and refugees.

7.2.1.2 Deforestation

Loss of aboveground vegetation leads to unstable topography/slopes. Most of the shelters and community structures are located on either slippery hillsides or flood-prone lands and are thus vulnerable to landslide, strong wind, and floods; an example of the congested, chaotic camps is shown in Fig. 7.2. The small ponds in and around camps become big lakes and pathways become muddy, which makes access challenging and has implications for humanitarian operations (Loy 2019). Collapse of one shelter can affect several other shelters which are in close proximity and on lower levels of slope. Rohingya refugees are highly dependent on external services,

Fig. 7.2 Congested camps made up of shelters located on fragile slopes in Kutupalong refugee camp, Cox's Bazar, Bangladesh, August 2018

and such conditions hinder access to humanitarian services. The elderly population of 60 years and above are at high risk of falling on unstable slopes, especially during the rainy season. A shelter located in the proximity of a waterbody at the bottom of a hill is at risk from flooding as the shelter gets submerged due to a rise in water level.

7.2.1.3 Vulnerable Population

Children represent 55% of all refugees, and 6100 unaccompanied and separated children were registered in October 2018. Females constitute 52% of Rohingya refugee populations. Female-headed households constitute 12% of the Rohingya refugee household population. Approximately 17.35% of the Rohingya refugee mothers are single parents. The illiteracy rate is 73 % among the refugees (Oxfam et al. 2018). The high level of illiteracy, female-headed households, children, people with disability, and aged group are classes particularly vulnerable to disasters. Income opportunities for women are limited compared to men in the camps which reduces female household's ability to meet their own basic needs. Women face constraints in accessing information as they are excluded from mosques, which are an important source of information in camps, preventing women from receiving first-hand information. This is a major challenge for women-headed households.

7.2.1.4 Human and Wildlife Conflict

An International Union for Conservation of Nature and United Nations High Commissioner for Refugees study (IUCN and UNHCR 2018) found that the refugee camp lies within an active corridor of elephants, which has been actively used for centuries. Refugee camps destroyed the wild elephant's habitat, posing a threat to the wild elephant's survival and to human security. The restrictions on elephants' free movement, the scarcity of their food, and new settlements in their habitats by refugees fueled human-elephant conflicts in the camp area, leading to deaths on both sides. Since September 2017, human-elephant conflicts have occurred on the edge of the refugee camps in the elephant corridors of Ukhiya subdistrict. Rohingya refugees, particularly adult females and girls who venture into nearby forest to collect fuelwood, are at high risk of physical attack from elephants (Rahman 2019b).

7.2.1.5 Limited Social Capital for Community-Based Response

Several studies have conceptually and empirically asserted the importance of social capital for disaster recovery, stressing that communities with stronger social capital are likely to achieve faster recovery (Aldrich 2012). Social capital is a set of social relations that provide access to resources (Tirmizi 2005). Community-based preparedness is the backbone of the globally recognized Cyclone Preparedness Programme of Bangladesh, which significantly reduced the loss of lives due to cyclone conditions. However, refugees who recently arrived in Cox's Bazar have limited knowledge of new terrains and neighborhoods. Considering the recent influx and context, the social capital is likely to be limited for the Rohingya refugee, which limits the community-based disaster response.

7.2.1.6 Safety Risk and Disaster Risk due to Fragile Shelters

The push to address urgent humanitarian needs, at the peak of the response and for monsoon preparedness, has meant that much of the bamboo used for shelters is poor quality with limited durability. Exacerbating this, nearly all structures have been constructed using untreated bamboo in direct contact with the ground, creating perfect conditions for pests and rot. The above factors mean that most of the bamboo within the camps will need to be replaced within the next 0–20 months. The structural assessment against wind loading categorized structures into four categories where the best one can withstand wind speed of 40–80 km/h, while the weakest one can withstand only 20–40 km/h (Shelter/NFI Sector 2018). Most shelters do have a natural windbreak protection such as trees. It is important to note that on the Saffir-Simpson scale, cyclones are ranked from category one through category five in order of increasing intensity. The wind speed of cyclone category one ranges from 119 km/h to 153 km/h (Iman et al. 2006). It can be inferred that existing shelters cannot withstand category one cyclone.

Female-headed households face additional safety risk of intrusion due to fragile construction materials of their shelters. These factors have a mental toll on female inhabitants as they are not able to sleep at night.

7.2.1.7 Congested Unplanned Camps

The spontaneity of mostly unplanned sites produces very dense areas (10 m^2 of land per person) in some of the refugee camps. The proximity of shelters creates fire hazards, protection threats and risks of communicable disease (SEG et al. 2018). The congested space also affects privacy of the refugees. Shelters are constructed of materials that are difficult to lock; thus inhabitants are vulnerable to intrusion constituting a safety risk.

7.2.1.8 Limited Cyclone Response Capacity

Bangladesh has a well-established and highly effective Cyclone Preparedness Programme (CPP). It involves a three-flag system: well-coordinated volunteer groups and a response mechanism. The system has been in practice since 1971 and is institutionalized. The Rohingya refugees are a relatively new arrival, and Bangladesh has expanded the scope of the CPP to cover the Rohingya refugee camps; however the expansion will take time to be instated. It is important to emphasize that the Cyclone Preparedness Programme has made a significant difference in saving lives from cyclone in Bangladesh.

7.2.1.9 Early Warning System

Early warning and its dissemination and actions following warning are an important component of the overall disaster response mechanism. In case of cyclone, tsunami, and other disasters, early warning is critical. The Great East Japan Earthquake and subsequent tsunami of March 2011 reemphasized the importance of an end-to-end early warning system, which is a system spanning all steps from hazard detection to community response. A Japanese government study found that only 58 % of people in coastal areas of Fukushima, Iwate, and Miyagi prefectures heeded tsunami warnings immediately after the earthquake and headed for higher ground. Of those who attempted to evacuate after hearing the warning, just 5 % were caught by the tsunami (Kumar 2012).

The effectiveness of warning systems in significantly reducing cyclone-related deaths in Bangladesh is well documented (UNDRR 2008). The early warning dissemination has been developed to cater the needs of the population of Bangladesh. However, the early warning message has limited reach among Rohingya due to cultural and language constraints, limited understanding of the CPP flag system, and a high level of illiteracy. Rohingya is the only spoken language that all refugees understand and prefer (TWB Communications 2018). There has been an effort to translate the key messages into the native language of the Rohingya refugees.

7.2.1.10 Lack of Social Considerations in Multipurpose Cyclone Shelters and Limited Capacity

Bangladesh government and partners have constructed approximately 4000 cyclone shelters, which can accommodate between 500 and 2500 individuals per shelter (Miyaji et al. 2017). Cox's Bazar was not a priority district under the Cyclone Preparedness Programme as it was earlier assumed that the local population would be able to evacuate into the local hills in the eventuality of a cyclone. Moreover, other structures have been identified for evacuation purpose during cyclone. The mapping of existing cyclone shelters identified by the government indicates that in Ukhiya, 75 structures are to be used as cyclone shelters with a capacity of 67,550 individuals, while in Teknaf 67 cyclone shelters have a capacity of 65,200 individuals (SEG et al. 2019). The existing capacity of cyclone shelters is insufficient for vulnerable Rohingya refugees and local populations. Approximately 80% of the Rohingya refugee population are women and children (UN Women 2020). People with disability face additional challenges during disaster response including evacuation. In Rohingya refugee camps, 11–13 % of the population are people with disability (CBM 2019). According to data collected by Humanity & Inclusion, approximately 3% of the children are mentally, intellectually, or developmentally disabled (Yan 2018).

7.2.1.11 Need of Coordinated Risk Governance in Camps

The government of Bangladesh has identified the Office of Refugee Relief and Repatriation Commissioner (RRRC) as the nodal agency for Rohingya refugees. RRRC appointed the Camp-in-Charge (CiC) officers for each camp. The Office of Deputy Commissioner is the nodal agency for disaster response and preparedness at the district level. The humanitarian agencies working in the camps are coordinated by the RRRC and are involved in disaster risk management. The Armed Forces also play an important role in disaster management especially response. The myriad of agencies calls for a coordinated response and preparedness, which is evolving.

7.2.2 Hazard Profile of Cox's Bazar District, Bangladesh

Cox's Bazar district is prone to multiple hazards including cyclone, floods, and landslide. It is one of the most flood-prone areas of Bangladesh. It receives 72% of annual rainfall in 4 months (June to September) of rainy season. The hazard seasonality is displayed in Fig. 7.3.

Ukhiya subdistrict, where most Rohingyas reside, is not prone to riverine flooding (overflowing of a river inundating adjacent land). However, heavy rains have been known to result in flash floods and landslides, which can affect the local Bangladesh community and the Rohingya refugee population (ACAPS 2018).

In Teknaf subdistrict, shelters of Rohingya refugees are on flatter land along canals beside the coast, which makes them prone to riverine flooding in addition to

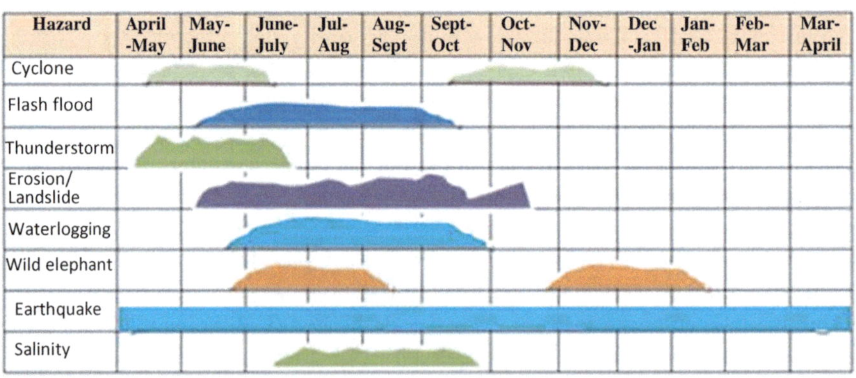

Fig. 7.3 Hazard seasonality of Cox's Bazar district, Bangladesh. (DDMC 2014)

flash floods. Both types of flooding and landslides are likely to result in damages in the sites and exacerbate the humanitarian needs of Rohingya people.

7.3 Understanding and Assessing Disaster Risk

Disaster risk has been defined as the potential loss of life, injury, or destroyed or damaged assets which could occur to a system, society, or community in a specific period of time, determined probabilistically as a function of hazard, exposure, vulnerability, and capacity (UNDRR 2017). Disaster risk management measures are based on the risk information; hence it is important to understand and assess risk of Rohingya refugee camps.

Dhaka University and development partners conducted a study to analyze the impact of floods and landslide on Rohingya refugees residing in Kutupalong registered camp and extension. For landslide, the impact area was calculated based on the degree of slope. A slope having more than 35 degrees has a risk of failure. For floods, the main river flood levels were calculated empirically, based upon field measurements taken of high-water levels as indicated from people who have lived in the area for over 20 years.

It is important to note that the Ukhia Upazila (subdistrict) disaster management plan and landslide risk assessment in high-risk location in district was conducted through the Comprehensive Disaster Management Programme (2005–2015). However, the Kutupalong-Balukhali refugee camp site was established in a protected forest area for which no risk assessment has been carried out (UNDP 2018b).

23,934 families or 102,036 individuals are at risk of being affected directly from landslides and floods. In-depth analysis identified 85,876 individuals at risk from floods, 23,330 individuals at risk from landslide, and 7170 individuals at risk from both landslide and floods (UNHCR 2018). The estimated figure of refugees at risk of direct impact by flood and landslide for all camps, beyond Kutupalong registered camp and extension, increased to 246,600 individuals in 57,424 households, of which 41,751 individuals were at highest risk of landslide and prioritized for relocation (SMSD 2019).

7.4 Measures to Reduce Vulnerability and Exposure to Cyclone and Flood Hazards

In consideration of the high level of risk in the Rohingya refugee camps, several risk reduction measures were initiated by the government of Bangladesh and development partners since early 2018. Risk reduction measures are ongoing.

7.4.1 Reducing Environmental Degradation and Restoration of Resource Base

As part of environmental rehabilitation efforts, 6,897,029 slip pieces of vetiver grass, shrub and grass, 696,200 gm of seeds, and 340,174 tree saplings were planted (SEG et al. 2018). A 3-year project started in 2017, which engaged refugee and host communities to rehabilitate deforested areas and protect water sources. It also aimed to provide technical assistance to host community agricultural groups to produce high-demand and nutritious food sources in the next 2–5 years complemented by nutrition education at the household level. The direct provision of materials for alternative fuel sources and facilities has been made to address the immediate and growing cooking fuel concerns of the camp population. The activities were implemented in a phased manner. By December 2018, a total of 17,481 (16,402 Rohingya and 1079 host community) households were provided with liquefied petroleum gas startup kits (cylinder, stove, regulator, and hose), though these need to be continually replenished (IOM 2018). The pipe water supply scheme for refugee camps has been initiated, which will reduce pressure on the depleted underground water level.

These interventions are likely to reduce environmental stress and will help in regeneration of flora; however impacts can be measured after 5–10 years. The interventions also relieve social implications as females go to nearby forest to collect cooking fuel and are at physical safety risk from antisocial elements and human-wildlife conflict.

7.4.2 Social Considerations Strengthening Cyclone Shelters

Development partners mapped existing cyclone shelters and facilities in Ukhiya and Teknaf subdistricts from April to December 2018 (REACH 2019a). These structures can be used for temporary evacuation during cyclone. Development partners improved and strengthened the existing facilities, which can be used for temporary evacuation. The existing cyclone shelters were strengthened in their structural and nonstructural aspects. Partitions were created in shelters for lactating mothers. Toilets were constructed for female evacuees. Iron grills were also installed in cyclone shelters for protection. Considering needs of people with disability, ramps were constructed, and provision for wheelchair access was made. The cyclone shelter management committee members were sensitized regarding Rohingya refugees, their roles and responsibilities, early warning system, and safe evacuation planning which has been carried out following guidelines (IOM 2018).

In 2019, several cyclone-resistant learning spaces were constructed within the camp boundaries, as part of ongoing improvements of 59 educational structures to serve as cyclone shelters and the rehabilitation of 60 existing shelters. The World Bank and ADB will construct 73 climate-resilient multipurpose structures, which will significantly improve the temporary evacuation facility (SEG et al. 2019).

Table 7.1 Relocation update as of 7 February 2019

Sr. no.	Relocation reason	Households	Individuals
1	Landslide risk	6003	26,844
2	Flood risk	1662	7333
3	New arrivals	2558	10,523
4	Infrastructure	721	3073
5	Others	101	442
6	**Total**	**11,117**	**48,508**

Source: ISCG et al. (2019)

7.4.3 Prioritized Relocation of Households at Risk of Landslide

To reduce landslide and flood risk and congestion, the government of Bangladesh has expanded the available land to a total of over 6500 acres. Government and development partners have prioritized 41,751 individuals/9660 households for relocation within camps or into newly extended camps. The relocation reasons include new arrivals of refugees, infrastructure development, and high risks from floods. Refugees at risk of landslide have been prioritized (ISCG et al. 2019). The breakdown of completed relocation is given in Table 7.1.

7.4.4 Structural Measures for Risk Mitigation

Access to refugee camps is critical to the humanitarian response especially during cyclones and floods. A multi-sectoral and multi-agency effort is essential to mitigate the risk. Efforts have led to formation of the Site Maintenance and Engineering Project (SMEP) office, which included the International Organization for Migration, the UN Agency for Refugee, and the World Food Programme. SMEP led risk mitigation activities in camps. Approximately 25 km of canals were dredged in the camps to reduce flooding risk, and 400,000 m^2 of slope were stabilized in 2018 (SEG et al. 2018). The slope stabilization interventions included plantation, covering slopes with tarpaulin, terracing, leveling of hills, sandbags on fragile slopes, pipe/box culverts, and others as seen in Fig. 7.4. New infrastructures including more than ten bamboo footbridges and six footpaths were created.

7.4.5 Strengthening Shelters

Recognizing the fragility of shelters, several measures were taken to strengthen the shelters. Tie-down kits were provided, which can resist the uplifting forces. Wall bracing kits were provided, which can increase the wind resistance by 50% in high

Fig. 7.4 Access road development in Rohingya refugee camps in Kutupalong-Balukhali, Cox's Bazar, Bangladesh, June 2018

wind exposure areas. Treatment of bamboo can significantly increase the life of shelters as existing bamboo shelters are in direct contact with the ground; 6199 Rohingya refugee households have benefited, recorded July 2019 (Shelter/NFI Sector 2019). Bamboo strengthening during construction is shown in Fig. 7.5.

7.5 Preparedness and Response to Cyclone and Flood Risk

7.5.1 Sector- and Camp-Level Emergency Preparedness and Response Plan

Camp-level disaster response is the first line of defense in case of a disaster. A guideline on camp-level emergency preparedness and response plan was shared with all camps. It was completed with technical assistance from the Site Management Support agencies of each camp. The plan was reviewed on key parameters, namely, camp-specific risk profile, trigger point for emergency response, response structure with camp-level volunteers, identification of extremely vulnerable individuals, prepositioning of stocks, food distribution plans, medical response, simulation at camp level, and cyclone warning dissemination plan (ISCG 2018a, d). A series of

Fig. 7.5 Shelter construction and strengthening in camp. (Chakmakul camp, Cox's Bazar, Bangladesh, October 2018)

tabletop exercises were conducted at camp level to test the plans. A Cox's Bazar-level emergency preparedness and response plan was prepared, which included sector-specific preparedness and response activities. The plan was updated using lessons learned from the monsoon season of 2018. It is a dynamic document and continues to be periodically updated. A 3-day simulation exercise for cyclone response was also conducted to check preparedness of Cox's Bazar district.

The management of high-tide incidents at Shamlapur refugee camp illustrates the importance of camp-level preparedness and response. Shamlapur camp number 23, in Teknaf subdistrict of Cox's Bazar, is located on the coast of the Bay of Bengal. Being coastal and one of the most low-lying Rohingya refugee camps, it is prone to high-tide sea levels. Between 12 and 15 July 2019, high-tide over 4 m was forecasted, which could have led to possible inundation of several refugee camps. ISCG communicated the high-tide dates and time to all stakeholders including camp-level/field-level agencies. The site management team reviewed the preparedness measures and put sandbags at vulnerable spots to reduce impact of tides. Safety volunteers were deployed to keep a vigil on tide/water level. The high-risk area within the Shamlapur camp and other low-lying areas were identified as the priority areas for evacuation. Shelters on high ground were identified for accommodating the evacuees. Communities were sensitized regarding do's and don'ts related to high tide. The detailed planning and proactive response measures led to no significant damage being incurred (ISCG 2019b).

7.5.2 Cyclone Preparedness Programme (CPP) Extended to Refugee Camps

Bangladesh government extended the CPP to refugee camps to improve cyclone preparedness. A minimum of 20 CPP volunteers from each Rohingya refugee camp, including women, have been trained on cyclone preparedness and early warning. Government aimed to increase the number of trained volunteers to 50 in each camp; a workshop preparedness training meeting facilitated by the CPP is shown in Fig. 7.6. The volunteer's skill was tested during depressions and cyclones formed in the Bay of Bengal in 2018, on more than one occasion. The preliminary findings related to Rohingya refugee performance as CPP volunteer indicated that refugees were fully engaged in the response exercise. There were several areas for improvement, which should be addressed through refresher training and mock drill exercises. The Cyclone Preparedness Programme took several years to mature as it

Fig. 7.6 Cyclone preparedness training in camp (Camp 8W, Cox's Bazar, Bangladesh, September 2018)

involves community, volunteers, and government. The CPP in Rohingya camps will also need time to mature; hence it is important to have a medium-term approach.

7.5.3 *Preparedness for Emergency Response*

Considering access issues in the aftermath of a cyclone or heavy rainfall, several volunteer groups were formed at the camp level. Also, porters were engaged before the rainy season to ensure supply of relief material in case of road blockage. These were proven to be effective measures during the rains of 2018. The experience of engaging volunteers during 2018 rainy season highlighted the need for comprehensive mapping of volunteers to avoid duplication and rationalization. The need to develop a protocol for coordination among different volunteer groups and their activation and deactivation were also realized. The standard minimum package has been identified and details are shown in Table 7.2.

7.5.4 *Awareness Generation*

Awareness of "do's and don'ts" related to disaster is a key component of community-level preparedness. Do's and don'ts related to disasters in pictorial form were distributed at community level. A website "Shongjog" was created, which contains messages, pictorial do's and don'ts, video, report, etc. related to cyclone, floods, landslide, and other hazards (http://www.shongjog.org.bd, Accessed 22 Nov 2020). The website is also a repository of information related to sectors such as health, sanitation, and communication. Radio listening groups were created across all camps for men, women, adolescent boys, and adolescent girls.

At camp level, approximately 90 information hubs have been created, which provide information on essential lifesaving needs to re-enforce access and utilization of basic humanitarian services as well as provide a platform for interacting with

Table 7.2 The standard initial emergency package (ISCG 2019b)

Shelter/non-food items	WASH (water, sanitation, and hygiene)	Food
Emergency shelter kit	Cyclone WASH kit is based upon 2-week supply/household	1 carton high-energy biscuit (5 kg) equivalent for 100 packets lasting for 1 week
1 × tarp	Aquatabs (33 mg) × 200	
1 × bundle of 6 mm rope	Bathing soap 100g × 5	
2 × floor mats	Jerry cans (10L) × 2 or Jerry can (10L) × 1 and bucket (10L) × 1	

communities' views and feedback. In case of emergency arising due to an oncoming cyclone or severe storm, the Info Hubs/IFCs re-enforce warning signs through loudspeakers/megaphones. Hence directions for safety and security such as guidance on where to take shelter have been provided (CwC 2019).

7.5.5 Information Management During Cyclone and Rainy Season

Information is the most valuable commodity during emergencies or disasters. It is what everyone needs to make decisions. It is necessary for rapid and effective assistance for those affected by a disaster (PAHO 2009). A "Daily Incident Report" was generated every evening by the ISCG, which captured any incident reported until 6 pm in the camp. The incident categories were landslide/erosion, wind/storm, flood, fire, and lightning. The affected population were categorized into total affected individuals and households, displaced, injured, dead, and missing. The immediate response measure across sectors was also captured. This report was shared electronically with the government of Bangladesh officials and all development partners. It was helpful in providing overall situation in camps every day and thus in quick decision-making. The format of the Daily Incident Report was revised in 2019.

Moreover, on extremely heavy rainy days, a short report on impact in camps was prepared and shared with the development partners during the day. It was prepared on ad hoc basis. A weekly report "Monsoon Emergency Response Update" was also prepared, which provided analysis of the impact of rain in camps. It also provided rainfall and wind data of the past week and a forecast for the coming week.

7.5.6 Rapid Needs Assessment

A Rapid Joint Needs Assessment system was developed in preparedness for disasters. The format for conducting assessment was developed, and several enumerators were trained on conduct of the assessment. It will deliver two outputs, one at the end of 24 h and the second at the end of 72 h. The trigger mechanism for activation of the assessment, geographical scope, and targeted population was also identified. The workflow of Rapid Joint Needs Assessment (JNA) including data collection, enumerators, and key informants (KI) is shown in Fig. 7.7, where "KoBo" refers to the digital tool used by humanitarian agencies (REACH 2019b). Simulation exercises were conducted to check the preparedness of assessment team.

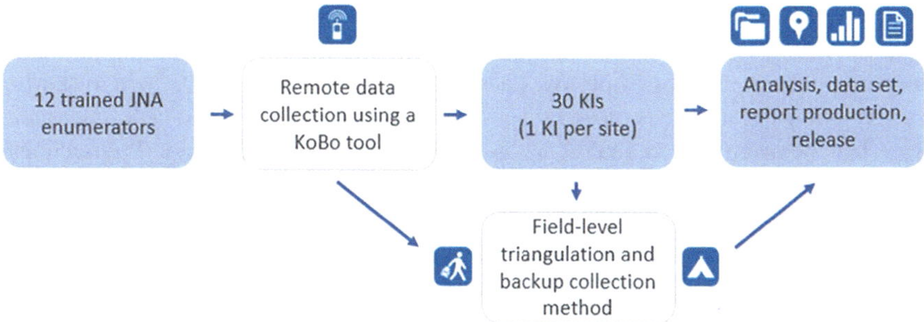

Fig. 7.7 Joint Rapid Needs Assessment workflow. (ISCG 2018c)

7.5.7 *Institutional Arrangements for Disaster Response*

A special coordination structure was created to ensure seamless coordination during disaster response for Rohingya refugees. For emergency level 3 (cyclone), the disaster response will be coordinated by an Emergency Control Room (ECR). It was led by the Office of Deputy Commissioner and included three other pillars, namely, Refugee Relief and Repatriation Commissioner (RRRC), Armed Forces Division, and Inter-Sector Coordination Group (ISCG). This arrangement was important as the Office of RRRC is the nodal agency for refugee camps, the Deputy Commissioner is in charge of the overall district, Armed Forces Divisions play a pivotal role in disaster response, and ISCG is the nodal body for coordinating the humanitarian response in Cox's Bazar. The ECR was activated on 25 July 2008 as more than 400 mm of rain were recorded in 24 h. The four-pillared ECR held meeting took several decisions including allowing access to camps beyond 5 pm by humanitarian actors and use of community facilities for evacuation purposes (UNDP 2018b). The ECR mechanism requires further strengthening as it is a relatively new concept. There is a need of capacity development of the support staffs to ECR on coordination, data processing, and information management.

7.5.8 *Emergency Preparedness and Response Task Force/ Emergency Task Force*

An Emergency Task Force was created to coordinate disaster preparedness and response of humanitarian agencies engaged in Rohingya refugee response in Cox's Bazar. The Emergency Task Force was earlier titled as Emergency Preparedness and Response Task Force. It was chaired by the Inter-Sector Coordination Group secretariat and had open membership, which included sectors, UN agencies, NGOs, Bangladesh Red Crescent Society, and donors. The task force facilitated preparation of the sector cyclone and monsoon emergency preparedness and response plan and its updating. It also facilitated preparation of the camp emergency response and preparedness plan. The weekly meeting provided a forum to discuss the flood/monsoon response in the preceding week and draw key lessons. It also helped in sharing the cyclone/rainfall forecasts and discussions of preparedness. The task force mandate has been reviewed and it is now titled as the Emergency Preparedness Working Group.

The constant review of disaster response and feedback into preparedness by the Emergency Task Force is illustrated through the fire incident case at one of the Rohingya refugee camps in Cox's Bazar. A fire incident occurred around 1:30 am on 5 September 2018 at Chakmarkul refugee camp, which engulfed 14 shelters and affected 30 families/126 individuals (ADRA 2018). The swift response by government, development partners, and refugee and host community volunteers led to zero casualty and zero injury. A special meeting of the Emergency Task Force was held on 9 September 2018 to draw key lessons and identify measures for improvement. The Bangladesh Army, Fire Services and Adventist Development and Relief Agency, and the Site Management Support agency of Chakmarkul camp were special invitees as they played key roles in the fire response. The meeting identified several short-term and medium- to long-term measures along with responsible agency for fire risk management. Some the measures include standard operating procedure for fire at camp level, re-pressurization of fire extinguishers, refresher course on fire safety for volunteers, fire exit route to be displayed in camps, fire risk to be included in future site planning, and space for water storage facility for fire response. The special meeting on the Chakmarkul fire incident reflects the importance of constant learning from events and feeding it into an ongoing response for improvement.

7.6 Challenges Faced in Implementing Risk Reduction Measures and Response to Cyclone and Floods

7.6.1 Medium-Term Risk Mitigation Needs vis-à-vis Short-Term Policy

Evidence from past crises illustrates that once someone is displaced for 6 months, they are likely to be in exile for years. Eight out of ten refugee crises last for 10 years or more, and one in five lasts for more than 30 years (Crawford et al. 2015).

The Bangladesh government has been generously hosting the Rohingya refugees and kept its border open; however the government is clear that the displaced Rohingya population should be repatriated to Myanmar and that Bangladesh is unprepared to host the displaced population over the mid-term or long-term (RSiS 2018). Disaster risk reduction requires medium- to long-term approach for substantial risk reduction as it requires correction/major adjustments in sectorial response planning and implementation such as shelter, sanitation, health, and education.

7.6.2 Sudden Influx Leading to Unplanned and Congested Settlement

Many parts of the refugee camps are accessible only by foot even in non-rainy seasons due to high congestion (as high as 10 m^2 of lands per person against minimum standard of 45 m^2). Most of the roads are not suitable for travel and get muddy, so vehicles cannot ply. Response agencies engage on foot volunteers to deliver relief. The congested and unplanned settlement makes swift and comprehensive disaster response a challenge in Rohingya refugee camps.

7.6.3 Evacuation in Case of Extreme Event

In the case of cyclones, especially those of higher category, making landfall in and around Cox's Bazar, mass evacuation of the refugees is necessitated. However, the number of refugees, restriction on movement of refugees, logistics for evacuation, and lack of infrastructure for evacuees make it extremely difficult to evacuate all vulnerable members. Thus, decisions regarding evacuation in the event of an extreme event are herculean tasks and have dimensions of maintenance of law and order. The decision process involves several questions such as which group (refugee or host community) should be prioritized and within each group which section (aged group, people with disability, children, female-headed household, lactating mother, and pregnant women) of community should be prioritized and the determination of a place of evacuation including allocation of cyclone shelters.

7.6.4 Limited Disaster Data and Risk Information

Risk information and disaster data of appropriate scale are extremely important for decision-making related to disaster risk management. The Rohingya refugee camps are in a small area, which was previously a forest/reserved area; disaster data and risk information at a microscale are not available. Also, the dynamic situation in camps changes the vulnerability and risk profile frequently. Decision-making related to disaster is a challenge due to limited data on camps. The one-time macro-level risk assessment exercise has limited usage in risk management planning.

7.6.5 *High Turnover of Actors Engaged in Refugee Crisis*

The effective coordination between different actors in this emergency response faced additional challenge due to high turnover rate of humanitarian staff, military officials, and government appointees such as the Camp in Charges. Humanitarian agencies mobilized surge capacity, drawing technical experts and field support teams to facilitate the coordination and planning for the emergency response. Surge teams operate on short-term rotations of 4–6 weeks. The military officials and the Camp in Charges appointed by the Ministry of Public Administration are deployed to the camps on short rotations (Cook and Ne 2018). The short duration deployment of government staffs and humanitarian workers in refugee crisis entails loss of precious resources and time in orientating newly arrived workers.

7.6.6 *Host Community vis-à-vis Refugee*

The Bangladeshi community around refugee camps are highly vulnerable to cyclone, floods, high tide, and landslides. In Teknaf and Ukhia subdistricts, 89 % of total houses are made of straw, bamboo, and tin sheds, which is fragile in strong wind and cyclone (DDMC 2014). A cyclone will badly impact both host community and Rohingya community. In such scenario, the Bangladeshi host community and Rohingya refugee require relief assistance and recovery support. In case of limited resources, prioritization of assistance will be important. The allocation of resources will require a fine balance as any bias, including perceived bias by refugee or host community, can have impact on social cohesion between host community and Rohingya refugees.

7.7 Key Lessons for Managing Disaster Risk in the Refugee Context

Every response and recovery in the aftermath of a refugee crisis or disaster are considered unique; however learning from the past can provide guidance to future policy makers and managers engaged in a refugee crisis or disaster. The overall aim of disaster risk management and response measures being implemented in the ongoing Rohingya refugee crisis is to avert a disaster within disaster. The key lessons from implementation of risk reduction and preparedness measures and response to floods and cyclone/strong wind in 2018 in the Rohingya refugee camps of Bangladesh are summarized in this section.

A cyclone or an earthquake is likely to affect refugee and host population unless host communities are highly capacitated or are far from the earthquakes' sphere of activity/epicenter. Disaster preparedness and responses only geared to refugee camps in the aftermath of a large-scale disaster can be counterproductive. It is important to plan preparedness and response in coordinated manner with the government, especially local authorities. This aids clear demarcation of roles and responsibility as well as in addressing needs of affected refugee and host community. For example, in the Rohingya refugee crisis, the four-pillared Emergency Control Room mechanism involving government and development partners was initiated. The four-pillared approach ensured coordination between the nodal government agency for disaster management, the nodal government agency for refugee, the government agency with maximum expertise and resources for disaster response, and the nodal development partners agency responsible for refugee response.

The influx of refugees on a large scale can have serious impact on the environment, including flora and fauna, in a matter of few weeks. It can take several years to rehabilitate the lost environment, including forests. The Rohingya refugee camps led to loss of close to 5000 acres of forest/reserved area since August 2017. More than 340,000 tree saplings were planted in camps and the most affected surrounding communities in 2018 to reverse the process of forest degradation. For 2019, it was targeted to cover/rehabilitate 100 acres through environmental restoration activities (SEG et al. 2019). It is important to consider environmental issues early in refugee response. Environmental issues shall be integrated into the settlement and infrastructure planning and implementation at initial stages. This helps in integrated planning and helps to avoid piecemeal approach.

It is important to have updated disaster risk information, of appropriate scale, of the refugee site. In case of non-availability of risk information, risk assessment should be conducted. The scale of risk information is important. In case of the Rohingya refugee crisis, flood and landslide risk assessment was conducted, and it helped to create a critical momentum that enabled significant risk mitigation efforts including development of additional sites and relocation of those most at risk (UNDP 2018b). However, discrepancies between the hotspots identified by risk assessment and site of actual landslide were observed during 2018. It is important to note that there can be several reasons including civil works at refugee sites after the risk assessment. It is recommended that updated risk information is important for risk mitigation and preparedness measures.

The influx of refugees is likely to create a dilemma in addressing short-term needs vis-à-vis medium-term needs. The Bangladesh government has been generously hosting the Rohingya refugee; however it is unprepared and unwilling to host the displaced population over the mid-term or long-term. The lack of policy level articulation hinders medium-term measures, which can help in reducing risk rather than only responding to disaster events. The entire Rohingya refugee population received basic emergency shelter kits to help them survive 2018s rainy season; however there is a need for more robust and safer shelters, which are secure in cyclone, landslide, and floods. This has implications on funding as many external funding sources cannot support medium-term risk mitigation measures through emergency funding windows.

The importance of community-based disaster risk management is well established. The Bangladesh Cyclone Preparedness Programme extended to refugee camps, and Rohingya volunteers were trained in three-flag cyclone warning systems. This helps in improved access of the early warning message as Rohingya is the only language that all refugees understand (Translators Without Borders 2018). The do's and don'ts related to cyclone, floods, and landslide have been translated into Rohingya language. Rohingya refugees played an important role in dousing midnight fire accidents in September 2018. It is important that the refugees are actively engaged in disaster risk management, and messaging should be in the local language. The engagement of refugee population has multi-fold benefits from the ownership of the intervention to sustainability of intervention. A community-based preparedness in refugee contexts is medium-term intervention as social capital is the backbone of any community-based program. In a refugee crisis, there is limited social capital as refugees are new to the geographical location and neighborhood.

It is important to plan for the worst-case scenario of natural hazard-induced disaster. A refugee situation is highly fluid in its initial months and years due to several factors including lack of data, changing built landscape, evolving refugee rehabilitation policies of host government, embryonic relationships between host community and refugees, availability of funds, and fluctuating refugee populations. The worst-case scenario may lead to additional use of resources, but protection of lives is critical. In the case of Rohingya refugee crisis, the emergency preparedness for monsoons was made for the worst-case scenario. It is recommended that in case of lack of data, it is important to plan for the worst-case scenario for a natural hazard induced disaster in refugee context. This may be a challenge for the worst-case cyclone and earthquake scenario as response planning will dramatically change from average case to worst case. In earthquake and cyclone scenario, it is important to plan for various scenarios based on resource availability and capacity of the host local government.

A refugee response program involves several dozens to hundreds of humanitarian agencies. Most of the humanitarian agencies work in their specific area of expertise such as shelter, sanitation, camp management, health, and logistics. These agencies may not necessarily have in-depth expertise in disaster risk management including weather forecast interpretation. The creation of a multi-agency or multi-sector platform for bringing humanitarian agencies together to discuss, plan, coordinate, and implement disaster preparedness is essential in refugee camps. The platform shall facilitate in developing a joint disaster response plan for coordinated effort. It is important to highlight that a dynamic setting like a refugee setting may throw new challenges in managing disaster risk. The platform should convene discussion after a disaster event to draw lessons, and it should feed into the response. In the Rohingya refugee crisis, an Emergency Task Force was established at Cox's Bazar district level. This platform convened the meeting of the humanitarian agencies every week to jointly review preparedness and response and feed the lessons into response. It also proved useful for dissemination of weather forecast with interpretation in simple language for humanitarian agencies. The platform facilitated to develop inter-sector and sector cyclone and monsoon preparedness and response plans. Future maintenance and creation of a continually updated multi-sector inter-agency platform for coordinating disaster preparedness and response in a refugee crisis remains an imperative for the safe development of displaced communities.

References

Akhter M, Uddin SMN, Rafa N et al (2020) Drinking water security challenges in Rohingya refugee camps of Cox's Bazar, Bangladesh. Sustainability 12(18):7325. https://doi.org/10.3390/su12187325

ACAPS (2018) Assessment capacities project: Rohingya crisis: pre-monsoon review summary report, March. Available via ACAPS. https://www.acaps.org/special-report/rohingya-crisis-pre-monsoon-review. Accessed Nov 2019

ADRA (2018) Adventist development and relief agency: fire incident at Chakmarkul Rohingya Camp 21, Power Point Presentation at UNHCR Campus, Cox's Bazar, Bangladesh in the Emergency Task Force meeting., 9 September

Aldrich DP (2012) Building resilience: social capital in post-disaster recovery. University of Chicago Press. https://doi.org/10.7208/chicago/9780226012896.001.0001

CBM (2019) Responding to the Rohingya Crisis, News section, 4 April. Available via Christian Blind Mission. https://www.cbm.org/news/news/news-2019/responding-to-the-rohingya-crisis/#:~:text=As%20of%20February%202019%3A,camp%20and%20the%20host%20community. Accessed Nov 2019

CwC (2019) Communications with Communities (CwC) Working group: standard operating procedures for information hub and information service center. Available via OCHA Services. https://www.humanitarianresponse.info/en/operations/bangladesh/document/standard-operating-procedures-sop-information-hub-and-information. Accessed November 2020

Cook A, Ne FY (2018) Complex humanitarian emergencies and disaster management in Bangladesh: the 2017 Rohingya exodus. S. Rajaratnam School of International Studies. https://doi.org/10.13140/RG.2.2.15958.75844

Crawford N, Cosgrave J, Haysom S, Walicki N (2015) Protracted displacement: uncertain paths to self-reliance in exile. Humanitarian Policy Group. https://doi.org/10.13140/RG.2.1.2154.3769

DDMC (2014) District Disaster Management Committee: Cox's Bazar, Bamgla-German Sampreeti & Ministry of Disaster Management and Relief, Bangladesh: District Disaster Management Plan District: Cox's Bazar, August. Available via Indonesia DOKUMEN. https://dokumen.tips/documents/dm-plan-coxs-bazar-districtenglish-version-2014.html. Accessed Aug 2019

Iman RL, Johnson ME, Watson CC (2006) Statistical aspects of forecasting and planning for hurricanes. Am Stat 60(2):105–121. https://doi.org/10.1198/000313006X106019

ISCG (2018a) Inter Sector Coordination Group: Emergency Preparedness Coordination Team. Power Point Presentation at Emergency Task Force Meeting at Ukhiya Upazila, Bangladesh on 22 November. Available via OCHA Services. https://www.humanitarianresponse.info/en/operations/bangladesh. Accessed (10 Feb 2021)

ISCG (2018c) Shelter/Non-Food Item Sector. Inter Sector Coordination Group: Structural Assessment Against Wind Loading, Power Point Presentation at the Emergency Task Force meeting, IOM Meeting Room, Cox's Bazar, Bangladesh on 16 September. Available via OCHA Services. https://www.humanitarianresponse.info/en/operations/bangladesh. Accessed (10 Feb 2021)

ISCG (2018d) Site Management and Site Development Sector: Monsoon Emergency Response Update: Rohingya Refugee Crisis. Cox's Bazar, Bangladesh (8–14 August 2018), ISCG, 15 August. Available via Reliefweb. https://reliefweb.int/sites/reliefweb.int/files/resources/iscg_monsoonemergencyupdate_15august2018.pdf. Accessed (Oct 2020)

ISCG (2019b) Inter Sector Coordination Group: Cyclone 72 Hour Response Plan: Coordination mechanisms, stocks and prepositioning, needs assessment protocols and planned response, June. Available via Humanitarian Response. https://www.humanitarianresponse.info/sites/www.humanitarianresponse.info/files/documents/files/cyclone-72hrplan_v1.0_2019june.pdf. Accessed 10 Feb 2021

ISCG, RRRC, UNHCR, IOM-NPM, Site Management Sector (2019) Bangladesh: Cox's Bazar Refugee Population as of 30 September 2019. 9 October. Available via Humanitarian Response. https://www.humanitarianresponse.info/sites/www.humanitarianresponse.info/files/documents/files/190930_cox_bazar_reference_map_with_pop_id0225.pdf. Accessed (Oct 2020)

IOM (2018) International Organisation for Migration: Rohingya Humanitarian crisis Response, Annual Achievements-2018, IOM Bangladesh. Available via IOM UN Migration. https://bangladesh.iom.int/sites/default/files/publication/Annual%20Report%202018%20copy%20 2019%2005%2006.pdf. Accessed Oct 2019

Kumar S (2012) Acting on disaster warnings: Don't miss the human factor. Available via Sci Dev Net. https://www.scidev.net/global/opinions/acting-on-disaster-warnings-don-t-miss-the-human-factor/. Accessed Nov 2020

Loy I (2019) Thousands displaced as heavy rains batter fragile Rohingya camps, The Humanitarian News, 8 July. Available via The New Humanitarian. https://www.thenewhumanitarian.org/news/2019/07/08/thousands-displaced-heavy-rains-batter-fragile-rohingya-camps. Accessed on 3 Dec 2020

Miyaji M, Okazaki K, Ochiai C (2017) A study on the Use of Cyclone Shelters in Bangladesh. J Archit Plan 82(737):1871–1880. https://doi.org/10.3130/aija.82.1871

NRC (2019) Torrential rains destroy shelters for displaced people in northeast Nigeria, 23 August. Available via Norwegian Refugee Council. https://www.nrc.no/news/torrential-rains-destroy-shelters-for-displaced-people-in-northeast-nigeria/. Accessed Oct 2020

Oxfam (2020) Up to 250 people per single water tap in refugee camps braced for arrival of coronavirus, 6 April. Available via Oxfam International. https://www.oxfam.org/en/press-releases/250-people-single-water-tap-refugee-camps-braced-arrival-coronavirus. Accessed on 23 Nov 2020

Oxfam, Save the Children and Action Against Hunger (2018) Rohingya Refugee Response Gender Analysis: Recognizing and responding to gender inequalities. Available via Reliefweb. https://reliefweb.int/sites/reliefweb.int/files/resources/rr-rohingya-refugee-response-gender-analysis-010818-en.pdf. Accessed Dec 2020

PAHO (2009) Pan American Health Organisation: Information management and communication in emergencies and disasters: Manual for disaster response teams. ISBN: 978-92-75-12993-7

Pinto GM, Attwood J, Birkeland N, Nordbeck HS (2014) Exploring the links between displacement, vulnerability and resilience, 4th International Conference on Building Resilience, Salford Quays, United Kingdom, 8-10 September 2014. Procedia Econ Financ 18:849–856. https://doi.org/10.1016/S2212-5671(14)01010-7

Rahman AT (2019a) Fighting Water Scarcity in Cox's Bazar Refugee Camps. 4 November. Available via UNHCR. https://www.unhcr.org/news/latest/2019/1/5c2fc16a4/fighting-water-scarcity-coxs-bazar-refugee-camps.html. Accessed (23 Nov 2020)

Rahman M (2019b) Rohingya refugee crisis and human vs. elephant (Elephas maximus) conflicts in Cox's Bazar district of Bangladesh. J Wildlife Biodivers 3(3):10–21. https://doi.org/10.22120/jwb.2019.104762.1057

Rajaratnam School of International Studies (RSiS) (2018) Complex humanitarian emergencies and disaster management in Bangladesh: The 2017 Rohingya exodus, July.

REACH (2019a) Bangladesh – Cox's Bazar: Cyclone Shelter Database. October. Available via REACH. https://www.humanitarianresponse.info/en/operations/bangladesh/infographic/bangladesh-coxs-bazar-cyclone-shelter-database-i-october-2019. Accessed (22 Nov 2020)

REACH (2019b) Participation of Rohingya Enumerators in Data Collection Activities: Findings of a Pilot Assessment in Cox's Bazar. Bangladesh, April. Available via Reliefweb. https://reliefweb.int/sites/reliefweb.int/files/resources/reach_bgd_brief_rohingya_enumerator_pilot_april2019_1.pdf. Accessed (3 December 2020)

SEG, Partners (2019) Strategic Executive Group and Partners: Joint Response Plan for Rohingya Humanitarian Crisis: January-December 2019. Available via Reliefweb. https://reliefweb.int/sites/reliefweb.int/files/resources/2019%20JRP%20for%20Rohingya%20Humanitarian%20Crisis%20%28February%202019%29.compressed_0.pdf Accessed (22 Aug 2020)

Shongjog (2020) Communication with communities in Bangladesh. Available via Shongjog. www.shongjog.org.bd/ Accessed 22 Nov 2020

Site Management and Site Development (2019), Cox's Bazar, Relocation Update, 7 February.

Sphere Association (2018) The Sphere Handbook. Available via Sphere. https://spherestandards.org/handbook-2018/ Accessed (Nov 2020)

Strategic Executive Group (SEG) and Partners (2018), Joint Response Plan for Rohingya Humanitarian Crisis: Mid-Term Review, March-December 2018. Available via Humanitarian Response. https://www.humanitarianresponse.info/sites/www.humanitarianresponse.info/files/documents/files/2018_jrp_mid_term_review_v28.pdf Accessed 15 Aug 2019

Tirmizi, ASN (2005) The contribution of levels of social capital to community development. Ph.D Thesis. Iowa State University. Available at Iowa State University Digital Repository. https://lib.dr.iastate.edu/cgi/viewcontent.cgi?article=2598&context=rtd. Accessed 24 Aug 2019

TWB (2018) Translators Without Borders Communications: The language lesson: What We've Learned About Communicating with Rohingya Refugees, Translators Without Borders, 4 December. Available via Translators Without Borders. https://translatorswithoutborders.org/the-language-lesson-what-weve-learned-about-communicating-with-rohingya-refugees/ Accessed on 11 Dec 2020

UN (2018) United Nations: Report of the United Nations High Commissioner for Refugees: Part II Global compact on refugees, 13 September. Available via UNHCR. https://www.unhcr.org/gcr/GCR_English.pdf. Accessed 16 July 2019

UNDRR (2008) Early warning systems can save lives when cyclones strike, 6 May. Available via United Nations Office for Disaster Risk Reduction. https://undrr.org/news/early-warning-systems-can-save-lives-when-cyclones-strike. Accessed 29 Nov 2020

UNDRR (2019) Global Assessment Report on Disaster Risk Reduction, Geneva, Switzerland. Available via United Nations Office for Disaster Risk Reduction. https://gar.undrr.org/sites/default/files/reports/2019-05/full_gar_report.pdf. Accessed 4 Dec 2020

UNDP (2018a) Impact of the Rohingya Refugee Influx on Host Communities, Bangladesh Country Office, November. Available via United Nations Development Programme. https://www.humanitarianresponse.info/sites/www.humanitarianresponse.info/files/2019/08/UNDP_Impacts-of-the-Rohingya-Refugee-Influx-on-Host-Communities_Nov2018.pdf. Accessed 6 Dec 2020

UNDP (2018b) Extreme weather and disaster preparedness in the Rohingya Refugee Response-Monsoon season 2018 lessons learnt. December. Available via Reliefweb. https://reliefweb.int/sites/reliefweb.int/files/resources/UNDP%2520DRR%2520-%25202018%2520-%2520Monsoon%2520lessons%2520learnt%2520report.pdf. Accessed (11 Dec 2020)

UNDRR (2017) Terminology on Disaster Risk Reduction, 2 February. Available via United Nations Office for Disaster Risk Reduction. https://www.unisdr.org/we/inform/terminology. Accessed 7 Aug 2019

UNGA (2016) New York Declaration for Refugees and Migrants, 3 October. Available via United Nations General Assembly. https://undocs.org/a/res/71/1. Accessed 18 July 2020

UNHCR (2013) Death toll in Thailand refugee campfire reaches 37; UNHCR aids victims, 26 March. Available via The office of the United Nations High Commissioner for Refugees. https://www.unhcr.org/news/latest/2013/3/51517c9e9/death-toll-thailand-refugee-camp-fire-reaches-37-unhcr-aids-victims.html. Accessed on 23 Nov 2020

UNHCR (2015) UNHCR, Displacement and Disaster Risk Reduction. Available via The office of the United Nations High Commissioner for Refugees. https://www.unhcr.org/5665945e9.pdf. Accessed 23 Nov 2020

UNHCR (2019) Figures at a glance, Statistical Yearbooks, 19 June 2019. Available via The office of the United Nations High Commissioner for Refugees. https://www.unhcr.org/figures-at-a-glance.html. Accessed 5 Aug 2019
UNHCR, IOM and ADPC (2018) Flood and Landslide - Impact Analysis. Rohingya Refugee Crisis, Bangladesh, 21 January. Available via OCHA Services. https://www.humanitarianresponse.info/en/operations/bangladesh/infographic/refugees-situation-cox%E2%80%99s-bazar-risk-analysis-kutupalong-balukhali. Accessed (11 Dec 2020)
UNHCR, IUCN (2018) Survey Report on Elephant Movement, Human-Elephant Conflict Situation, and Possible Intervention Sites in and around Kutupalong Camp, Cox's Bazar, 22 February. Available via The office of the United Nations High Commissioner for Refugees. https://www.unhcr.org/5a9946a34.pdf. Accessed 2 Dec 2020
UN Women (2020) Promoting the Women. Peace and Security Agenda in the Rohingya Refugee Crisis in Cox's Bazar, Bangladesh, 27 October. Available via Reliefweb. https://reliefweb.int/report/bangladesh/promoting-women-peace-and-security-agenda-rohingya-refugee-crisis-cox-s-bazar. Accessed (11 Dec 2020)
World Bank (2018) Brief: Bangladesh Disaster Risk and Climate Resilience Program. 25 July. Available via The World Bank. https://www.worldbank.org/en/country/bangladesh/brief/bangladesh-disaster-risk-climate-change-program Accessed (10 Aug 2019)
Yan W (2018) Help is scarce for Rohingya children with intellectual disabilities. 22 August. Available via Global Post. https://www.pri.org/stories/2018-08-22/help-scarce-rohingya-children-intellectual-disabilities. Accessed (19 August 2019)

Chapter 8
Risks for the Environment, Biodiversity, Humankind, and the Planet

Kelly Swing

Abstract Factors tied to the environmental ills of the planet are explored primarily in connection with human attitudes and behavior. It would appear that some very basic aspects of our nature, at least partially founded in survival instincts, are at the root of the complications that loom darker with each passing day. The simple drive to reproduce and then to acquire sufficient resources for the next generation to do precisely the same has led to a monstrous human population and exorbitant exploitation of natural resources. Combined with our extraordinary capacity as ecosystem engineers, there seem to be no limits to our incessant occupation of more space and conversion of the landscape for our own benefit with little regard for other organisms or our own future needs. The outcome is a downward spiral of worsening conditions for ourselves as well as cataclysmic consequences for millions of other species. Although these problems have been recognized for some time, the majority of humans remain rather complacent about our highly predictable, and lamentable, pathway. Perceptions of proportional individual sacrifice versus community benefit will require serious modification if we are to successfully navigate the building storm on the horizon.

Keywords Environment · Risks · Historical perspective · Ethical considerations · Human behavior

8.1 Introduction

In every sense, we are living at a crucial moment in the trajectory of humankind as well as the planet and its biota. Scientists have directly monitored the day-to-day of global conditions in highly reliable ways for more than a century. Meanwhile,

K. Swing (✉)
Tiputini Biodiversity Station, College of Biological and Environmental Sciences, University of San Francisco de Quito, Quito, Ecuador

Tropical Ecology Program, Boston University, Boston, MA, USA
e-mail: kswing@usfq.edu.ec

J. N. Furze et al. (eds.), *Earth Systems Protection and Sustainability*,
https://doi.org/10.1007/978-3-030-85829-2_8

innovative technologies have allowed us to look back across millennia for the purpose of gaining broader perspective on Earth's history, especially in the realms of geology, biology, evolution, and climatology. Understanding the past can allow comprehension of the present and provide possibilities for extrapolation into the future, thereby simplifying decision-making. For far more than half a century, well-masticated evidence of negative anthropogenic impacts at every scale on every continent has been made available to the ever-increasing human population. With this cascade of scientific facts laid before us, some countries have responded with relevant policies and actions to remedy deteriorating conditions, while billions of other humans go on living and reproducing as though nothing has changed one iota. Inventors and theoreticians have continually offered reasoned strategies to overcome most of the related challenges, but so far, inaction has been our greatest response. Considering the grim predictions for the coming decades, especially in light of stereotypical human behavior, an assessment of pertinent risks – past, present, and future – is well overdue.

> "We're not gonna make it, are we? I mean humans." (John Connor)
> "It's in your nature to destroy yourselves." (Terminator)
> Prophetic dialog from the 1991 Arnold Schwarzenegger movie *Terminator 2: Judgement Day*

8.1.1 How Did We Get Here?

It took a colossal amount of time – more than 4.5 billion years – for the first humans to appear on the planet (Mallick et al. 2016). From the moment of becoming "human," it took a couple hundred thousand years to reach a total population of one billion individuals, a figure some bacteria could attain in a matter of several hours (Todar 2017). Although our reproductive capacity is quite limited in comparison to most of the smaller organisms in existence, our exceptional intelligence combined with social support systems has led us to being particularly successful at leaving offspring in subsequent generations. Dedicated parental care no doubt helps to overcome many complications of early survivorship (Klug and Bonsall 2014), but it is almost certainly the surprising capacity to modify our surroundings for our own benefit (building shelter, growing crops, domesticating animals) that gave us unusual opportunities to disproportionately increase our numbers. After reaching that first billion mark around 1800, it only took about one century to surpass the second and then only half a century to get to three (Smil 2011). From there we went into overdrive with the rate of a billion more per decade for a couple of scores and are now headed toward nearly 10 billion by mid-century (Garfield 2016).

> We have a finite environment – the planet. Anyone who thinks that you can have infinite growth in a finite environment is either a madman or an economist. (Sir David Attenborough, 1926–)

There is no escaping the fact that each and every one of us occupies space, consumes resources, and produces waste. There is also no escaping the fact that humans represent a ponderous burden on the all-encompassing Gaia system. The Gaia concept suggests that the planet Earth, as a single, living, integrated organism, can only provide her plenitude of essential ecosystem services under certain conditions. Given that we indeed wish to enjoy the full bounty nature has to offer (survive and thrive comfortably with access to resources and functionality), Gaia, although quite resilient, must be treated with respect and dignity, not assailed with constant exploitation and pollution. Settling the score for all our past transgressions will not be resolved by 10% of the total population sorting their trash in order to feel good about themselves (UNEP 2017; Wilkins 2018). Going right on as though the scale of the current situation does not impose unprecedented urgency is thoroughly nonsensical. We must also consider that an exponential push for more affluence across nearly all societies, accompanied by greater materialism and consumerism, means that our summed impacts are increasing far faster than our absolute numbers (Shah 2014). The foreboding idea of carrying capacity (Daily and Ehrlich 1992; Hardin 1960) inevitably springs to mind, logically indicating that there is truly a limit to how much life can exist or thrive in any given area. Though our creativity allows for a modicum of stretching the envelope, all reason would insinuate that there is most definitely an endpoint which is real. Slowly but surely, we have elbowed our way from being small, rare clans of hunched-over savannah roamers to dominating every habitable part of the planet (Diamond 1999). At this point, humankind, combined with all our domesticated animals and livestock, possibly represents over 98% of all mammalian biomass on the planet (Bar-On et al. 2018; Carrington 2018). Logically, it is implied that we have "competitively displaced" (Rees 2017) the vast majority of wild mammals, and animals of all sorts, across the globe (Smil 2011). Harvard emeritus professor E.O. Wilson (2016) has argued that our only chance of maintaining a healthy planet that includes human civilization "as we know it today" is to keep at least 50% of global ecosystems intact. Sadly, we blew past that figure decades ago, meaning that we now face the chore of not only dialing down the rate of population growth but actually reversing it, not only curtailing our exploitation of nature but giving her space and time to recover on the large scale. Paul and Anne Ehrlich had said basically the same thing in *The Population Bomb* a half century ago when our numbers were less than half of what they are today (Ehrlich 1968). Instead of getting a response in the form of a plan of action, the Stanford professor was promptly labeled an alarmist and his ideas are dismissed for the most part. Though his predictions did not come to pass as quickly as projected, the general line of reasoning remains valid all these years later, essentially confirming, by and by, a large portion of the thought processes Thomas Malthus originally published in 1798.

> If men could learn from history, what lessons it might teach us! But passion and party blind our eyes, and the light which experience gives is a lantern on the stern, which shines only on the waves behind us! (Samuel Coleridge, 1831)

During modern times, especially the last two-thirds of the last century, we have almost incessantly heard that we must change our ways to save the planet. Rachel Carson (1962) sounded the first powerful alarm of the modern era, relating more to our behavior than our absolute numbers, pointing out that our indiscriminate use of toxic chemicals (including insecticides, pesticides, fungicides, weed-killers, and fertilizers) to rule the world was silencing nature and would soon have grievous implications for human health and well-being. More than a century earlier, American painter Thomas Cole, through his series, "The Course of Empire," was already bemoaning the conversion of landscapes into municipal/industrial complexes (Comegna 2016). Now, during the first score of the Third Millennium, abundant pleas in the name of pandas and polar bears, elephants, rhinos and giraffes, cheetahs, monarch butterflies, whales and sharks, honeybees, and pangolins have increasingly gone out. Repeated advisories nowadays get minority attention from members of the choir, but a larger portion of the human population goes on ignoring all the indications that our world is indeed going to hell in a handbasket (Steffen et al. 2018). Yet another bleak projection?

> How can we be so arrogant? The planet is, was, and always will be stronger than us. We can't destroy it; if we overstep the mark, the planet will simply erase us from its surface and carry on existing. Why don't we start talking about not letting the planet destroy us? (Paulo Coelho, 1947–)

Greater numbness, rather than a definitive plan for today and into the future, has been the response in far too many cases. The naive masses continue to think that the world beyond their everyday domain is independent of themselves and their personal actions (Rees 2017). They seem to believe that we are somehow impervious to what happens over the backyard fence.

> Every person takes the limits of their own field of vision as the limits of the world. (Arthur Schopenhauer, 1788–1860)

For every doomsayer, there is an equally outspoken denier who justifies his/her position with arguments that inevitably signal either a woeful lack of valid information conflated with distorted interpretations or a profound misunderstanding of how science actually works. Religious perspectives currently play a larger role than in other recent periods; faith rather than physics appears to be the justification for many environmental policy decisions among a dogged and formidable conservative sector in the USA (Vox 2017). Dead reckoning, nonetheless, implies that someone must be right and someone must be wrong. In the modern world with nearly limitless communication, it should be a simple matter to dismiss the possibility that people in positions of power do not have access to relevant, reliable information. Consequently, given the preponderance of scientific evidence confirming a potentially catastrophic future (Skidelsky 2015; Steffen et al. 2018; Goodell 2018; Mooney 2018), for the perspective of skepticism and rejection to genuinely represent a problematic situation at the political level, deliberate manipulation (Baumrind 1985; Franta 2018; Temple 2018) or blind delusion of the voting population would be necessary. That appears to be precisely the pernicious scenario that we are living.

> Do you know how [the people] make it through each and every day? They believe against all odds and all evidence that tomorrow will be better than today. (Mayor of the town of Dirt, from the animated movie Rango, 2011)

Currently, a phenomenon known as the Titanic mentality (Battles 2001) seems to pervade society. Scientists, nearly in consensus, say we have all the requisite evidence to confirm that disaster is imminent. Being comfortable on board an "unsinkable" vessel, a 4.6 billion-year-old planet, many continue their lives as though nothing is wrong, instead of responding in a reasonable way to the oncoming challenges by trying to understand what virtual iceberg lies ahead, making efforts to change our course to avoid the collision altogether or at least to soften the blow. Were the drive focused on saving ourselves instead of the planet, would our response be more substantive?

It would definitely be easier and cheaper, in the short term, to do absolutely nothing, going on with "business as usual." The downside is that such a choice could easily put us into a devastating situation within less than half a century (Franta 2018; Goodell 2018). At present, however, a few individuals with tremendous power and position dominate microphones and airwaves insisting that the world citizenry has en masse been duped by the scientific community; the wheels of "progress" simply must continue to grind out greater mountains of money; the economy must grow unhampered or the people will undisputedly suffer even more; and untapped resources must all be exploited without exception to keep any possible competitive edge in the rat race that is life in modern times. As a consequence, are we guilty of sacrificing the Earth's species on the altar of economic growth and material wealth?

> Our task must be to free ourselves by widening our circle of compassion to embrace all living creatures and the whole of nature and its beauty. (Albert Einstein, 1879–1955)

By narrowing our gaze to include little more than the bottom line, have we unwittingly chosen an inexorable path toward ecosystem collapse? (Diamond 2005; Franta 2018) Simplistic explanations and facile solutions, even when they are clearly contrary to historical perspective, or offer little in the way of evidentiary support, once coupled with fear mongering and xenophobia, are jet fuel for populism, making for easy manipulation of the unconscious and uneducated mobs.

> Those who can make you believe absurdities can make you commit atrocities. (Voltaire, 1694–1778)

The classic "argument of the beard" (a.k.a. the continuum fallacy) should invade our thoughts. This is the idea that if 100,000 hairs are enough to make a "good beard," then certainly 99,000 would suffice. The argument continues progressively, always reducing the number of hairs but maintaining the conclusion, to the absurd endpoint where a few scattered hairs on a face can still be considered a "good beard" (Fallacy Files 2021). Are we using the same logic to justify what is acceptable in the way of how few trees it takes to be called a forest, or whether being recognized as a forest only involves an assessment of the number of trees as opposed to all members of the forest community (Suárez 2018), or how little of nature must be left in order to produce "enough" oxygen in a breathable atmosphere, potable water, and so on?

Regrettably, the gift of charisma affords a disproportionately influential voice to certain individuals – whether they are knowledgeable or ignorant; true visionaries or blind to realities; whether they depend upon substantiated evidence or intentionally contort the truth – who are experienced or novices, however well-intentioned or evil, caring or indifferent, generous or selfish, be they humane or malevolent, be they prejudiced or fair and just, be they ethically upstanding or be they dedicated swindlers, be they innately responsible or totally unaccountable in word and deed, regardless of having been democratically elected to occupy high office or having violently taken control as a recognized tyrannical dictator.

> Everyone is entitled to his own opinion, but not to his own facts. (US Senator Daniel Patrick Moynihan, 1927–2003)

For people in the here and now to be dismissive of science or of its importance in everyday life and in commerce is more anomalous than ever in the past. Innovation founded in scientific principle has been the mother of essentially all human advances since the invention of the atlatl and the wheel. Modern conveniences – ranging from comfortable shoes to computers, from laparoscopic interventions and intercontinental flight to Internet communication systems, from hurricane warnings to seismically resistant buildings, from laundry detergent to disposable contact lenses – are all tangible results of technology and engineering coming together to convert the underlying science into practical products. Disregarding the sheer volume of trade associated with scientifically derived products represents an unforgiveable level of ignorance in any modern leader or decision-maker.

To take it a step further, self-interested willful ignorance on the part of those in power is at once intentionally abusive and reprehensible. Not responding quickly and decisively to abundant scientific evidence related to the climate puts us into an ever-deepening quagmire (Goodell 2018) with negative implications in nearly every aspect of human life and existence. Not investing in the scientific education of young minds so that they may have greater access to the tools needed to compete in a world more heavily dependent upon science each year will relegate entire generations to the backseat of future economies. On the other hand, expanding scientific knowledge and functionality in young generations can improve the outlook for the future in relation to everything from exploration to discovery, from agriculture and healthcare to manufacturing and transportation, and from human rights and justice to effective conservation of nature and sustainable management of natural resources.

> Understanding the laws of nature does not mean that we are immune to their operations. (David Gerrold, 1944–)

8.2 Protecting Our Own Interests by Protecting Our Surroundings

> The environment is humanity's first right. (Kenule Saro-Wiwa, 1941–1995)

The Environmental Protection Act (1970) combined with associated legislation in the USA, including the Clean Water Act, the Clean Air Act, and the Endangered Species Act, came together to make a huge impact on the way societies address their relationship with nature and natural resources by taking legal responsibility for our actions and our future. We should point out that these laws/policies came into being in a time when the mass consumption of leaded fuels was yet to be amply recognized as a serious health threat; planting flowers along major US roadways was widely considered an outstanding benefit for nature. During the 1970s, through television spots lamenting the littering of the American countryside in what had once uniformly been indigenous territories, "Iron Eyes" Cody, portraying the dubious plight of ancestral lands and cultures, seemed to provoke some of the first noticeable impacts on broad public awareness.

Since its inception, the Environmental Protection Agency (EPA) has been somewhat controversial. This had nothing to do with the biological/chemical evidence showing that we were already poisoning ourselves to death. Although some early datasets were still rather limited and needed more elaboration, there were plenty of dismal indications for all to see. Common sense suggested that something simply had to be done. The Great Smog of London in December of 1952, by killing thousands of individuals, spawned the first legislative responses to contamination in the world, eventually leading to the 1956 Clean Air Act in Great Britain (Brimblecombe 2006). Even a president of the USA so reviled by history as Richard Nixon could understand that "Oh say, can you see?" (excerpted from the US national anthem) presented unsafe implications for the skylines of major cities like New York, Chicago, Los Angeles, and Pittsburg. Smog had grown so severe in some places that there were daily warnings not to go outside due to health risks. In the USA, a river, the Cuyahoga in Cleveland, Ohio, was so polluted it actually caught fire in 1969, leading to an environmental movement that produced the Clean Water Act of 1972 (Latson 2015). Even so, there was always an underlying stewing perspective that simply too much consideration was being given to nature, wild animals, native plants, ecosystems, wilderness, and natural resources (Benohr and Lynch 2018). England's Clean Air Act of 1956 "remains a seminal piece of legislation because it created a belief that a better environment was possible and worthwhile despite the fact that at times it would restrict our individual freedom" (Brimblecombe 2006). Fear of precisely how to word or enforce the intentions in such philosophies made for lengthy discussions and even led to political skirmishes in some regions.

Nevertheless, by the 1980s, mentalities were changing as the EPA was having noticeable impacts in the real world. Many toxic compounds (EPA/Toxic Substances Control Act website) had been eliminated from everyday usage, standards had been set for acceptable levels of multiple chemicals in the workplace (US Department of Labor Occupational Safety and Health Administration, founded 1971), and several endangered species had been recognized and were receiving much needed attention. Things were far from perfect, but we were off to a strong start. There was reason for a degree of optimism in spite of a long list of distractions from what could only be considered a laudable goal for all of humanity. Some specific complaints, usually from self-interested parties, would undoubtedly always be a part of the process of sorting out what is permissible and what is simply too deleterious to life, human or otherwise.

As a result, some iconic endangered species – American alligator, brown pelican, bald eagle, and humpback whale – eventually recovered enough to be taken off the list (US Fish and Wildlife Service Environmental Conservation Online System). While this is cause for hope and celebration, it has the rueful side effect (Dye 2017) of leading the public to believe that there is no longer any need to provide support for these species, which once again places an air of doubt over their futures. In most cases, delisting has been the result of painstaking, relentless endeavors on the part of dedicated teams over decades (Jacobson 2013). Those directly involved in these successful outcomes know all too well that a short lapse in vigilance could put these same species right back on the brink of demise and into truly unrecoverable status (Sieswerda 2018). If a powerful and wealthy nation like the USA had allowed its national symbol, the bald eagle, to go extinct within its territory, what would that imply about the potential futures of all other species – endangered at the moment or not – across the globe? And what would that imply about a nation's basic values?

> If this [the conserving of nature] is not done, future ages will certainly look back upon us as a people so immersed in the pursuit of wealth as to be blind to higher considerations. (Alfred Russel Wallace, 1823–1913)

8.3 What Indeed Is the Basic Role of Government?

Although governments, along with the laws they produce and the policies they espouse, are involved in many aspects of our lives at different levels, most of us probably think of them primarily as serving to tell us what we cannot do and that there will be grave consequences if we dare violate established norms – much like a strict parent. It turns out that having some basic standards for acceptable behavior founded in simple, straightforward principles that most humans readily accept and support is precisely what separates civilization and order from chaos and fear, thereby allowing us to be productive as individuals, communities, and societies. This concept is reflected by the fact that most cultures and religions proclaim some version of what Westerners refer to as the "Golden Rule" – "Do unto others as you would have them do unto you." At some level, governments are meant to protect us from one another, and from ourselves. How we come up with systems for monitoring behavior, setting norms, and doling out punishments has varied across cultures since the first rules were made at societal levels, but the underpinning principles are ideally founded in concepts of fairness and functionality.

In most modern systems, it would appear that governments indeed have served to lessen or overcome many challenges, but recently, shortfalls in the realm of the environment, especially for climate concerns, pollution, and mass extinctions, have been far too common. In efforts to compensate for lack of funds or commitment or action on the part of the state, numerous NGOs have stepped in to take on important conservation roles across the globe. Although certain governmental agencies in many countries have played admirable parts in the past, overall performance records during the current century have mostly missed the mark while the stakes rapidly increase in tandem with the galloping rate of accumulated impacts. Scornfully, far too many games and manipulations, broadly referred to as "politics," get in the way of an already cumbersome process. While "politics" can mean everything from legitimate arguments and negotiations to ethical trade-offs, we must recognize that other considerations such as special favors, pay-offs, kick-backs, and so forth (commonly known as corruption and regularly including nepotism) too often come into play.

8.4 An Ethical Exploration: What If?

Let's follow up on the theme of governments serving to protect us by contemplating one of the most widely accepted of all philosophical approaches to life, one included among the Bible's Ten Commandments – "Thou shalt not kill" (Exodus 20). It is quite easy to see why we all like the resulting protection for ourselves and accept that it applies to and benefits all fellow human beings. If the rule is violated, then some punishment is prescribed. In most cultures, however, there are valid exceptions. What if someone is killed due to an accident? Laws typically have special allowances for such cases. What if the accident happened because someone was doing something absolutely stupid near the victim? That may be categorized as "negligent homicide" (technical language certainly changes between systems), and the consequences would usually be distinct from a planned execution ("premeditated murder"). There would probably, however, need to be some procedural evaluation of intent. What if someone sets out to kill you and you are obligated to use deadly force to save yourself? Or your family? Or, in the military, members of your own platoon? (Here, we must recognize the gaping loophole of war, in which formal declaration of someone as your enemy means they, by default, lose most of their rights within your system.)

What if someone is intentionally poisoning his neighbor so as to get access to his property or his spouse? Logically, this would be easy to categorize as intentional, premeditated homicide as well. What if we don't know the precise motive? It would still be murder, right? What if the poisoning happens due to oblivion as opposed to being part of a defined plan? Is responsibility or culpability the same? What if the individual exposing his neighbor to poison has no intention to commit murder but knows he is engaged in something that results in someone downstream being threatened chemically? What if the one releasing the toxin is completely unaware of his/her impacts? Is there an obligation to know or to act responsibly? What if someone knows fully well they're releasing deadly compounds, are they guilty of premeditated murder? What if the neighbor isn't killed outright but suffers because of a chronic, non-lethal dosage? Is killing someone or causing them harm under these circumstances still a punishable offense? Of course, it is. What if the level of harm cannot be readily quantified? Or a connection is not easily drawn between cause and effect? What if the one doing the poisoning is entirely conscious of what is going on but continually makes decisions based on how much money he/she can make given various scenarios in which neighbors die quickly, suffer for years, or only have minor discomfort all the time? What if such a business owner only considers the bottom line, always takes the cheap alternative, and goes right on poisoning his/her neighbors wholesale with complete knowledge of what's happening? What if the business makes piles of money and employs dozens of workers whose families depend upon the resulting income? What if the local economy benefits from cash flowing from the precise industry poisoning its neighbors? What if 100 employees and their families benefit from income associated with this operation but another 100 individuals are severely poisoned by its effluents? What if the proportions change? What if the balance is 1000:1 or turns out to be the reverse, 1:1000?

> No man may poison the people for his private profit. (Theodore Roosevelt, 1858–1919)

Is our perception of the poisoning ameliorated philosophically or legally because of the production of cash flow? What if the business owner has powerful connections in the legislative or judicial branch of government and influences regulations or their interpretation to make exceptions so the business can increase profit margins? What if the offending party is best friends with the chief executive? What if "the system" favors a particular industry simply because it makes money? Or in response to donations along party lines and, as a consequence, ignores the rights of neighbors to not be killed or harmed?

If you are personally being well paid, would it matter that your own children are being poisoned? What if a company offered you generous sums of money for license to "frack" in your neighborhood? Which matters more for this decision? Your personal knowledge of this extractive process regarding environmental impacts and public health concerns or how much money flows to you or your region from the resulting hydrocarbons?

What if the situation isn't quite so clear? What if the victims are only mildly poisoned and it is difficult to assign responsibility? What if the impact affects something difficult to measure in the first place? What if the parameter in question were intellectual capacity, as in the historical cases of lead in paints and fuels (Hemberg 2000), lowering the IQ of victims by 10 or 20 points?

What if everyone downstream from a company's drain outlet suffers "inexplicable" maladies, but no one else does? What if everyone who lives in that area is poor? Or belongs to an oppressed minority with miniscule economic or voting power? Or they are a small population of dark-skinned indigenous people? What if a similar problem were quickly identified and remedied because it impacted a wealthy senator's neighborhood? What if the same problem were never addressed because it "only" affected people of a particular culture or religion? What if only the most vulnerable (the elderly, small children, early first-trimester embryos) were affected?

What if those affected are other species? What if those species are emblematic megafauna? What if, on the other hand, they are lowly arthropods, or mollusks, or fish? What if we're talking about commercially valuable species like shrimp, scallops or oysters, tuna or salmon? What if an activity leads to mercury accumulation in swordfish (Lyons 1970; US Food and Drug Administration website) which would poison everyone who consumes it but makes money for fishermen and restaurants along the way? What if a governmental administration, beholden to a specific industry, ignoring all scientific and medical evidence, preferentially allows that industry to release toxic substances or sell merchandise known to be hazardous (like certain pesticides)? In a practical sense, is this very different from despotic regimes that have intentionally used chemical weapons against their own people? If the level of suffering is a consideration, dying within minutes of an explosion would almost certainly be preferable to chronically debilitating exposure or languishing with cancer for years.

In the USA, legislators have moved toward excluding testimony from unbiased experts as a standard part of "fact gathering" hearings. Instead, nowadays, "expert" witnesses are specifically chosen solely to share their perspectives, comments, and opinions in accordance with their partisan positions (Devins 2005). Without doubt, this implies insidious intentionality through selective inclusion/exclusion of evidence, scientific or otherwise. Should this be considered a form of "witness tampering"?

Given someone has knowledge that could prevent a disaster, would it be negligent not to share that information? Would it also be criminally negligent not to receive, understand, or apply said information? What if those in power actively exclude the most knowledgeable or experienced individuals from conversations? Suppose a member of the Titanic crew had seen that iceberg in the North Atlantic and said nothing, would he be directly responsible for the loss of 1500 lives? Had no individuals seen the dangerous looming chunk of ice, is no one culpable? Was not having a proper lookout equally negligent? Could willfully ignoring such information be considered criminal negligence? Suppose the helmsman had knowingly, intentionally directed the "unsinkable" vessel into that iceberg due to a sense of arrogant invulnerability, what would society's response be?

Do legislators sincerely believe that their choosing ignorance, or arrogance, absolves them of all accountability? Is the ostrich with its head in the sand while its hatchlings are devoured by a jackal not responsible? Given the condition "ignorance of the law is no excuse," then intentionally ignoring the facts must be seen in a similar light.

> Where knowledge is a duty, ignorance is a crime. (Thomas Paine, 1737–1809)

8.5 Relevance of Basic Human Rights Policy to Average People

One of the basic human rights recognized by the United Nations is the right to live in an environmentally sound setting. Principle 1 of the 1992 Rio Conference on Environment and Development stated that all humans "are entitled to a healthy and productive life in harmony with nature." Article 24 of the African Charter on Human and Peoples' Rights (Banjul, June 26, 1991) states that "All peoples shall have the right to a general satisfactory environment favorable to their development."

> The environment is where we all meet, where all have a mutual interest; it is the one thing we all share. (Lady Bird Johnson, 1912–2007)

Surprisingly perhaps, many governments, local to national, choose not to adopt laws that reflect this stance, effectively ignoring the indisputable connection between environmental health and public health, often due to political or economic pressures. There are, however, exceptional cases that take the theme one massively important step further.

By 2008, the South American country of Ecuador had ratified a new constitution that proclaimed the rights of nature, thereby becoming the first country to recognize legal status for "the integrated existence and maintenance of nature as well as the regeneration of life cycles, structure and evolutionary processes" (Ecuadorian Constitution, Article 71). Most scientists would agree that evolution is classically an ongoing and often slow process; its specific mention as part of that statement would imply a truly long-term commitment. A corollary reflected in that same document is "the right for the population to live in a healthy and ecologically balanced environment" (Ecuadorian Constitution, Chapter 2, Section 2, Article 14) which is parallel to basic human rights recognized by the United Nations.

> What we are doing to the forests of the world is but a mirror reflection of what we're doing to ourselves and to one another. (Mahatma Gandhi, 1869–1948)

The following observation linking the condition of nature to considerations regarding human rights was certainly an appropriate way to begin the new millennium.

> Human rights cannot be secured in a degraded or polluted environment. The fundamental right to life is threatened by soil degradation and deforestation and by exposures to toxic chemicals, hazardous wastes and contaminated drinking water. Environmental conditions clearly help to determine the extent to which people enjoy their basic rights to life, health, adequate food and housing, and traditional livelihood and culture. It is time to recognize that those who pollute or destroy the natural environment are not just committing a crime against nature but are violating human rights as well. (Klaus Toepfer, Executive Director of the United Nations Environment Programme, 2001, 57th Session of the Commission on Human Rights)

Words are relatively easy; actions are more challenging. The humane philosophy expressed by Mr. Toepfer has not been applied equitably in remote areas or to all strata of society. Examples abound across the globe and are frequently explained away as a necessary compromise.

> I don't understand why when we destroy something created by man, we call it vandalism, but when we destroy something created by nature, we call it progress. (Ed Begley Jr., 1949–)

Arguments for "the greater good" have yielded death and destruction for all who might stand in the way of "progress." The way the modern "civilized" world treats peoples still living ancestral lifestyles amounts to even less respect than that afforded nature, or even pets and livestock, in most cases. We don't only tell these groups that they must give up their homes but also their way of life – because some people they don't even know want something from beneath their feet, something they may have never heard of. What if a group of total strangers showed up at your doorstep saying that they had found some magic liquid (crude oil) or pretty rocks (gold, diamonds) under your house and that you have no choice but to give up your home and land because this product can make these intruders rich?

> Genocide, the physical extinction of a people, is universally condemned, but ethnocide, the destruction of a people's way of life, is not only condoned, it is universally celebrated as part of a development strategy. (Wade Davis, 1953–)

What if these strangers make all kinds of promises to get into your backyard? You are told that you can have access to modern healthcare, formal education, jobs, steady income, electricity, running water, a "real" house, and "stability," including security in this life and the next. They are told so many wonderful things by the outsiders that only a fool would object. Too good to be true? What they never hear upfront, however, is that, along with the rose-colored fantasy, they should also expect contaminated water, incessant industrial noise replacing the natural sounds of the forest, more illness through exposure to diseases introduced from other parts of the world, the kinds of jobs that no one would seek out (back-breaking menial labor such as working with a machete for 8–10 hours per day in the tropical sun) for less than minimum wage, the poorest excuse for education ever offered, relegation to the lowest rungs of the newly joined society (through active exclusion related to prejudice and racism), less than appropriate medical attention, and so on (Swing et al. 2012). What if the stories about opportunity are not real because the converts will never be able to truly fit into the new society and their children can never be as self-sufficient as their grandparents were?

> Do unto those downstream as you would have those upstream do unto you. (Wendell Berry, 1934–)

8.6 Why Should We Care?

> One of the penalties of an ecological education is to live alone in a world of wounds. (Aldo Leopold, 1949 Sand County Almanac)

Although most of the risks associated with extreme human population and our current myopic behavior toward the environment are broadly familiar (Diamond 1999), we will enumerate a few here below. In light of the well-known threats, however, our responses have been far less than satisfactory. This is a one-time global experiment without historical precedent, there can be no absolute certainty in terms of human or economic impact, but we can easily predict that losses will potentially exceed any measure that can be rationally categorized as acceptable, affordable, or readily manageable. Complacence is not a viable option. On the world timeline, the window of opportunity for proactive maneuvers closed decades ago (Rich 2018). At best, realistic possibilities for reactive mediation are also running out quickly.

8.6.1 Risk 1

We continue to wait with our arms folded while permitting a tiny fraction of self-interested politicians to disconnectedly drive the discussion with their unfounded skepticism instead of taking decisive action based on the best available scientific evidence. Out of some perverse sense of short-term convenience, we ignore long-term reality. When precisely does patience become negligence?

By continually electing officials who have little or no expertise in science (Miller 2002), we have allowed what is at its core a scientific challenge to become an overwhelmingly political issue. At present, it looks as though we may be forced to go on waiting until worsening environmental conditions (related to human population, pollution, climate change, and the extinction pulse) become desperately moral issues before deciding to do much about them. Every passenger in a plane would prefer a pilot who responsibly pays attention all the time, not one who must desperately pull out of near crash situations all the time. Anticipating challenges and taking definitive action to avoid predictable problems are always part of a logical strategy, but concrete decisions continue to be complicated by distractions of far less consequence. By doing nothing, we move ever closer to "a point of no return" in which the climatic trajectory of the planet moves into another phase, one which has no possibility of being reversed through human efforts (Steffen et al. 2018).

> Science is what we do to keep from lying to ourselves. (Richard Feynman, 1918–1988)

8.6.2 Risk 2

Our inaction directly or indirectly provokes untenable levels of competition and conflict due to changing climate and the resultant redistribution of traditional production of agricultural commodities along with many other consumable goods. As weather patterns shift, historical origins of numerous resources will move faster than sociopolitical adjustments can accommodate them. Many regionally stereotypic crops will come to be incompatible with local temperature and rainfall regimes. Everything from water to watermelons, from rice paddies to hamburger patties, from automobile factories to economic centers will likely present points of contention as both availability and transportation become modified or exasperated (Goodell 2018). All that will translate into heated arguments and even wars over the vital liquid we must imbibe or use for irrigation, over items as simple as fruits and vegetables, over every form of staple grains, animal protein of any kind, manufacturing and the labor force and investments, as well as all associated cash flow.

Thermal comfort zones for people and cultures will also move; ecosystems will suffer the same pressures. Human migrations in every part of the planet are already problematic considering legal, economic, and social concerns. This overall scenario is essentially guaranteed to worsen with time (Alden 2018).

> Human history becomes more and more a race between education and catastrophe. (H.G. Wells, 1866–1946)

8.6.3 Risk 3

We lose valuable resources forever. The current rate of species extinction is exponentially greater than anything documented across previous millions of years of history. Without question, the Sixth Mass Extinction is real and is being caused by human activity (Kolbert 2014). By calling these losses "extinctions," we suggest with our language that these are isolated events without any particular cause or that entire species simply decline and inexplicably disappear without provocation. Arguing that species go extinct as part of a natural cycle is similar to saying that JFK died of natural causes. Unfortunately, the application of this term "extinction" conveniently allows us to sidestep culpability through misdirected euphemism. We should call these losses what they are – exterminations. In all honesty, humans are likely responsible, directly or indirectly, for 99.9% of the current wave of species-wide deaths. Just because you are not personally out poaching elephants and rhinos, harpooning the last whale, or cutting the throat of a rare sea turtle while she is nesting, that does not serve as exculpatory evidence. "Extinction" means something has been extinguished, put out like a flame, literally smothering its last spark of life. These disappearances are the result of active occurrences, not passive ones – cumulative acts committed by real perpetrators, even if they are as nebulous as "society." Extinction is not something that "just happened" while we weren't looking; modern humans are nearly uniformly part and parcel of this story. As such, we are as likely to be eventual victims as many other species.

Considering that only a few of the millions of extant species have been explored for potential benefits to ourselves, there is ample probability, a guarantee even, while lounging in our sea of oblivion, of throwing away tremendous treasures from nearly every realm before they are identified.

> We should preserve every scrap of biodiversity as priceless while we learn to use it and come to understand what it means to humanity. (E.O. Wilson, 1929–)

8.6.4 Risk 4

By doing nothing, or very nearly nothing, we miss an enormous economic opportunity and get stuck in the past (Temple 2018). We try to hold onto old, outdated technologies because of their former relevance and historical money-making capacity, thereby forever getting left behind in the modern world. As time progresses, humankind is being forced to move toward energy sources other than fossil fuels. Imagine countries had made a move on innovation in this arena when the issue originally came to be a public concern, in the 1970s and 1980s (Rich 2018); those visionary nations would now be enjoying staggering market dominance in relation to globally applicable technologies. A few private corporations were aggressive in related endeavors, but without state support, especially in the form of tax breaks, finance was difficult to impossible in the early years. In the last decade, several countries, mostly in Europe but also including China, have made serious policy decisions to make this energy sector transition, and they have become the current trendsetters.

> The Stone Age didn't end for lack of stone… (Ahmed Zaki Yamani, 2005)

8.6.5 Risk 5

We forget valuable lessons that have been repeatedly experienced in nearly every corner of the globe; we ignore historical perspective altogether; we embrace amnesia of the last 200 years and allow it to rule the foreseeable future. In the developed world, we seem to have completely lost sight of what our surroundings were like before the industrial revolution and, more importantly, the last half century. We have become blind to, or have intentionally chosen to ignore, the many environmental improvements made over the last several decades due to the visionary interventions of members of previous generations. Some skeptics are profoundly and mistakenly dismissive of the value of the achievements of the American EPA and its influence across the globe. As obvious examples, we have lost awareness of many former cases of concentrated air pollution associated with specific cities. Many do not currently believe that general living conditions, air or water quality (Fortuño 2017) and so forth, can easily return to what we experienced in the 1970s and 1980s or become worse than ever before, especially in relation to availability of certain resources like fisheries and forests. Being unaware of a situation and the reality of its consequences presents a palpable risk for ourselves and everything that shares the Earth.

> Facts do not cease to exist because they are ignored. (Aldous Huxley, 1894–1963)

If visionary leaders, public and private figures, had not made courageous efforts to set aside important expanses of land, rivers, and sea in the form of national parks, wildlife refuges, and wilderness reserves, we would have long ago lost virtually all intact representations of nature as well as the services they provide gratis. Had we depended exclusively upon the hungry masses to leave some base reproductive populations so as to maintain access to renewable resources, the plight of the Dodo would have been much more contagious. We can be thankful that we still have a large proportion of our natural heritage to celebrate and to consider as potential resources, but our gratitude must be accompanied by reasoned respect and scientifically coherent management (Myers 1992; Curtis et al. 2018; Fritts 2018).

> Anti-intellectualism has been a constant thread winding its way through our political and cultural life, nurtured by the false notion that democracy means that my ignorance is just as good as your knowledge. (Isaac Asimov, 1920–1992)

8.6.6 Risk 6

We underestimate the importance of long-term data collection as well as the institutions involved in documenting and analyzing pervasive events and trends. A global system of field stations (data collection sites, "Earth observatories") representing a variety of ecosystems on all continents and in all seas at a wide range of latitudes should be maintained and even expanded to quantify changing environmental and climatic conditions on local, regional, and planetary scales while monitoring real-time responses of biodiversity to ensure availability of pertinent information necessary for reasoned management strategies.

> An investment in knowledge always pays the best interest. (Benjamin Franklin, 1706–1790)

8.6.7 Risk 7

We fail to fully absorb the fact that environmental health is human health, that our well-being is inexorably tied to the quality of our surroundings. We do not quite comprehend that all we put into the environment, intentionally (including pesticides, herbicides, and fertilizers) or unintentionally (sewage, solid waste, and agricultural byproducts), has potential to harm ourselves as well as species upon which we depend. We continue to think that plastics in our environment *only* represent an annoying eyesore on land and a tangling risk in the seas but remain ignorant of the health risks associated with the hormone-mimicking compounds that plastics emanate. Anyone who understands anything about chemistry or physiology recognizes that because humankind has continuously pumped so many toxins into our medium, especially since WWII, it is absolutely impossible that our own bodies are not affected by the never-ending inundation of pollutants in our air, water, and food supplies. As the poisoning came on slowly at first, and because all individuals are not equally vulnerable to the abhorrent chemical cocktail loosed on the world, we simply accepted exposure as an unavoidable part of modern life, often completely unaware of its extent or significance. Just because a minor amount of toxins in our lives may be bearable, that does not mean that a little more every day ad infinitum can be managed physiologically or should ever be considered acceptable. Although the industrial/governmental complex may have convinced the masses that there is no alternative, nothing could be further from the truth.

> It is easier to fool the people than it is to convince them that they have been fooled. (Mark Twain, 1835–1910)

It is abundantly clear that a much larger proportion of the population must have access to education in the fundamentals of biology so that people in all walks of life can understand that any assault on nature results, directly or indirectly, is a rebounding assault on ourselves. Incorporation of this concept by the masses would improve every aspect of individual and public health across the planet every day and could diminish healthcare costs by billions of dollars per country every year.

Such basic information in the hands of the masses, despite socioeconomic discrepancies, could likely have made a huge impact on how the COVID-19 crisis has been handled – from the White House to the schoolhouse, from the ivory towers of New York and London to the crowded slums of Mumbai or Lagos. During the pandemic, widespread comprehension of the science-based principles of health, hygiene, and medicine could have preemptively saved hundreds of thousands of lives, millions of jobs, and trillions of dollars because more people would have acted quickly; they would have known how to respond properly and why. Basic knowledge could provide the human population with the capacity to apply critical thinking to populist rhetoric, thereby avoiding the waste of precious time on distracting myths and lies. The ultimate result would be a very different mindset, one of innately cooperative "herd immunity," one based on sense of community, good citizenship, informed decisions, and rational, mindful behavior, not distorted perceptions.

8.6.8 Risk 8

We remain unaware of the personal value of national parks, wildlife refuges, wilderness areas, and state reserves. Tracts of intact land and sea certainly help to maintain natural resources and heritage, but they also provide countless invaluable ecosystem services such as production of oxygen, absorption of carbon dioxide, recycling of water, and stabilization of climate. Beyond all that, public lands are the property of every citizen despite nearly universal ignorance of that fact. As the masses are unaware of this patrimony, they tend to be lax in their management and oversight. Most individuals care much more about the car parked in their driveway than they do about their own expansive lands held in trust by the state. Disappointingly, some governmental administrations are willing to take advantage of the ignorance of all those stakeholders and readily pass along the value of resources to "friends" in industry. This transfer of patrimony from the general population specifically to benefit a select few individuals or corporations that are loyal to the party in power under the guise of some "greater good" is relatively simple; such changes in status are much more frequent than most people know (Golden Kroner et al. 2019). By exploiting human nature, and the old adage "out of sight, out of mind," the loss goes unnoticed by most stakeholders, especially if they live hundreds or thousands of kilometers away. For an analogy, if you have a family heirloom locked away in a safety deposit box, you don't forget that it is valuable just because you don't see it every day. There is nearly unbounded risk that the "owners" of natural gems in the form of parks and refuges, set aside for us all by early visionaries, never notice the importance or value of their property and that it can easily be lost to posterity through indifference and neglect. Distance and distraction make us susceptible to this pilfering every day.

> You cannot protect the environment unless you empower people, you inform them, and you help them understand that these resources are their own, that they must protect them. (Wangari Maathai)

8.6.9 Risk 9

We continue to altogether disregard the struggle and bloodshed, individual and collective, that has repeatedly yielded abundant resources only to have the fruits of those efforts and sacrifice go to waste through mass myopia induced by the most primordial of human weaknesses: self-centered greed. Throughout history, humans have fought fervently (massacred multitudes, eradicated entire cultures, desecrated holy sites, tortured countless captives, destroyed swathes of infrastructure, and burned crops) in order to gain access to the resources held by enemies and rivals. Once the bounty has been acquired, the prudent strategy would be to make a reasoned assessment of the new inventory and then proceed to plan how to best exploit it without overdoing it. Logic would have humankind, like all creatures great and small, competing for life-sustaining requirements, but uniquely utilizing the greatest asset of all – our intellect – to chart a sustainable, functional course into the future. As it turns out, however, one of the biggest surprises is that the supposed only thinking species on this planet does not put that massive brain to full use but, instead, has repeatedly chosen, almost without exception, a phase of dedicated plunder, driven primarily by primitive desires for instant gratification (usually measured in money), ignoring the judicious minority visionaries who would conscientiously opt to follow the elementary lessons of "The Little Red Hen" (Williams 1918).

A monumental example would be the management of oil reserves in the USA over the last several decades. As this powerful nation has typically produced less than half its fuel needs, there is an ongoing and pervasive concern for national security related to energy availability due to its direct bearing on productivity and economic strength. Nonetheless, each and every day, the USA exports around 6 million barrels of its irreplaceable oil, some to countries categorized as adversaries. Ironically, the country concurrently imports 10 million barrels of the same stuff every day. Instead of recognizing that keeping their own oil in the ground would serve to foment energy independence into the future, they have chosen to continuously sell off a portion of their sovereign assets for short-term economic gains. This, in combination with the fact that oil production is heavily subsidized within the USA, confirms that cash flow for Big Oil takes precedence over energy security for the nation. Were economic gains directed specifically to relieve any ongoing or future offshore reliance through investment in locally available sustainable or renewable energy sources like solar, wind, and hydroelectric power, this might be considered a visionary transition strategy, but most of the profits end up in the pockets of industry giants and their stockholders. In a parallel way, the potential value of genetic resources related to native flora and fauna was entirely dismissed by the Trump administration, while access to resources of known value like crude oil, coal, and natural gas took full priority. As usual, choices between easily accessible short-term profitable resources outweigh potential long-term benefits. In this particular case, the stunning revelation that fossil fuels are part of an already obsolete system based on antiquated and dirty technology that will eventually be replaced is certainly playing a role in the drive to "get while the getting's good."

8.6.10 Risk 10

Women continue to be systematically excluded from education, management, and legislative processes at every level of society. If governments, corporations, and entire cultures insist on ignoring or suppressing the voices, ingenuity, rights, intelligence, creativity, and caring nature of the majority of human beings based on gender, the outcome simply cannot be the best one possible. The wholesale exclusion of women from education in many parts of the world effectively undercuts the potential for all facets of civilization. Not educating girls and women nearly guarantees uncontrolled population growth and absolutely all the concomitant socioeconomic ills that can, in no way, be avoided if we indeed reach a total population of 10 billion. Furthermore, discrimination, exclusion, and oppression based on gender, ethnicity, religion, or other sociocultural differences represent an abomination to the principles of human rights and provoke these same kinds of losses to everyone.

8.7 Conclusion

Were it possible that the human race could even moderately live up to our specific epithet, *sapiens*, we should have plenty of capacity for predicting many aspects of the future and reasonably working toward conditions that are both comfortable and maintainable, physically and fiscally, over the long haul. An important question is whether we might have the ability to choose wisely. Can we muster the determination to keep something for the future or will we squander it all? Or will we lazily criticize the evolutionary process for selecting survival traits in us that have surreptitiously gotten out of hand?

The challenges are enormous, especially the challenge to see beyond populist and partisan politics so that modern technologies and strategies may be applied in efficient ways. As long as leaders and individual citizens put more effort into looking for a place to lay the blame for current circumstances than they put into unifying the masses to reach sustainable solutions for the future, few substantial advances will be made. In the end, the most profound risk for the environment and for humanity might lie in not having leadership that can be trusted to make the hard decisions between immediate personal political gains and the true greater good for posterity.

> Nobody makes a greater mistake than he who did nothing because he could do only a little. (Edmund Burke, 1729–1797)

References

Alden S (2018) UN Scientific paper suggests capitalism has to die in order for the planet to be saved. Available via August 28 GritPost https://gritpost.com/un-paper-capitalism/. Accessed 18 Mar 2021

Bar-On YM, Phillips R, Milo R (2018) The biomass distribution on Earth. PNAS 115(25):6506–6511. http://www.pnas.org/content/early/2018/07/31/1810141115#page. Accessed 19 Mar 2021
Benohr J, Lynch PJ (2018) Should Rivers Have Rights? A growing movement says it's about time. Available via Yale Environment 360. https://e360.yale.edu/features/should-rivers-have-rights-a-growing-movement-says-its-about-time. Accessed 18 Mar 2021
Battles JB (2001) Disaster prevention: lessons learned from the *Titanic*. Proc (Baylor) Univ Med Cent 14(2):150–153
Baumrind D (1985) Research using intentional deception: ethical issues revisited. Am Psychol *40*(2):165–174
Brimblecombe P (2006) The Clean Air Act after 50 years. Weather 61(11):311–314
Carson RL (1962) Silent Spring. Houghton Mifflin, New York
Carrington D (2018) Humans just 0.01% of all life but have destroyed 83% of wild mammals – study. Available via The Guardian. https://www.theguardian.com/environment/2018/may/21/human-race-just-001-of-all-life-but-has-destroyed-over-80-of-wild-mammals-study. Accessed 18 Mar 2021
Comegna A (2016) Art as ideas: Thomas Cole's the course of empire. https://www.libertarianism.org/columns/art-ideas-thomas-coles-course-empire. Accessed 18 Mar 2021
Curtis PG, Slay CM, Harris NL et al (2018) Classifying drivers of global forest loss. *Science 361(6407):1108–1111*. https://doi.org/10.1126/science.aau3445
Daily GC, Ehrlich PR (1992) Population, sustainability, and earth's carrying capacity. Bioscience 42(10):761–771. https://doi.org/10.2307/1311995
Devins N (2005) The Academic expert before congress: observations and lessons from Bill Van Alstyne's Testimony. Duke Law J 54:1525-1554. https://scholarship.law.duke.edu/cgi/viewcontent.cgi?article=1259&context=dlj. Accessed 18 Mar 2021
Diamond J (1999) Guns, germs and steel. The fates of human societies. Norton & Co, New York
Diamond J (2005) Collapse: how societies choose to fail or succeed. Viking Press, New York
Dye L (2017) The dilemma of delisting species. Available via ABC News https://abcnews.go.com/Technology/story?id=120117&page=1. Accessed 18 Mar 2021
Ehrlich PR (1968) The population bomb. Ballantine Books, New York
Fallacy Files (2021) The fallacy of the heap. http://www.fallacyfiles.org/fallheap.html. Accessed 18 Mar 2021
Fortuño M (2017) La economía del agua: El futuro se avecina complicado. Available via World Economic Forum. https://es.weforum.org/agenda/2017/03/la-economia-del-agua-cada-vez-sera-mas-importante?utm_content=buffer5182f&utm_medium=social&utm_source=facebook.com&utm_campaign=buffer. Accessed 18 Mar 2021
Franta B (2018) Shell and Exxon's secret 1980s climate change warnings. The Guardian. https://www.theguardian.com/environment/climate-consensus-97-per-cent/2018/sep/19/shell-and-exxons-secret-1980s-climate-change-warnings?CMP=fb_gu. Accessed 18 Mar 2021
Fritts R (2018) What's causing deforestation? New study reveals global drivers. Mongabay. https://news.mongabay.com/2018/09/whats-causing-deforestation-new-study-reveals-global-drivers/ Accessed 18 Mar 2021
Garfield L (2016) The world's population is growing faster than we thought. Available via Science Alert. https://www.sciencealert.com/the-world-s-population-is-growing-faster-than-we-thought-new-report-finds. Accessed 18 Mar 2021
Golden Kroner RE, Qin S, Cook CN et al (2019) The uncertain future of protected areas and waters. Science 364(6443):881–886. https://doi.org/10.1126/science.aau5525
Goodell J (2018) The melting Artic is a real-time horror story – why doesn't anyone care? Available via Rolling Stone. https://www.rollingstone.com/politics/politics-news/arctic-ice-melting-716647/. Accessed 18 Mar 2021
Hardin G (1960) The competitive exclusion principle. Science 131(3409):1292–1297
Hemberg S (2000) Lead poisoning in a historical perspective. Am J Ind Med 38(3):244–254

Jacobson L (2013) Only 1 percent of endangered species have been taken off list, says Cynthia Lummis. Available via Politifact. https://www.politifact.com/truth-o-meter/statements/2013/sep/03/cynthia-lummis/endangered-species-act-percent-taken-off-list/. Accessed 18 Mar 2021
Klug H, Bonsall MB (2014) What are the benefits of parental care? The importance of parental effects on developmental rate. Ecol Evol 4(12):2330–2351. https://doi.org/10.1002/ece3.1083
Kolbert E (2014) The sixth extinction: an unnatural history. Holt, New York
Latson J (2015) The burning river that sparked a revolution. Available via Time. https://time.com/3921976/cuyahoga-fire/. Accessed 18 Mar 2021
Lyons RD (1970) Mercury found high in 89% of Swordfish Tested. Available via The New York Times. https://www.nytimes.com/1970/12/24/archives/mercury-found-high-in-89-of-swordfish-tested-mercury-content-is.html. Accessed 18 Mar 2021
Mallick S, Li H, Lipson M et al (2016) The Simons genome diversity project: 300 genomes from 142 diverse populations. Nature 538:201–206. https://doi.org/10.1038/nature18964
Malthus TR (1798) An essay on the principle of population. Johnson, London
Miller JD (2002) Civic scientific literacy: a necessity in the 21st century. J Feder Am Sci 55(1). Available via Federation of American Scientists. http://www.fas.org/faspir/2002/v55n1/scilit.htm. Accessed 18 Mar 2021
Mooney C (2018) At this rate, Earth risks sea level rise of 20 to 30 feet, historical analysis shows. Available via Washington Post. https://www.washingtonpost.com/energy-environment/2018/09/20/antarctica-warming-could-fuel-disastrous-sea-level-rise-study-finds/?utm_term=.785028bfcaa6. Accessed 18 Mar 2021
Myers N (1992) The primary source: tropical forests and our future. Norton, New York
Rees WE (2017) What, me worry? Humans are blind to imminent environmental collapse. Available via The Tyee. https://thetyee.ca/Opinion/2017/11/16/humans-blind-imminent-environmental-collapse/. Accessed 18 Mar 2021
Rich N (2018) Losing earth: the decade we almost stopped climate change. Available via New York Times Magazine. https://www.nytimes.com/interactive/2018/08/01/magazine/climate-change-losing-earth.html?smid=fb-share. Accessed 18 Mar 2021
Sieswerda P (2018) The Trump administration vs. N.Y.'s marine life: offshore drilling will put whales and other creatures in deep jeopardy. http://www.nydailynews.com/opinion/ny-oped-trump-administration-vs-ny-marine-life-20180816-story.html. Accessed 18 Mar 2021
Shah A (2014) Consumption and consumerism. Available via Global Issues. http://www.globalissues.org/issue/235/consumption-and-consumerism. Accessed 18 Mar 2021
Skidelsky R (2015) Is western civilisation in terminal decline? Available via The Guardian. https://www.theguardian.com/business/2015/nov/17/is-western-civilisation-in-terminal-decline. Accessed 18 Mar 2021
Smil V (2011) Harvesting the Biosphere: The Human Impact. Popul Dev Rev 37(4):613–636. http://vaclavsmil.com/wp-content/uploads/PDR37-4.Smil_.pgs613-636.pdf. Accessed 18 Mar 2021
Steffen W, Rockström J, Richardson K et al (2018) Trajectories of the earth system in the anthropocene. PNAS 115(23):8252–8259
Suárez E (2018) Biodiversity and the society of superlatives (commentary). Available via Mongabay. https://news.mongabay.com/2018/08/biodiversity-and-the-society-of-superlatives-commentary/. Accessed 18 Mar 2021
Swing K, Davidov V, Schwartz B (2012) Oil development on traditional lands of indigenous peoples; coinciding perceptions on two continents. J Dev Soc 28(2):257–280
Temple J (2018) How the science of persuasion could change the politics of climate change. Available via MIT Technology Review. https://www.technologyreview.com/s/610719/how-the-science-of-persuasion-could-change-the-politics-of-climate-change/. Accessed 18 Mar 2021
Todar K (2017) The growth of bacterial populations. Available via Online Textbook of Bacteriology. textbookofbacteriology.net/growth_3.html. Accessed 18 Mar 2021
UNEP (2017) Our planet. Available via United Nations Environment Programme. http://web.unep.org/ourplanet/December-2017. Accessed 18 Mar 2021

United States Department of Labor Occupational Safety and Health Administration. https://www.osha.gov/. Accessed 19 Mar 2021

United States Environmental Protection Agency. https://www.epa.gov/. Accessed 19 Mar 2021

United States Fish and Wildlife Service. https://ecos.fws.gov/ecp0/reports/delisting-report. Accessed 19 Mar 2021

United States Food and Drug Administration. https://www.fda.gov/Food/FoodborneIllnessContaminants/Metals/ucm115644.htm Accessed 19 Mar 2021

Vox L (2017) Why don't Christian conservatives worry about climate change? God. Available via The Washington Post. https://www.washingtonpost.com/posteverything/wp/2017/06/02/why-dont-christian-conservatives-worry-about-climate-change-god/?utm_term=.3287c8fc404e. Accessed 19 Mar 2021

Wilkins M (2018) More recycling won't solve plastic pollution. Available via Scientific American. https://blogs.scientificamerican.com/observations/more-recycling-wont-solve-plastic-pollution/. Accessed 19 Mar 2021

Williams FW (1918) The little red hen: an old English folk tale. Available via Project Gutenberg. https://www.gutenberg.org/ebooks/18735. Accessed 18 Mar 2021

Wilson EO (2016) Half-Earth: Our Planet's Fight for Life. Norton & Co., 272pp

Part III
Nepal and India Community-Facilitated Disaster Management

Fig. 1 Community members walking along the flooded Triyuga river, Fattepur, Saptari, Nepal, July 2019. Increasing settlements along the rivers are increasing flood risk in countries like Nepal
Fig. 2 Two elderly people evacuating during a July flood in Jhanduwa, West Champaran, India, July 2020. Regular monsoon flood has greater impact on the poor and elderly people, with limited resources for preparedness, response and recovery
Fig. 3 Manual shallow tube well drilling in Balara, Sarlahi, Nepal, July 2018. Drought spells are increasing in the Terai area of Nepal, affecting the livelihood of marginalized farmers. Low cost shallow tube well has proven to be a good source of groundwater irrigation
Fig. 4 Task force members of community disaster management committees disseminating early flood warning to the communities in Saptari, Nepal, July 2018. Engaging communities in early warning communication and dissemination has helped in saving lives and properties
Fig. 5 Community discussion for gender issues in early warning and flood risk reduction and management in Saptari, Nepal, July 2020. Community engagement and gender perspective in local-level disaster risk reduction planning enables developing inclusive environments for vulnerable groups
Fig. 6 Gathering community feedback for climate change adaptation in Nepal, July 2020. Engaging communities' knowledge and experience in adapting climate change helps in better planning and project development
Fig. 7 School students participating in an awareness rally organized to mark the international day for disaster risk reduction (DRR) at Saptari, Nepal, October, 2019. Engaging youths in DRR is crucial for preparedness, response and recovery
Fig. 8 Training local partners (non-governmental organizations) in Nepal, March 2018. Training local partners in resilience, climate change adaptation and disaster risk reduction is crucial in communicating scientific knowledge to target communities
Fig. 9 Community disaster management committee (CDMC) representatives creating COVID-19 awareness on precautions and safety measures in Hanumannagar Kankalini municipality, Saptari, Nepal, April 2020. The CDMCs are playing crucial roles in mainstreaming disaster risk reduction management at the community level
Fig. 10 Road construction in Nepal opposite the Banepa Bardibas Highway, July 2020. Road construction without proper environmental management has created environmental problems as well as increasing landslide risk in the hill slopes
Fig. 11 Farmers utilizing zero till in West Champaran, India, June 2019. Small-scale subsistence farmers are adopting new technologies and practices to adapt with climate change
Fig. 12 Women farmers cropping an improved variety of brinjal (eggplant) in Saptari, Nepal, January 2020. Access to improved seed varieties and market helps rural farmers in adapting to climate change and provides their earnings
Images were provided by the Lutheran World Relief, Nepal in 2020. Further details of disaster risk reduction management in Nepal are given in chapter 9.

Fig. 1

Fig. 2

Fig. 3

Fig. 4

Fig. 5

Fig. 6

Fig. 7

Livelihood CAPITALS
RESILIENCE ATTRIBUTES
HUMAN CAPITAL
ECONOMIC CAPITAL
SOCIAL CAPITAL
NATURAL CAPITAL
PHYSICAL CAPITAL
SCALE
LEARNING
DIVERSITY
RAPIDITY
FLEXIBILITY
ROBUSTNESS
SELF-ORGANIZATION
EQUALITY EQUITY
CHANGE AGENT
GENDER INTEGRATION
SHOCKS
NATURAL DISASTER
STRESSORS
MARKET DOWNTURN
FOOD INSECURITY
CLIMATE CHANGE

Fig. 8

Fig. 9

Fig. 10

Fig. 11

Fig. 12

Chapter 9
Disaster Risk Reduction and Management Policy in Nepal: A Centralized-Decentralized Dichotomy

Gyawali Narayan, Shah Rakesh Kumar, and James N. Furze

Abstract The disaster risk reduction and management (DRRM) policy framework in Nepal provides a glass half-full or half-empty, depending on one's viewpoint. With the caveat that the division of responsibilities among the three layers of government in new federal structure is a work in progress, it is unclear how previous acts apply to a changed system of government. The over-arching DRRM framework emphasizes the importance of prevention, preparedness, and mitigation, rather than just response. The current chapter includes detail of governmental policy and field information from three local government municipalities.

The capacity of the Nepal government to meet well-intentioned policies is low; disaster management responsibility of the local government officials is one among many and less prioritized than other responsibilities. The institutional setup for budgetary allocations continues to prioritize response over preparedness, despite the cost-savings associated with the latter and regardless of policy frameworks. There are claims that the extent of disaster response is politically determined. Further, the implementation and interpretation of policy is generally more important than the policy itself. The existing policies generally discuss the higher-level approach, such as institutionalizing DRRM but do not set out specific points on how this should be done to benefit the target communities. Nepal has made significant progress in

G. Narayan (✉)
Agriculture and Forestry University, Rampur, Chitwan, Nepal
e-mail: gyawalin@gmail.com

S. R. Kumar
Lutheran World Relief, Kathmandu, Nepal
e-mail: rakeshkumarshah2007@gmail.com

J. N. Furze
Royal Geographical Society (with the Institute of British Geographers), London, UK

Laboratory of Biotechnology and Valorization of Natural Resources,
Faculty of Sciences-Agadir, Department of Biology, Ibn Zohr University, Agadir, Morocco

Control and Systems Engineering Department, University of Technology-Iraq, Baghdad, Iraq
e-mail: james.n.furze@gmail.com; jamesfurze@hotmail.com

J. N. Furze et al. (eds.), *Earth Systems Protection and Sustainability*,
https://doi.org/10.1007/978-3-030-85829-2_9

DRRM policies in recent years. However, institutionalization and the capacity of the officials and stakeholders need to be strengthened for operationalizing DRRM policy framework to strengthen community resilience.

Keywords Disaster risk reduction · Early warning · Policies · Localization

9.1 Introduction

Nepal falls in the top 20 list of the most multi-hazard-prone countries in the world and is ranked 4th, 11th, and 30th in terms of vulnerabilities to climate change, earthquake, and flood risk, respectively (UNDRR 2019). This has established Nepal as one of the most disaster-prone countries in South Asia with more than 80% of the total population of the country living under the risk of disaster (MoHA 2018c). Nepal has invested substantially in improving disaster management mechanisms in recent years: disasters have quickly evolved from being uncontrollable "Acts of God" to events that can be prepared for and managed. Still, as the 2015 Nepal earthquake demonstrated, disasters "greatly set back hard-won development gains, particularly in low income countries like Nepal" (Shrestha et al. 2014). Until relatively recently, disasters were treated as random occurrences, and policy toward disasters focused on the response after the disaster had struck.

The federal government of Nepal is shifting from a reactive to a proactive approach toward disaster risk management, strengthening legal frameworks, planning, and institutions for disaster risk management. The Disaster Risk Reduction and Management (DRRM) act endorsed in 2017 emphasizes the shift from a disaster response approach toward a multilayered approach stressing risk reduction. The government of Nepal is committed to implementing the new Sendai Framework for Disaster Risk Reduction 2015–2030 aiming to strengthen the focus on disaster risk reduction (MoHA 2018a). While there has been a shift witnessed in, for instance, the Sendai Declaration, budgetary allocations in relation to disasters are skewed toward response rather than preparedness.

Considering the importance of disaster risk reduction and management, the Constitution of Nepal has clear provisions for disaster management functions to be operated in all the three levels of government in the new federal structure. DRRM is included in Schedule 7, Schedule 8, and Schedule 9, implying that DRRM falls under the sole authority of local government, along with shared authority between federal, provincial, and local levels (Hayes et al. 2020; MoHA 2016, 2018b). The constitution places the responsibility of disaster management with local governments and is also on the concurrent list for all three jurisdictions (IOM 2019). The Government of Nepal from federal to local level has formulated different plans, policies, acts, and regulations for DRRM and is in process of setting up institutional arrangements for mainstreaming DRRM in local level planning process. However, limited capacity and resources and the traditional approach for DRRM prioritizing response over mitigation and preparedness are some of the hurdles that need to be overcome, for mainstreaming DRRM in development planning from the federal to the local level in Nepal.

Nepal has recently politically restructured with three tires of political systems – federal, provincial, and local government. There are adequate DRRM plan and policy formulations in process for all three of these government authorities. Provincial and local governments (municipalities) are the implementers of plans and policies. However, they have limited capacities, awareness, understanding, and knowledge to formulate and implement DRRM plans and policies. They need additional capacity conducting trainings and orientations, both technical and managerial in understanding the policies and procedures and practicing it.

This chapter includes the methodology, major DRRM plan and policies, DRRM policy commitment, localizing the DRRM act and policies at municipal level, implications of the policies at implementation, and stakeholder engagement in DRRM and gives a conclusion detailing policy formulation and requirements for future strategical integration in Nepal and elsewhere. The remainder of this chapter is structured with Sect. 9.2 detailing methodologies considering DRRM in Sect. 9.3; Sect. 9.4 lays out existing DRRM policy trends in Nepal; Sect. 9.5 indicates the DRR policy commitments, and Sect. 9.5.1 qualitatively summarizes focus group discussion and informant interview evidence blending localized realities with documented indications; Sect. 9.6 discusses the implications of policy in its implementation, and Sect. 9.6.1 indicates primary evidence of implementation problems; Sect. 9.7 notes stakeholders' engagement in DRRM, and Sect. 9.8 gives concluding remarks.

9.2 Methodology

The investigative study of this chapter entailed an in-depth review of the published documents and DRR policies of government of Nepal and is accompanied with subsequent primary field data collection.

In the review of data, qualitative sources were collected and analyzed following seven steps:

Step 1: Exploring facts and evidence
Step 2: Initiating the search and exploration
Step 3: Storing and organizing information
Step 4: Selecting/deselecting information
Step 5: Expanding the search and exploration to include one or more "MODES" (Media, Observation(s), Documents, Expert(s) and Secondary Data)
Step 6: Analyzing, synthesizing, and comparing information
Step 7: Presenting and summarizing the conclusion

These seven steps are multidimensional, interactive, emergent, iterative, dynamic, holistic, and synergistic, being fundamental tenets of social science research (Onwuegbuzie et al. 2010) (Fig. 9.1).

The study was carried out in three communities of Saptari district and three communities in Nawalparasi district in Nepal as listed in Table 9.1 with their respective latitude and longitudinal coordinates.

Fig. 9.1 Map showing the study areas across Nepal. Nawalparasi district is indicated on the left; Saptari district is indicated on the right

Table 9.1 Description of study areas and communities of Saptari and Nawalparasi

District	Municipality	Community	Coordinates
Saptari	Saptakoshi	Sakhubani	86.95693°E; 26.72921°N
	Hanumannagar Kankalini	Bisanpur	86.80099°E; 26.45501°N
		Gobargaraha	86.87519°E; 26.46067°N
Nawalparasi	Susta	Narsahi	83.834728°E; 27.417191°N
		Susta	83.869586°E; 27.358502°N
		Ratangunj	83.850313°E; 27.379677°N

9.2.1 *Primary Data Collection*

9.2.1.1 Focus Group Discussion

Visits to Nawalparasi and Saptari districts of Nepal were made to collect primary data and information regarding disaster risk reduction and management at local levels.

During the visits, 6 focus group discussions (FGDs) of 8 to 12 community people were carried out. The age group of the participants ranged between 20 and 70 years with an average 70% male and 30% female participation. Semi-structured discussions of about 2 hours were held regarding vulnerabilities of the communities, their mitigation, preparedness, response and recovery efforts, support from local government with regard to disaster risk reduction and management, change in the disaster

Fig. 9.2 Focus group discussion with the people of Narsahi in Susta, Nawalparasi (August 2019). The people representing the Community Disaster Management Committee (CDMCs) are leaders for community leadership and engagement in disaster preparedness, mitigation, response, and recovery

governance in the federal structure of the country and establishment of municipalities, appropriateness, and efficiency of the government efforts in disaster risk reduction and management and the impact of disaster governance (Fig. 9.2).

9.2.1.2 Key Informant Interviews

Eight key informant interviews (KIIs) were carried out with five community leaders and contained one female and four males, one disaster risk reduction (DRR) expert male, and two male government officials. The interviews were held over an hour and a half. The FGDs and KIIs were semi-structured and were transcribed with support of a note taker.

9.3 Disaster Risk Reduction and Management Policy Trends in Nepal

Originally formulated in 1982, the Natural Disaster Relief Act (NDRA) also known as the Natural Calamity Relief Act (NCRA) was the first DRR policy in Nepal and paved the way for DRR policy (Jones et al. 2014; Nepal et al. 2018). With changes in governance systems, increased knowledge in DRR and needs for addressing different aspects of disaster other than relief are required. The government of Nepal has formulated a number of acts, regulations, plans, policies, and frameworks that have been directly or indirectly supportive in DRRM. Thus, DRRM has evolved through the following:

- Natural Calamity (Relief) Act, 1982
- National Action Plan for Disaster Risk Management 1996
- Local Self Governance Act, 1999
- National Strategy for Disaster Risk Management in Nepal, 2009
- National Strategy for DRM, 2009
- Three Year Interim Plan, 2007–2010
- National Disaster Response Framework, 2013
- Disaster Risk Reduction and Management Act, 2017 (replaces Natural Calamity (Relief) Act, 1982)
- Local Government Operation Act, 2017
- National Policy for Disaster Risk Reduction, 2018
- Disaster Risk Reduction National Strategic Plan of Action, 2018–2030
- National Disaster Risk Reduction Policy, 2018

The Disaster Risk Reduction and Management Act 2017 replaced the 1982 Natural Calamity Relief Act, which did not cover the broader spectrum of hazard mitigation and disaster risk reduction and management. Formed a few days before the 2015 Gorkha Earthquake, the bill covered a range of disasters including health emergencies, famine, industrial accidents, and pollution, as well as weather-related disasters and earthquakes. It also involves pathways for creation of more powerful institutional arrangements to deal with disasters. This stems in part from lessons learned after the 2015 earthquake, which constituted a particular lack of coordination between different arms of government. The act is intended to address the entire cycle of disaster management shown in Fig. 9.3, as well as a diverse range of disasters faced.

The DRRM act 2017 has clear provisions for all the stages of disaster management through its definition of terms described as:

- "Disaster" – meaning a natural or unnatural disaster that creates emergency situation at any place whereby life or property is lost, or livelihood and environment is adversely affected
- "Disaster Risks Reduction" – meaning analysis and assessment of risks before disaster, disaster prevention or reduction of harms to be caused by disaster, and works concerning minimization of disaster risks in development activities
- "Disaster Counteraction" – meaning works of search, rescue, and relief to be carried out immediately after occurrence of an incident of disaster and the words also include preparedness of actions for countering disaster
- "Disasters Management" – meaning entire activities concerning disaster risk reduction, disaster counteraction, and disaster recovery

The disaster cycle of Fig. 9.3 can traverse either clockwise or anticlockwise based on its nexus with development. The nexus has three separate but interrelated

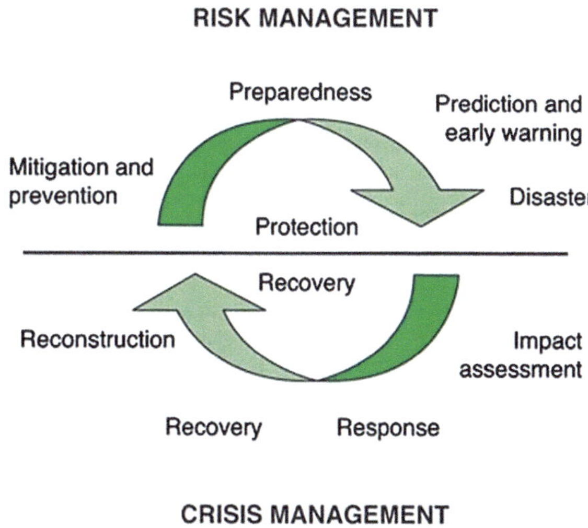

Fig. 9.3 Disaster risk reduction management cycle (Chakrabarti 2017)

dimensions (Chakrabarti 2017), gains of development are eroded by disasters, and the risk of disaster is created by a deficit in development. Development creates new risks of disasters that compound existing layers of risk. Thus, DRRM plans and policies should not be in isolation but rather mainstreamed into the development process.

Localization of DRRM in Nepal has plans and policies for disaster management rather than just relief and recovery from the federal to provincial to local (municipal) level. The newly formulated DRRM act and regulation can be considered an exemplary step taken by the government of Nepal for its commitment toward disaster risk reduction and management in Nepal (MoHA 2016). The federal level plans and policies are almost formulated and endorsed. The provincial and local governments (municipalities) are also in process of formulating local acts, policies, and plans for localizing DRRM (IOM 2019). However, following localization evidences of DRRM such as those collected within this study, there is a risk that localization plans and policies could be more on paper and less in practice. Governments fail to put resources in place to enable satisfactory implementation and absence of strong regulatory provisions (Scott and Tarazona 2011).

9.4 DRR Policy Commitment

The government of Nepal is an active member of the international DRR community, which has committed and endorsed different international commitments, policies, frameworks, decisions, protocols, and agreements for DRRM (MoHA 2018c). Nepal participated and presented a national action plan on disaster management in the first world conference on disaster risk reduction in Yokohama, Japan, in 1994.

Table 9.2 Hyogo Framework goals and priorities

Hyogo Framework strategic goals	Hyogo Framework priorities for action
1. More effective integration of disaster risk considerations into sustainable development policies, planning and programming at all levels, with a special emphasis on disaster prevention, mitigation, preparedness and vulnerability reduction 2. Development and strengthening of institutions, mechanisms, and capacities at all levels, in particular at the community level, that can systematically contribute to building resilience to hazards 3. The systematic incorporation of risk reduction approaches into the design and implementation of emergency preparedness, response and recovery programmes in the reconstruction of affected communities	1. Ensure that disaster risk reduction is a national and a local priority with a strong institutional basis for implementation 2. Identify, assess and monitor disaster risks and enhance early warning 3. Use knowledge, innovation and education to build a culture of safety and resilience at all levels 4. Reduce the underlying risk factors 5. Strengthen disaster preparedness for effective response at all levels

The National Action Plan on Disaster Management 1996 was based on the Yokohama strategy. Similarly, Nepal continued to adhere to its international commitments through endorsing the three goals and five priorities (Table 9.2) of the Hyogo Framework for Action (HFA) between 2005 and 2015 (MoHA 2015).

Nepal has adopted the Sendai Framework for Disaster Risk Reduction (SFDRR) 2015–2030 and integrated the four priorities and guiding principles into its planning and policy making (MoHA 2017; UNISDR 2015). The four priority areas of the SFDRR are as follows: (i) understanding disaster risk, (ii) strengthening disaster risk governance to manage disaster risk, (iii) investing in disaster risk reduction for resilience, and (iv) enhancing disaster preparedness for effective response and to rebuild more effectively through recovery, rehabilitation, and reconstruction.

The DRRM Act 2017 and the Disaster Risk Reduction National Strategic Plan of Action 2018–2030 are in line with the SFDRR. Aside from these, the government of Nepal has endorsed the Sustainable Development Goals of 2015–2030, through the periodic plan of 2016/2017–2018/2019 that mainstreamed and internalized the 2030 Agenda (NPC 2020) Paris Agreement on Climate Change 2015. This was implemented through submission of its first Nationally Determined Contribution (NDC) in 2016 that outlined Nepal's planned contribution to achievement of the goals outlined in the Paris Agreement (MoFE 2018). Commitments were made for participation in different regional and international conferences and workshop to share Nepal's progress in mainstreaming DRR in development planning (MoHA 2018d). Enduring commitment and continual progress have assisted the government gaining financial and technical support for DRRM.

9.5 Localizing the DRRM Acts and Policies

Since the early 1990s, it has been apparent that Nepal requires a proactive disaster management act and policy to address different aspects of disaster management. This led to the drafting of the Disaster Management Bill and Policy in 2007 in consultation with different stakeholders (MoHA 2018c). Finally, the DRRM Act was passed by the parliament in 2017 followed by the National Disaster Risk Reduction Policy of 2018. The DRRM Act 2017 mandates the federal, provincial, and the local government with roles and responsibilities in disaster risk reduction and management along with institutional setup like Provincial Disaster Management Committee (PDMC), District Disaster Management Committee (DDMC), and Local Disaster Management Committee (IOM 2019). There is an urgent need for localization of DRRM acts and policies as the impacts of disasters are most immediately and intensely felt at the local level, local actors are the first responders to a disaster, and the local level is where the governments and communities can engage with each other for mitigation, preparedness, response, and recovery (UNISDR 2018).

The Local Government Operation Act 2017 has further defined the roles and responsibilities of local governments in DRRM (Hayes et al. 2020). The Constitution of Nepal 2015 also has clear provisions for localizing the DRR policies at local level as it has defined DRRM as the authority of the local government. The newly formed federal structures and provincial and municipal governments do not have adequate plans, policies, acts, regulations, guidelines, institutional framework, resources, and technical or managerial capacities to practice the constitutional rights at local level referred to as localizing of DRR plans and policies (Khanal 2020). Localization is defined as the implementation of supranational policy into projects at the appropriate subnational level to ensure the service delivery to the appropriate level population, and broadly speaking, localization is the process of making something local in character or restricting it to a particular place (Patole 2018). Localization prioritizes subnational planning and resource allocation by local and regional governments in a specific sector based on subnational variations (Lucci 2015). Thus, in Nepal once DRRM localization is complete, the local government or municipality will have all the necessary plans, policies, regulations, institutional framework, trained human resources, financial arrangements, and capacity for delivering disaster preparedness, mitigation, response, and recovery services to the target population.

Local governments are in the process of localizing DRRM by formulating municipal act, policy, and action plans for DRRM in addition to numerous other local acts, policies, and operational guidelines (IOM 2019). However, since DRRM localization is planned and being implemented with a top-down approach from federal to provincial to local, the process is "centralized decentralization."

9.5.1 Focus Group and Key Informant Evidence

During the focus group discussion with local leaders, local government officials and DRR experts from Susta municipality in Nawalparasi and Hanumannagar Kankalini and Saptakoshi municipalities from Saptari gave evidence in response to localization of DRRM. The key informants informed the researchers that the municipality executive boards comprising of the elected members are aware that local policy, acts, and action plans are to be formulated for localizing DRRM.

Municipalities did not complete the task for various reasons such as they did not have adequate information and guidelines as what has to be done exactly. The municipalities also did not have adequate technical human resources, knowledge, and capacities for carrying out risk assessments; the ministerial and departmental guidelines or directives were not clear in response to the delegated authority, procedures, contents, timeline, and elimination of overlapping roles and responsibilities.

Full implementation of the newly endorsed DRRM Act is challenging as local governments do not have trained human resources, adequate financial support, and technical inputs (Pradhan and Chauhan 2020; Scott and Tarazona 2011). The requirements and capacity assessments of 14 rural and urban municipalities for disaster risk reduction and management in Nepal carried out by the IOM in 2019 found that only 2 out of 14 municipalities had developed local DRRM Act. However, all three municipalities of Saptari have formed Local Disaster Management Committees and are taking the necessary decisions and actions for disaster response and recovery as well as mitigation and preparedness. This clearly shows that the long practice of policy for disaster response and recovery is still in practice though the new acts and policies have prioritized prevention, preparedness, mitigation, and risk reduction.

The three municipalities of each district did not carry out detailed risk assessment and investment plans for risk reduction, which is crucial for strengthening community resilience. The three municipalities have established Local Emergency Operation Centers (LEOCs) as mandated by the DRRM Act. However, the LEOCs do not meet all the minimum requirements for operationalizing as stated in the NEOC Standard Operating Procedure 2014 (IOM 2019). Target E of the Sendai Framework aims to "Substantially increase the number of countries with national, and local disaster risk reduction strategies by 2020." The United Nations sustainable development goal (SDG) 15 states it is necessary to "Take Urgent Action to Combat Climate Change and Its Impacts" (UNDRR 2020). As Nepal is a signatory country for the Sendai Framework for both DRR and SDGs, it is of paramount importance for the local government to prepare local disaster risk reduction strategies and take urgent actions to meet the commitments and targets. Further the DRRM Act of 2017 has provision for establishing local DRR funds and preparing its operational guideline.

The studied municipalities have formulated DRR fund operational guidelines and allocated budget for disaster preparedness, mitigation, response, and recovery. The government officials during KII agreed that the DRR fund has to prioritize investments in all the stages of disaster. In practice, the fund is targeted to be used more in relief.

9.5.1.1 Primary Feedback Indicating the On-Ground Reality of Mismanagement of Policy and Consequences of Flood Risk

During the focus group discussion, the communities of Saptakoshi and Hanumannagar Kankalini municipalities stated that the government response during floods of 2019 was not timely; as a consequence it was difficult and time-consuming for them to get relief provided by the local government. The municipalities did not adopt or adhere to clear identification and classification guidelines for the flood-affected households which led to bias in selecting the beneficiaries for relief and recovery.

During 2019 floods in Saptari, 537 houses were damaged in Saptakoshi municipality and 1188 households in Hanumannagar Kankalini (DCA 2019). Sharing the experiences of 2019 flood response and relief, the respondents of FGD in Saptakoshi informed that some of the flood-affected households received relief material related to sanitation and hygiene in September, 3 months after the disaster, which was not timely or appropriate. The DRR expert informed that unavailability of adequate resources for relief with the local government; longer time required due to disputes and conflicts in selection criteria; questions in the validity of the data for finalization of flood-affected household list at the community, wards, and municipality and district level; and external agencies' support as per their resource availability are some of the major causes of untimely relief support.

The DRR expert and local leaders agreed that the municipalities have allocated local resources (budget) of the DRR fund. However, the budget is not timely due to the administrative and financial compliances and also not properly utilized in disaster preparedness, response, and recovery activities. A large portion of the fund goes into regular development activities. This clearly shows that formulation of local policies and guidelines still needs revisions and amendments to address the concerns and problems of the vulnerable people and intended objectives.

In Sakhubani, the communities had carried out plantation over around 500m^2 of land for the protection of riverbank erosion in June, as per the plan that the earthen embankment and spur diversion upstream would be repaired before the onset of monsoon. However, due to delay in the repair of the embankment and diversion, the flood not only destroyed the plantation area but also inundated nearly 30000m^2 of cultivable area destroying crops (paddy and vegetables) and caused siltation. Mistiming of development projects in regard to disaster risk compound and amplify the vulnerability and losses to the communities.

Figure 9.4 shows an example cultivable area which was destroyed by mistimed actions and little or no actions to prevent the disaster. Losses to crops, cultivable area, animal, and the communities ensued, due to overflow from surrounding water bodies. Besides losses in the human communities, water resources were also lost.

Fig. 9.4 Damaged crop and land resources in Sakhubani due to the lack of preparedness (Sakhubani, Nepal, July 2019)

The DRRM Act 2017, Local Government Operation Act 2017, and the National Policy for Disaster Risk Reduction 2018 have mandated the local government for establishing, operating, and strengthening Disaster Information Management System (DIMS) and multi-hazard early warning system (EWS) at the local level. However during the field visit and observation, it was found that none of the three municipalities had instigated a DIMS, with the exception of Saptakoshi in Saptari where an external agency supports the municipality. The system was not operational. Municipalities were not directly involved in accessing flood early warning and disseminating it to the communities.

The flood EWS and the local hydrology and meteorology stations are managed by the Department of Hydrology and Meteorology (DHM). Flood EWS is crucial in saving lives and properties, but the ownership and activity of the local government in early warning system operation and management is very low. The DHM has prepared a standard operating procedure (SoP) which states that the local government needs to provide financial and administrative support for community-based early warning system and can request federal government for technical assistance and additional resources (DHM 2018). There is a gap in the delineation of clear roles and responsibilities and resource sharing among the federal and local government, as well as in the development agencies supporting/strengthening the community-based EWS.

The federal structure has devolved considerable authority for delivering EWS services to municipal governments which should result in improved communication of warnings and enhanced flood response capabilities. However, whether municipal governments will have adequate funds and technical capacity to do this remains a critical challenge and query (Chinaporn et al. 2019). The local government representatives in both Saptari and Nawalparasi explained that the local government does not have technological access or trained human resource for managing EWS at the municipal level. The municipalities are involved in information dissemination and response. The municipal chair during the KII expressed his dissatisfaction over unclear roles and responsibilities during the 2019 flood response. There is an urgent requirement for more efficient structural organization in Nepal.

9.6 Implications of Policy at Implementation

The devolution of administrative powers and authority from central government to local governments has taken place in the federal government of Nepal. The transition from a unitary to a federal system of governance remains challenging in the absence of several laws to be formulated and enacted, institutional arrangement at all government levels is yet to be finalized, and policies and guidelines are yet to be prepared. There are substantial overlaps, duplication, and ambiguities among the three tiers of government (Shrestha 2019).

The provincial and local government (municipalities) have authority to develop acts, plans, policies, and regulations as per their requirement toward localizing development plans and policies. Further, there is devolution of authority and ownership in disaster governance. The acts, policies, and plans define the roles and responsibilities of the federal, provincial, and local government in disaster risk reduction, preparedness, response, and recovery. However, disaster governance in federal structures is not smooth due to implications of policy at implementation stages.

The Constitution on Nepal 2015 refers to spheres of governance and service deliveries as (i) exclusively federal functions, (ii) exclusively provincial functions, (iii) exclusively local functions, (iv) concurrent functions of federal and provincial governments, and (v) concurrent functions of federal, provincial, and local governments (NPC 2017). It has stipulated the DRRM is a sole authority of local government and also as a shared authority among federal, provincial, and local governments (Hayes et al. 2020).

The vision of disaster governance in the federal structure of Nepal is that the local government takes the lead in disaster risk reduction and management as far as local capacity and resources can support. In case the DRRM requires additional capacity and resources beyond the capacity of local government, the provincial and federal governments have outlined that they will back up or lead accordingly. However, there is a gap in clear delineation of the roles, responsibilities, and accountability in the federal structure as shown in Table 9.3.

Table 9.3 clearly shows disaster management is under the jurisdiction of all the three tiers of government with shared responsibilities. This has created confusion, dissatisfaction, conflict, and grievances for localizing DRRM with serious consequences. Schedule 1 lists preparedness, rescue, relief, and rehabilitation as concurrent powers of federal and state or provincial government. Schedule 2 of the constitution lists disaster management as the local level power, and in Schedule 3 of the constitution, disaster management is also in the list of concurrent powers of federal, state, and local level. This has created uncertainty in clear responsibility and delineation as to how power is shared.

Table 9.3 Delineation of power by the Constitution of Nepal for disaster management to the three tiers of government

Schedule in the constitution	Scope
1. Concurrent Powers of Federation and State	Early preparedness for rescue, relief, and rehabilitation from natural and manmade calamities
2. List of Local Level Power	Disaster management
3. List of Concurrent Powers of Federation, State and Local Level	Disaster management

Shared authority between federal, provincial, and local governments and the extent to which the tiers should exercise their authority are not delineated (Hayes et al. 2020). The government has realized this, and the first interprovincial council meeting organized in Dec 2019 for Federalism Implementation Action Plan recommended to "amend the Disaster Risk Reduction and Management Act clarifying the structure and roles of the provincial government and local level and designated Ministry of Home Affairs for it" (AYSPS 2019).

The DRRM Act and policies do not adequately address the gap but serve to create further bureaucracy which ultimately leads to loss of resources. Consequently, confusion and dispute occur during disaster response and recovery. Similarly, there is no standard communication protocol between the chairs of federal, provincial, and local government committees and CDOs which hinders information sharing and risks informed decision-making (The Asia Foundation 2019).

As per the DRRM Act and Local Government Operation Act, the Local Disaster Management Committee (LDMC) leads disaster management at local level. There is also a District Disaster Management Committee (DDMC) at district level for disaster management. Two authorities with similar mandates on disaster management within each district create confusion and dispute, which results in delayed response, relief, and recovery. Security forces are mandated with the search and rescue responsibilities under the command of the Chief District Officer (CDO), chair of DDMC.

Local government has to be dependent on the district for search and rescue. Similarly, though the local government is established as an independent body and mandated with disaster management, such multiple institutions and structures create complications, impeding quick response and decision-making. There is also duplication of resource allocation in the absence of effective communication mechanisms.

9.6.1 Primary Evidence of Implementation Problems

The government of Nepal has adopted a one-door policy for disaster response, relief, and recovery support. However, counterproductively, external support agencies are required to take permission from local government to support relief and recovery. The three municipalities of each district formed similar decisions in that this approach is inadequate.

9.6.1.1 Saptari District

During the focus group discussion of the current study with the flood-affected communities in Saptari, the communities responded that the relief distribution was not systematic and did not benefit the affected people most. The relief and recovery were biased and based on political affiliations, influencing power and reach to the decision-makers. The DRR focal person of the municipality FDGs too accepted the fact that relief and recovery could not be managed well in the absence of a robust information management system providing the response, relief, and recovery activities carried out within the municipality as well as the district and identifying the beneficiaries.

9.6.1.2 Bisanpur and Gobargaraha in Hanumannagar Kankalini Municipality

During the FGDs the communities of Bisanpur and Gobargaraha in Hanumannagar Kankalini municipality informed that despite several objections and complaints, about 25% of the beneficiaries of aid were least affected households, not the vulnerable groups who were in dire need of the relief support. In the absence of documentation requirement, not receiving information and notice on time, and unclarity about where and whom to contact, some of the affected households are deprived of the relief support. Duplication of relief is also rampant as most of the organizations prefer to distribute relief items to vehicle access road heads and do not coordinate with the municipality or ward offices for relief distribution.

The DRRM Act has integrated all key components of disaster risk reduction and management including measures for risk assessment, investments for risk reduction, strengthening disaster risk governance, and preparedness for effective response, recovery, rehabilitation, and reconstruction (IOM 2019). However, currently local governments do not have adequate capacity, skills, and resources for disaster risk reduction and management as envisioned by the constitution, DRRM Act, policies, and LGOA. Some of the key findings from the need and capacity assessment of 14 rural and urban municipalities on disaster risk reduction and management in Nepal carried out by IOM in 2019 were that there is a limited understanding on comprehensive DRRM; local acts and policies are exact copies of the sample and guidelines, DRRM plans and policies are less prioritized than others, multi-hazard risk assessment and plans are not available, and emergency response is ad-hoc and flat for all irrespective of the losses or damage suffered. This holds true for the municipalities of the current study.

9.6.1.3 Hanumannagar Kankalini Municipality

During the KII, the ward chairperson from Hanumannagar Kankalini Municipality admitted that he and other colleagues in the neighboring wards do not have adequate scientific knowledge on DRRM and are not provided with adequate trainings or information for legislative changes that take place after the local elected representatives take their responsibilities. This is also justified from the response of the communities during FGD where the two communities from Hanumannagar Kankalini municipalities responded that the local government representatives have very low level of DRR expertise. For DRR planning and policy making, personal perception and past events are taken as evidence. The municipality has not prioritized in adopting the risk and vulnerability assessment and disaster information management system, "Building Information Platform Against Disaster" (BIPAD); though such systems are already available at the national level, the Ministry of Home Affairs has a policy to have an updated DIMS at local level. In the absence of risk and vulnerability assessment, quantifying risk and risk informed planning and policy making, there is always a chaotic environment during disaster response compounding the risk which ultimately leads to delayed and disputed response and recovery, and preparedness and mitigation are least prioritized. This is evident from the fact, too, that none of the municipalities in the study districts, Saptari and Nawalparasi, have updated DIMS-BIPAD.

Bringing DRRM plans and policies in practice through localization requires risk categorization and ranking that will provide intelligent guidance for preparedness and ultimately risk prevention. It also requires identifying current and future risk and understanding the drivers of risk. This has also been identified as a challenge for localization. The local government can move ahead with knowing and understanding current future risk by (i) risk modeling and scenario building to consider climate change and or any other future threats for prospective risk management, (ii) comprehensive hazard, vulnerability, and exposure assessments based on past events and historical trends for corrective risk management, and (iii) loss and damage calculation, sectoral impacts for realized risk for compensatory risk management (UNISDR 2018).

The municipal council is responsible for the division of the local budget and resources to the different wards for disaster risk reduction based on the needs for DRRM. However, the community during the focus group discussion and the DRR expert during KII clearly stated that allocation of budget to the wards is based on political reach or flat for all wards. This has increased the vulnerability of the marginalized people in one hand and the misuse of the DRR fund on the other. Similarly, the DRR focal person agreed that drafting and endorsing DRRM laws and plans at local level are behind sectoral and other operational priorities as the municipality is more focused on meeting the compliances for receiving government funds.

Seventy-seven percent of the local government have incorporated DRM into their organizational structures (AYSPS 2019). However, the localization of DRRM acts and policies has several limitations and implications. It would not be wise to expect completion of localizing DRRM within a few years in the new federal structure of the country. It is good to have acts, plans, and policies that envision localization and have mandatory provisions and institutionalization for it. However, these are not enough to localize the DRRM. The three tiers of government should have a clear vision regarding the essence of localization, roles and responsibilities delineation, resource allocation, institutional setup, and institutionalization of DRRM from local to provincial to federal structure. Sectoral policies and plans such as the Land Use Policy 2015, National Urban Development Strategy (2017), and National Climate Change Policy 2019 have also incorporated climate and disaster risk management issues. Operationalization of DRRM at local levels requires translating such plans and policies to benefit the people (UNDRR 2019). The necessity for risk assessment and quantification and localization of DRRM was reinforced in the current study, given that only 14 % of respondents strongly agree that they have protection from disaster which is least among access to other basic services (MoFAGA 2019).

9.7 Stakeholders' Engagement in DRRM

Different local, national, regional, and international development partners are engaged directly or indirectly in DRRM advocacy, policy influencing, policy formulation, implementation, monitoring, review, and research from the very beginning. UN Agencies like United Nations Development Program (UNDP) and World Food Program (WFP); the International Federation of Red Cross and Red Crescent Societies; bilateral donors Japan International Cooperation Agency (JICA), US Agency for International Development (USAID), Department for International Development (DFID), European Commission Humanitarian Aid (ECHO), Asian Development Bank, and World Bank; and international and national NGOs have implemented their development programs. Further, they have come together under the leadership of the government of Nepal to establish the Nepal Risk Reduction Consortium (ADB 2015). The stakeholders have supported the government technically and financially in mainstreaming DRRM from central to local level and in preparedness, response, and recovery.

There are concerns that the consortium reduces the government's incentives to centralize DRR issues as a core priority which results in lessening the allocation of resources (budget) in national planning process (Taylor et al. 2013). The emergence of multi-stakeholder governance landscape for DRR is very much evident in Nepal and is unsurprising given the rise of NGOs, and in some areas this has given rise to duplication and inefficiency (Jones et al. 2014). The government has adopted measures like "the one-door policy" for DRRM program, response, relief, recovery, as well as formation of different committees, guidelines, information-sharing portals and consortiums, but limited capacity of the government and gaps between the development agency remits and government in joint effort for DRRM have made establishing a strong partnership between government and stakeholders in DRRM challenging.

The DRRM Act 2017 and the Local Government Operation Act 2017 have made provision for partnership with private and development agencies in localizing DRRM from federal to provincial to local levels. However, in the absence of a clear road map, the stakeholders' engagement in localizing DRRM at local level has not made any significant progress.

In the study area of Saptari and Nawalparasi districts, the DRRM projects supporting the local communities for risk reduction, capacity building, and strengthening their resilience to monsoonal flooding were/are financially and technically supported by bilateral donors or international development agencies through local nongovernment organizations. The representatives of all the three municipalities detailed that the municipalities and LDMCs do not have a major stake in the DRRM programs implemented. The local government representatives are contacted through introduction, discussion, and recommendation meetings for project introduction, approval, participatory, monitoring, and evaluation. There is generally a gap of engaging local government and communities in assessment, planning, and designing of projects.

The municipalities of Susta in Nawalparasi and Saptakoshi and Hanumannagar Kankalini municipalities in Saptari did not have an updated database and monitoring system to track the donor agencies, different DRR projects, supporting international non-governmental organizations (INGOs), beneficiaries, and projects related to DRR. In the FDG of the current chapter, in response to the question "does the community have access to external flood response and recovery services?", two communities, Sakhubani and Narsahi, responded that they believed there was a low portion of external services, and the four communities responded that they believed there was very low portion of external flood response and recovery services. Similarly, all the six communities responded that the quality of the external flood response and/or recovery services provided were ineffective. Also, the study communities did not know if there was a dedicated budget for flood response and recovery, and there was always uncertainty as they will receive support or not during a disaster. This clearly shows the gaps of community engagement by the local government and external agencies.

A simple computer-based information management system (Microsoft excel or other software platforms) can easily provide this service at the municipality level managed by the IT personnel. A web sharing service would ease the data entry for the development agencies remotely. Similarly, sharing of disaster contingency plans and information sharing on available resources for response and recovery will also help in building trust among the communities, local government, and external agencies for effective response and recovery. Inter-agency sharing and coordination among DRR donors and INGOs is also lacking at local level. There is a need for stakeholder mapping, assessment, and creation of a network for communication and sharing of information at the municipal level. The municipality must strengthen its collaboration with stakeholders for technical and financial support in localizing DRRM.

9.8 Conclusion

The current chapter concludes that the DRRM Act, policies, and plan are more focused to relief and response than preparedness; this may be akin to a top-down approach which merely accentuates hierarchical structure in Nepal and leads to unfair allocation of resources. A bottom-up approach of dealing with DRRM is purportedly favored by the communities, the Nepalese national, and local government. This premise aligns the need for equal or greater bias to be given to mitigation, preparedness, and response.

In the current organizational structure imposed upon the political borders of Nepal, the roles, responsibility, and scope of the three tiers of government are not clear and have duplication. There are very limited or no practical roles of provincial governance found in this study considering the implementation of DRRM Act and plan at the field. Local government is crucial because it deals with the community and faces all kinds of disasters.

Disaster resilience should be considered across sectors like health, agriculture, education, etc. integrated in all development programming, and discussed during annual planning and budgeting. This would refine preparedness and mitigation needs in countries which are subject to high levels of risk due to their topological, geographic, and demographic diversity. Federal government should have a clear road map for institutionalization of DRRM from community to national level. The road maps should include capacity building, institutional setup, resource allocation, and management with clear-cut roles and responsibilities. DRR databases are very poor in all tiers of the government of Nepal; this contravenes the localized approach that is purportedly favored. Web-based database needs to be strengthened for evidence-based DRR planning and effective decision-making processes.

Nepal has passed a series of acts, policies, strategies, and plans relating to disaster risk reduction and management. The government is dedicated to localizing the DRRM Acts, plans, and policies through technical backup, institutional setup, and directives but is still unclear how the acts and policies will transfer to the new federal system. Most local government institutions in Nepal have not made significant progress and achievements to satisfy this regard. Several issues remain uncertain given the introduction of a new federal system of government as well as the recent promulgation of the 2017 DRRM Act. Under the 2017 act, the roles and responsibility of three-tiered government is still not clear enough. The policy formulation and institutional setup does not give expected output unless there are ability and competence to operationalize the intent of the relevant acts and policies (Nepal et al. 2018).

Different research and studies have also highlighted the need of technical capacity building of the local government, resource allocation, and finances for formulation of local DRRM plan and policies (Hayes et al. 2020; IOM 2019; The Asia Foundation 2019). Policy formulation and institutional setup alone do not give the expected output unless they are complemented by the ability and competence to operationalize the intent of relevant acts and policies, in preparing, mitigating, rehabilitating, and responding the consequences in the event of disaster. There is a need to invest in government's ability to respond in addition to giving incentives at the community level that will indeed further preparedness (Molden et al. 2016).

Lastly there are enough DRR policies, regulations, plans, acts, and strategies; however the major point is these need to be promulgated to all three tiers of governments with clear roles and responsibilities. More technical and management supports are needed to local government as the implementer. Federal government support to the provincial government, and in turn provincial government should strengthen the capacities of the local government. Policies should be flexible and revised, based on learning through a bottom-up process. Integrated rule structures in governance with risk and vulnerability assessment, quantifying risk, database for vulnerable (or affected) households, and mapping and tracking of available external supports are required. The latter should augment coordinated approaches in preparedness, mitigation, response, and recovery and also planning and preparing for worst-case scenario which will certainly enable the local government in better DRRM and its localization.

Local government should develop a strategic action plan for addressing identified gaps and challenges such as technical capacities, human resources, financial availability, improvements to EWS, increased risk communication, and database and lead in all the stages of the disaster management cycle for effective service delivery to the target people. Resolving the bureaucracy and mismanagement of resources identified in governmental processes and local municipalities identified in this chapter ultimately leads to more effective use of our land resources amid harsh environmental change, the severity of which is only likely to increase in the future. Additionally we suggest a solution orientated approach for all stakeholders and indeed local community members. "Unit"-led or well-coordinated training of selected community individuals in essential mapping and database processes and instigating more regular community meetings would be a proactive start to reinforce the inevitability of future losses and diffuse any conflict between federal and local members.

Acknowledgment The authors would like to thank the Community Disaster Management Committees (CDMC) members of Narsahi, Susta, and Ratangunj of Nawalparasi district and Sakhubani, Bisanpur, and Gobargaraha of Saptari district for participating in the focus group discussions and providing their valuable information and inputs to this study. Similarly, authors are also thankful to the Susta Rural Municipality Chairperson, Susta ward number 4 ward chairperson, DRR focal person of Saptakoshi municipality, ward number 14 chairperson of Hanumannagar Kankalini municipality, chairperson of Koshi Victim's Society, Transboundary Citizen Forum chairperson of Nawalparasi, and Narsahi Community Disaster Management Committee secretary for providing their time and information regarding DRRM policies and localization at the municipal level. The authors express their gratitude to the reviews and all the people of Nawalparasi and Saptari who provided their support and coordination to complete the study.

References

ADB (2015) Disaster risk management earthquake emergency assistance project. Available via Asian Development Bank. https://www.adb.org/projects/documents/nepal-earthquake-emergency-assistance-project-rrp. Accessed (15 Feb 2021)

AYSPS (2019) Capacity Needs Assessment for the Transition to Federalism. Andrew Young School of Public Policy, Georgia State University and the Nepal Administrative Staff College. Government of Nepal, World Bank and UNDP. Available via World Bank. http://documents1.worldbank.org/curated/en/503181593615554741/text/Nepal-Capacity-Needs-Assessment-for-the-Transition-to-Federalism.txt. Accessed 10 Feb 2021

Chakrabarti D (2017) Mainstreaming disaster risk reduction for sustainable development: A guidebook for the Asia-Pacific. United Nations Economic and Social Commission for Asia and the Pacific (UNESCAP). Available via UNESCAP. https://www.unescap.org/sites/default/d8files/knowledge-products/publication_WEBdrr02_Mainstreaming.pdf. Accessed 15 Feb 2021

Chinaporn M, Wilkinson E, Lovell E et al (2019) The governance of Nepal's flood early warning system opportunities under federalism. Available via Overseas Development Institute. https://www.odi.org/publications/11418-governance-nepal-s-flood-early-warning-system-opportunities-under-federalism. Accessed 15 Feb 2021

DCA (2019) Dan Church Aid: Situation Report 7: Nepal Flood – Saptari District. Available via Reliefweb. https://reliefweb.int/sites/reliefweb.int/files/resources/DCA%20Saptari%20SitRep%2026%20July%202019.pdf. Accessed (15 Feb 2021)

DHM (2018) Standard Operating Procedure (SOP) for flood early warning system in Nepal. Government of Nepal, Ministry of Energy, Water Resources and Irrigation. Department of Hydrology and Meteorology. Available via Flood Resilience Portal. http://repo.floodalliance.net/jspui/handle/44111/2900. Accessed (15 Feb 2021)

Hayes P, Bhandari D, Neupane S, et. al (2020) Disaster risk reduction and management in Nepal: Delineation of roles and responsibilities. Oxford Policy Management Limited. Available via OPML Publications. https://www.opml.co.uk/files/Publications/a1594-strengthening-the-disaster-risk-response-in-nepal/delineation-of-responsibility-for-disaster-management-full-report-english.pdf?noredirect=1. Accessed 15 Feb 2021

IOM (2019) Needs and capacity assessment of fourteen rural and urban municipalities on disaster risk reduction and management in Nepal. International Organization for Migration, Nepal, Kathmandu Nepal. Available via Reliefweb. https://reliefweb.int/report/nepal/needs-and-capacity-assessment-fourteen-rural-and-urban-municipalities-disaster-risk. Accessed 15 Feb 2021

Jones S, Oven KJ, Manyena B et al (2014) Governance struggles and policy processes in disaster risk reduction: a case study from Nepal. Geoforum 57:78–90. https://doi.org/10.1016/j.geoforum.2014.07.011

Khanal BN (2020) Nepal: a brief country profile on disaster risk reduction and management. Available via Asian Distaster Reduction Centre. https://www.adrc.asia/countryreport/NPL/2019/Nepal_CR2019B.pdf. Accessed (15 Feb 2021)

Lucci P (2015) 'Localising' the Post-2015 agenda: what does it mean in practice? Available via Overseas Development Institute. https://www.odi.org/sites/odi.org.uk/files/odi-assets/publications-opinion-files/9395.pdf. Accessed 15 Feb 2021

MoFAGA (2019) Provincial and Local Governance Support Program (PLGSP) Programme Support Document. Government of Nepal, Ministry of Federal Affairs and General Administration. Available via PLGSP. https://plgsp.gov.np/sites/default/files/2019-12/PLGSPProgramDocument.pdf. Accessed (15 Feb 2021)

MoFE (2018) Nepal's National Adaptation Plan (NAP) process: reflecting on lessons learned and the way forward. Government of Nepal, Ministry of Forests and Environment. Available from NAP Global Network. https://napglobalnetwork.org/resource/nepals-national-adaptation-plan-nap-process-reflecting-on-lessons-learned-and-the-way-forward/. Accessed 15 Feb 2021

MoHA (2015) National progress report on the implementation of the Hyogo Framework for Action (2013–2015). Government of Nepal, Ministry of Home Affairs, Kathmandu, Nepal. Available via United Nations Office for Disaster Risk Reduction Prevention Web. https://www.preventionweb.net/english/hyogo/progress/reports/v.php?id=41755&pid:223. Accessed 15 Feb 2021
MoHA (2016) Disaster risk reduction in Nepal: achievements, challenges and ways forward. Available via DRR Portal. http://drrportal.gov.np/uploads/document/892.pdf. Accessed 15 Feb 2021
MoHA (2017) Disaster risk reduction through implementation of Sendai Framework Experiences and Lessons from Nepal. Global Platform on Disaster Risk Reduction 22-26 May, 2017 Cancun, Mexico. Available via Experiences and Lessons from Nepal - Nepal DRR Portal. http://drrportal.gov.np/uploads/document/893.pdf. Accessed (15 Feb 2021)
MoHA (2018a) Nepal Disaster Report 2017. Road to Sendai. Government of Nepal, Ministry of Home Affairs, Kathmandu, Nepal. Available via Government of Nepal, Ministry of Home affairs. http://drrportal.gov.np/uploads/document/1321.pdf. Accessed (15 Feb 2021)
MoHA (2018b) Disaster Risk Reduction National Strategic Plan of Action 2018 – 2030. Government of Nepal, Ministry of Home Affairs, Kathmandu, Nepal. Available via United Nations Office for Disaster Risk Reduction Prevention Web. https://www.preventionweb.net/english/professional/policies/v.php?id=64695. Accessed 15 Feb 2021
MoHA (2018c) National Position Paper on Disaster Risk Reduction and Management in Nepal. Available via Global Platform for Disaster Risk Reduction, 2019 Geneva. http://drrportal.gov.np/uploads/document/1514.pdf. Accessed 15 Feb 2019
MoHA (2018d) National Policy for Disaster Risk Reduction. Ministry of Home Affairs, Nepal. Available via United Nations Office for Disaster Risk Reduction Prevention Web. https://www.preventionweb.net/english/professional/policies/v.php?id=64693. Accessed 15 Feb 2021
Molden D, Sharma E, Acharya G (2016) Lessons from Nepal's Gorkha earthquake 2015. In: Singh SP, Khanal SC, Joshi M (eds) Lessons from Nepal Earthquake for Indian Himalayas and Gangetic Plains Central Himalayan Environment Association, pp 1–14
Nepal P, Khanal NR, Sharma BPP (2018) Policies and institutions for disaster risk management in Nepal: A review. Geogr J Nepal 11:1–24. https://doi.org/10.3126/gjn.v11i0.19546
NPC (2017) Nepal's sustainable development goals, Baseline Report, 2017. Government of Nepal, National Planning Commission, Kathmandu, Nepal. Available via National Planning Commission. https://www.npc.gov.np/images/category/SDGs_Baseline_Report_final_29_June-1(1).pdf. Accessed 15 Feb 2021
NPC (2020) Nepal national review of sustainable development goals. Available via United Nations Sustainable Development. https://sustainabledevelopment.un.org/content/documents/26541VNR_2020_Nepal_Report.pdf. Accessed 15 Feb 2021
Onwuegbuzie AJ, Leech NL, Collins KM (2010) Innovative data collection strategies in qualitative research. The Qualitative Report 15(3):696–726
Patole M (2018) Localization of SDGs through Disaggregation of KPIs. Economies. https://doi.org/10.3390/economies6010015
Pradhan C, Chauhan R (2020) Disaster resilience for sustainable development in Nepal. Disaster Risk Reduct. Methods Approaches Pract. An Interdisciplinary Approach for Disaster Resilience and Sustainability. https://doi.org/10.1007/978-981-32-9527-8
Scott Z, Tarazona M (2011) Decentralization and Disaster Risk Reduction, Study on Disaster Risk Reduction, Decentralization and Political Economy Analysis prepared as UNDP's contribution to the Global Assessment Report on Disaster Risk Reduction 2011. Available via United Nations Office for Disaster Risk Reduction Prevention Web. https://www.preventionweb.net/english/hyogo/gar/2011/en/bgdocs/Scott_&_Tarazona_2011.pdf. Accessed 15 Feb 2021
Shrestha R (2019) Governance and institutional risks and challenges in Nepal. Available via Asian Development Bank. https://www.adb.org/publications/governance-institutional-risks-challenges-nepal. Accessed (15 Feb 2021)
Shrestha M, Kafle S, Gurung M et al (2014) Flood early warning systems in Nepal; a gendered perspective. ICIMOD Working Paper 2014/4 Kathmandu ICIMOD 1–66

Taylor G, Vatsa K, Gurung M et al (2013) Review of the Nepal Risk Reduction Consortium (NRRC). NRCC, Kathmandu. NRCC, Kathmandu, Nepal. Available via Humanitarian Outcomes. https://www.humanitarianoutcomes.org/publications/review-nepal-risk-reduction-consortium. Accessed (15 Feb 2021)

The Asia Foundation (2019) The roles of local governments in disaster management and earthquake reconstruction. The Asia Foundation, Kathmandu, Nepal. Available via Democracy Resource Centre. https://asiafoundation.org/wp-content/uploads/2019/08/Nepal_Role-of-Local-Government-in-Disaster-Management.pdf. Accessed (15 Feb 2021)

UNDRR (2019) Disaster risk reduction in Nepal: status report 2019. United Nations Office for Disaster Risk Reduction (UNDRR), Regional Office for Asia and the Pacific. Available via United Nations Office for Disaster Risk Reduction. https://www.undrr.org/publication/disaster-risk-reduction-nepal. Accessed 15 Feb 2021

UNDRR (2020) Local leadership for disaster resilience: Profiles from Asia and the Pacific. c. United Nations Office for Disaster Risk Reduction – Regional Office for Asia and the Pacific. Available via United Nations Office for Disaster Risk Reduction. https://www.undrr.org/publication/local-leadership-disaster-resilience-profiles-asia-and-pacific. Accessed 15 Feb 2021

UNISDR (2015) Sendai framework for disaster risk reduction 2015–2030, In: Proceedings of the 3rd United Nations World conference on DRR. Sendai, Japan pp 14–18. Available via United Nations Office for Disaster Risk Reduction. https://www.unisdr.org/files/45069_proceedingsthirdunitednationsworldc.pdf. Accessed 15 Feb 2021

UNISDR (2018) Implementation guide for local disaster risk reduction and resilience strategies. A companion for implementing the Sendai Framework target E. 2018 - Public consultation version. United Nations Office for Disaster Risk Reduction (UNDRR), Regional Office for Asia and the Pacific (UNISDR)

Chapter 10
Expanding Loops in Sustainable Intelligent Driven Markets in Zimbabwe

Farai Nhakwi and James N. Furze

Abstract Issues of resource use and misuse provide serious concerns in both developed and developing economies. In order to provide a structure for safer development and provide solutions rectifying irrational resource use, electronic or e-commerce is discussed as a sustainable way of transacting and growing an economy. E-commerce is used to implement green concepts within market development of a range of products and is effective in curbing greenhouse gases. The activation of resource efficiency by the use of an e-commerce platform enables the use of a paperless system. Minimal capital and monthly expenses are required, while the community may be engaged to enable the success of the project. Hence commercial and economic development is linked via a loop with direct integration with the community. The use of e-commerce platforms and big data benefits are expatiated in this chapter. Zimbabwe is still in its infancy in e-commerce use. A case study of "Blue Chip" was used to relate the use of green concepts to the modern-day world in Zimbabwe in an intelligent, sustainable loop. Enhanced resource efficiency and its management through big data platforms will guide development of Zimbabwe for generations to come.

Keywords E-commerce · Sustainability · Big data · Resource efficiency · Community inclusivity · Low emissions

---

F. Nhakwi (✉)
Faculty of Finance and Economics, Zhejiang Gongshang University, Zhejiang, China

School of Business Sciences, Chinhoyi Universitu of Technology, Chinhoyi, Zimbabwe
e-mail: fnhakwi@gmail.com

J. N. Furze
Royal Geographical Society (with the Institute of British Geographers), London, UK

Laboratory of Biotechnology and Valorization of Natural Resources,
Faculty of Sciences-Agadir, Department of Biology, Ibn Zohr University, Agadir, Morocco

Control and Systems Engineering Department, University of Technology-Iraq, Baghdad, Iraq
e-mail: james.n.furze@gmail.com; jamesfurze@hotmail.com

J. N. Furze et al. (eds.), *Earth Systems Protection and Sustainability*,
https://doi.org/10.1007/978-3-030-85829-2_10

10.1 Introduction

The twenty-first century has seen an unprecedented accelerated growth of human populations accompanied by increasingly scarce resources. Thus efficient methods using minimal resources are emerging for use, given constraints of specific countries' development. E-commerce (electronic commerce) has been fundamental in furthering the implementation of "green" economies in both developing and developed countries. A green economy involves the production of goods and services with minimal damage on the environment and is effectively implemented through the pillar of e-commerce. E-commerce involves the trade of goods and services through the use of electronic platforms. This chapter outlines how, as part of a green economy, e-commerce is instrumental in the achievement of the "Sustainable Development Goals" (SDGs). The objective of this study is to illustrate the link between e-commerce and sustainability. Finally, given the current minimal use of e-commerce in Zimbabwe, implementation risks and solutions provide the main factors which must be considered to enable sustainable development in Zimbabwe.

10.2 Defining the Green Economy

The United Nations Environment Programme (UNEP) defines a green economy as one that results in improved human well-being and social equity while significantly reducing environmental hazards and organic insufficiencies (Sheng et al. 2011). In simple terms, a green economy is framed by three pillars, namely, low carbon emission, resource efficiency, and community inclusivity.

The world's population is approximately 8 billion and is expanding at a high rate with the expectation that it will hit the 9 billion milestone by 2050 (UNDESA 2019). A green economy ensures there is a reduction of the carbon footprint while production levels are kept optimal. To ensure a balance in current and future use of resources, green economy policies and standards have been instated, notably through operation of e-commerce. The use of green economy policies and standards provides a pathway to a safer future through sustainable development that minimizes adverse impacts on society while fostering economic growth in individual nations and the world at large.

It is essential that Zimbabwe adopts an environmentally ethical economy and facilitates the upscaling of "clean" technologies in agriculture, manufacturing, transactions, and consumption, thereby maximizing opportunities for upgrading the economy sustainably in alignment with more developed countries. As part of the global arrangement, the country agreed to a 33% reduction in gas emissions with a specific focus on energy production; however Nationally Determined Contributions (NDCs) have to stretch across sectors in order that they are achieved by the set deadline of 2030 as set by the Paris Accord (UNFCCC 2015). This will be achieved with assistance from the Global Climate Fund, which has the mandate to assist developing countries in embracing green economies. Further, the Zimbabwean ministry of environment, water, and climate change management department has the mandate to ensure that the economy embraces all forms of green technologies in the management of business structures and processes.

10.2.1 E-commerce

E-commerce is considered a multi-billion global industry that has emerged through the information technology revolution. E-commerce has enabled the global economy to integrate and companies to communicate on a worldwide scale. E-commerce is by definition the use of online transactions and is growing rapidly in emerging markets including those in developing economies (Uddin and Khan 2016). There are two main types of e-commerce: business-to-business (B2B) and business-to-consumer (B2C). Kaplan and Sawhney (2000) assert that B2B commerce through the Internet generates a lot of interest as it enables the bringing together of a huge number of buyers and sellers while fully automating transactions and reducing transaction costs for all players. In B2C e-commerce, the shopper can access different suppliers and investors through comparison shopping across domestic and international regions. Companies gain competitive advantage by not being physically constrained to operate in specific geographic areas, therefore lowering, dispersing, or completely removing storage costs.

The use of e-commerce not only promotes ease of producer-consumer interaction in business; it also contributes toward reducing environmental pollution and degradation, by encouraging paperless transactions and minimal use of fuel. The full utilization of e-commerce aids the promotion of a green economy through resource efficiency, community integration, and low carbon emissions (Oláh et al. 2019). E-commerce in Zimbabwe is still in early stages, but there is evidence that the rapid progression of online transactions has not only contributed to the performance of business but also to a form of sustainability.

10.2.2 Sustainable Development Goals Premise and Application in Development

After the millennium began, the United Nations developed sustainable development goals. The focus of the SDGs go beyond the current generation; the goals accentuate economic development while providing a strong foundation for the future generations. The SDGs are set out in Fig. 10.1 below as disseminated by the UN.

A green economy is complemented by the direct effects of e-commerce's increased efficiency. Sustainable approaches are at the center of a green economy.

Fig. 10.1 Sustainable Development Goals of the United Nations. (UN 2021)

Implementation of e-commerce increases compliance within supply chains, accelerates economic growth, and hence fosters green economies.

10.3 The Effect of E-commerce on the Sustainable Development Goals in Zimbabwe

E-commerce has already produced huge benefits in Zimbabwe in stimulating growth within the safe provisions as guided through the SDGs shown in Fig. 10.1. The current section comments on the use of e-commerce to facilitate the goals and remarks their specifics in Zimbabwean context.

10.3.1 Eradication of Poverty and Hunger (SDG 1 and 2)

E-commerce enables ordinary citizens to trade easily with only essential or minimum costs. Transaction costs are minimized as businesses operate in paperless systems and therefore have no need for a physical infrastructure to be put in place. Booth (2018) asserts that 72% of Zimbabweans live below the poverty datum line, in which the country has been rated as the 22nd poorest country in the world. A paperless business with limited physical infrastructure increases the number of local people who may establish successful business ventures. E-commerce enables local businesses to market their products and services online, accessing markets that are beyond their geographical location. This gives small to medium enterprises (SMEs) the equal opportunity of being able to compete on international platforms, thus promoting export opportunities with other countries. The successful implementation of e-commerce in business-to-business correspondence and business-to-client negotiation will go a long way in reduction of poverty in the country.

10.3.2 Good Health and Well-Being (SDG 3)

E-commerce has a huge impact on the achievement of the third SDG of good health and well-being. E-commerce decreases carbon emissions dramatically, which directly affects overall health of a nation. A green economy supplements community and individual health as businesses pursue cleaner, cheaper sources of energy and make use of online platforms in the purchase and supply of goods and services. Massive industrialization and the use of nonrenewable resources cause a decline in the general health of Zimbabwe, particularly through respiratory diseases in citizens. Logically, upon reducing the level of carbon emissions, the well-being of ordinary citizens is increased.

The reduction of business owners traveling in between locations for the purposes of transacting reduces carbon emissions of vehicles. Where companies make use of online platforms, there is a smaller carbon footprint in the energy deployed, as energy is only used when goods are being delivered. A healthier nation becomes more productive and ensures that future generations are free from the crippling effects of carbon emissions.

10.3.3 Affordable and Clean Energy (SDG 7)

Green economies enable the achievement of affordable and clean energy for national wide usage, meeting SDG 7. In a green economy, emphasis is on the use of renewable, clean sources of energy such as solar, hydropower, and bio-gas. At current Zimbabwe is for the most part dependent on nonrenewable (in particular coal)-sourced power. This is unsustainable and is highly polluting in both extraction of the fuels and emissions; the large carbon footprint leads to extreme environmental degradation. Energy is not affordable for average citizens; consequently most resort to the use of alternative fuels such as wood. This is detrimental to the environment in terms of emissions and disturbance and has led to deforestation of large portions of the country's forests. Only 40% of the country is forested, which is rapidly falling (Conservation International 2010). The use of solid wood fuel is unsustainable. In essence, a green economy (with e-commerce leveling the market) will ensure that the country has energy sources that can be accessed by every citizen and with minimal negative impact to the environment. The use of energy modeling is suggested to assist efficiency (Fonteneau and Ernst 2017; Nguyen et al. 2021), ensuring future generations' needs as nonrenewable and unsustainable fuels eventually run out.

10.3.4 Decent Work and Economic Growth (SDG8)

A green economy ensures that citizens have "decent" work and economic growth is assured. It is not enough that citizens simply have jobs; there is need for productive and generative sources of income, if possible of a circular (supply chain linked) nature (Sanguino et al. 2020). Ethical supply chains in Zimbabwe are in their infancy if present at all. A stimulus/incentive to local communities and businesses is required to produce growth. The use of e-commerce platforms in trading ensures that businesses can grow, resources required in production are reduced, while quality and price of goods sold are not compromised. Thus, profit margins become higher and businesses are able to expand. An effective green economy has the potential to triple the bottom line as resource efficiency is maximized. Economic growth can be achieved as a positive balance of trade which can be achieved as local businesses market their products to international markets. This increases export potential and further supplements foreign exchange trading. The combination of economical benefits will increase the gross domestic product (GDP) of Zimbabwe as a country. In the case that the GDP of a nation grows, other development goals are more accessible as capital gain facilitates.

10.3.5 Reduced Inequalities (SDG 10)

There has been a significant gap between the affluent and the poor in Zimbabwe which has made it difficult or impossible for SMEs to compete with large corporations. Gyimah-Brempong Kwabena (2002) asserts that the gap between the rich and poor increased dramatically between 1990 and 2005. Income inequality rose in more than two-thirds of countries particularly in Africa. Traditional methods of marketing goods and services require large capital injections, which may be disproportionate across sectors. E-commerce has leveled the playing field for all businesses. E-commerce means that SMEs can market their products online and thus access bigger markets without the necessity of investing vast financial or human resources. Local businesses are free to purchase supplies from cheaper service providers from other countries. The most popular source of raw materials for Zimbabwean businesses has become China and neighboring South Africa. Transactions have been made easier through the use of electronic means. Inequality between large corporations and SMEs is slowly being reduced as both compete on the same platform. Company equality should continue to lessen individual inequality, although there is a long way to go. The building of green business structures and compliance in supply chains will further ease the transition.

10.3.6 Responsible Consumption and Production (SDG 12)

Responsible consumption of resources and production can be achieved through green economies using e-commerce. The basic need to consume resources responsibly ensures that future generations are not put to a disadvantage and must be emphasized to local, national governmental, and community members in Zimbabwe. E-commerce is an obvious solution to increase resource efficiency in every sector. The rate at which resources inputs are converted to outputs is high in the e-commerce approach. Minimal resources are used to produce high-quality goods; thus the carrying capacity of the planet is not compromised. E-commerce increases benefits during production including reduced transport costs, human resources, time, and raw materials. Other resources such as business administration can easily be achieved through online platforms that connect several partners simultaneously. In essence, the cost of production is reduced drastically, while quality goods and services are provided to local and international clients.

10.3.7 Climate Action (SDG 13)

E-commerce within a green economy ensures the achievement of climate action/improvement. Climate change occurs largely as a result of an increase in carbon emissions since industrialization of the global economy. Zimbabwe's high industrialization rate (around 40%) due to economic challenges makes the country the potential key third world implementer of a green economy. Though the country's carbon footprint is much smaller than other countries, there is room for improvement. Through the use of e-commerce, the country can contribute to the fight against climate change as reforestation is feasible in a paperless economy. Further reforestation improves the aesthetic of the country, enriches diversity, and enhances ecological restoration. The emission of poisonous gases from nonrenewable energy such as petroleum fuels is reduced through reductions in transport costs. E-commerce is a fundamental pillar in a green economy and will encourage the use of less toxic materials in trade.

10.4 E-commerce Benefits to the Green Economy in Zimbabwe

E-commerce benefits the economy in reaching low carbon emissions, raising resource efficiency, and increasing community inclusivity. These benefits are core targets of the SDGs.

10.4.1 Low Carbon Emissions

Zimbabwe has the potential of fully capitalizing on renewable energy as it is strategically positioned for solar power in the sub-Saharan region of Africa; this should help greatly with a sorely needed reduction in the use of fossil fuels. With all the resources and capital, the e-commerce platform acts as a bridge for investors to find suitable partners as well as facilitates B2B and B2C transactions. The implementation of e-commerce policies and practice leads to a decarbonized economy low on carbon and low on the output of greenhouse gases. Handled with professionalism a low carbon system through e-commerce can be a major driver for job creation. Seventy-two percent of the total population of Zimbabwe is in poverty. Yet, the economy is empowered through the safeguarding of the environment and the reduction of poverty. Low emissions are key in business attractiveness, contributing in making Zimbabwe a favorable investment destination.

10.4.2 Resource Efficiency

Resource efficiency is the use of natural resources without compromising the quality of goods and services rendered or the resource supply, thus effecting a circular economy/sustainability. The use of e-commerce can eliminate the use of transactional paperwork which reduces deforestation while creating employment and lowering carbon emissions. In the future, e-commerce platforms will be able to release data to facilitate learning of the economic growth pathway as they serve vast amounts of big data. Importantly e-commerce platforms will save farmers from unnecessary costs. Farmers and businesses are saved costs in marketing, given the platform has designated provision of data for this; transportation networks; safe electronic transfer of funds and safe payment in the comfort of their farm or office at their convenience; and linking investors to possible investment partners.

10.4.3 Community Inclusivity

E-commerce is a "one-stop-shop" for goods, services, and investment thus offers a vast array of opportunities for community integration at both the local and international level. The Zimbabwe stakeholder community, locally and globally, has access to local goods and services online without geographical constraints or expenditure. This is an important step in Zimbabwean economic development as there are over three million Zimbabweans living in the diaspora (Pasura 2010). E-commerce promotes small businesses as they are given access to local and international markets without wasting resources. Figure 10.2 shows a local small business product of a solar-powered borehole in a rural area of Zimbabwe. This sustainable and

Fig. 10.2 Solar-powered boreholes integrating community and business needs for safe development. Source: Blue Chip Engineering Ltd. Zimbabwe (2021)

innovative approach integrates the community and future development needs of water and energy; it was set up within an e-commerce approach.

Full details of Blue Chip Ltd. are given in the case study of Sects. 10.5 and 10.5.1.

10.5 Case Study: Establishing a Green Engineering Sector in Zimbabwe

In recent years a company guided by the SDGs and green ethics has emerged in Zimbabwe – "Blue Chip Engineering" – which is owned by three patriotic Zimbabweans and has been set up with the vision of improving the lives of Zimbabweans, by making basic human needs accessible to all average citizens within the country. The company seeks to provide average Zimbabweans with clean sources of water that can be used for both domestic and commercial purposes using clean energy.

The company operates within strict green initiatives, ensuring that resources are conserved at all times.

Blue chip services include:

- Geological surveying
- Borehole drilling and casing
- Pump supply
- Electrical installation

- Solar installation
- Borehole flushing

Blue chips' mission is to become a leading service provider in water surveying, borehole drilling, and installing electrical and solar facilities, with follow-up engineering equipment supply and support. The company holds the values of quality assurance, honesty, punctuality, creativity, professionalism, and integrity.

Accentuated by the use of e-commerce, the key objectives of the company are to efficiently deliver to all customers regardless of geographical location or terrain; to ensure quality service provision that assures customer satisfaction; to remove bureaucracy as a hindrance to product delivery through coordinated and integrated management systems; to narrow the supply-demand gap; and to benefit the community through mentorship and apprentice programs.

Blue Chip has a commitment toward ensuring that citizens have access to clean water sources in their domestic and commercial localities using clean energy sources. This is achieved by providing quality drilling services at affordable rates for the average citizen. The company also provides solar installation as a sustainable means of pumping water for the client as it is both a renewable source of energy and is cost-efficient in the long run. This means that as a company, Blue Chip provides clients with a guarantee of water for everyday use without incurring the high cost of water from the national service provider. The green initiative is a motif within business operations which means that there is minimal resource wastage and environmental degradation. At the heart of service provision is the preservation of scarce resources of the country.

Clients benefit from Blue Chip as they are assured of a durable and clean source of water for both domestic and commercial purposes as the national service provider is highly unreliable, provides unclean water, and charges high rates. Clients are assured of quality products as the company prides itself in providing products and services of high quality that guarantee customer loyalty in the long run. Individuals enjoy their water without contributing to the degradation of the environment. Blue Chip serves domestic clients, corporate entities, and the farming and mining sectors to ensure water and energy needs are met. Larger clients such as corporations are also able to make use of battery storage of solar-generated energy in order to sustain their additional development needs. Naturally, the correct protocols for battery storage are followed and casing/storage guidance is provided (Manimekalai et al. 2013; Akinyele et al. 2017). The case study continues in Blue Chips' use of e-commerce in the following section.

10.5.1 E-commerce and Marketing for Company Startup

The company makes use of a strong marketing infrastructure through the use of social media and an interactive website. These platforms ensure that the company is able to reach clients beyond the local geographic location. This is in line with the company's vision to service the entire country rather than one locality. Using electrical dissemination marketing their service provision, Blue Chip reaches across Zimbabwe, and additionally clients from the diaspora community can communicate online and view pictures and videos that act as empirical evidence of the company's operations.

In Fig. 10.3 the requirements of the community drive the development of the green business. The company offers services (inputs) which act to facilitate development (outputs) of Zimbabwe. The loop of Fig. 10.3 acts intelligently and is set up as a self-regulating system. Economic development pressure acts on the loop causing potential contraction; however as the demographic of Zimbabwe increases and the economy suffers, competition imposed by lower pricing of borehole companies acts in favor of Blue Chip who outsources this process, leading to expansion of the loop of Fig. 10.3.

The outsourcing of input (b) of Fig. 10.3 encourages development of the economy. The submersible pumps used in input (c) of Fig. 10.3 contain an automated system which ensures continual supply of water for the community or corporate clients. Within the services provided by the company, several supply chains operate. Internationally certified (Hunan Puqi Geologic Exploration Equipment Institute, China) equipment is used to identify the underground water source and produces geological insight. Hence with further outsource to civil engineering experts, Blue Chip identifies above- and below-ground recommendations to ensure that when a borehole is drilled, double pipe casing may be used to prevent collapse of the borehole in the future. Materials used (casing pipes, drilling experts, and gravel packing) are outsourced from the cheapest available sources. Within electrical installation the submersible pump, pressure tank, piping, garden tap, and other fittings including a water tank and tank stand are sourced from Zimbabwean production, South Africa, and China. Solar installation also involves materials sourced from Zimbabwe, South Africa, and China including solar panels, a solar pump, and controller. As the cost of the payment is removed from the customer, in the long term the supply and use

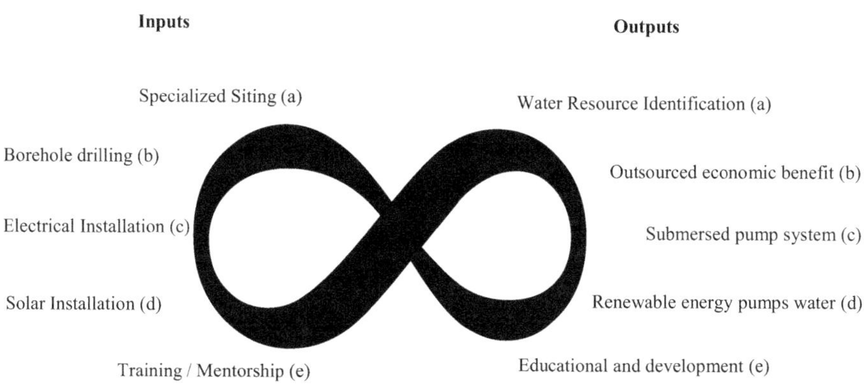

Fig. 10.3 Operations, requirements, and benefit loop used by Blue Chip Ltd. in Zimbabwe

of these materials have a beneficial effect to the economy, especially given the increased efficiency of e-commerce orchestrated supply chains.

To ensure standards are maintained and resources are not wasted, the pump size is determined by a 12 hour test prefitting in relation to the nature (flow rate/volume) of the water source. Further the water purity is certified by WHO (2011), upheld by ZINWA (Zimbabwe National Water Authority), ensuring safe domestic and commercial use. Additional functions of Blue Chip such as Fig. 10.3 Input (e) have an obvious benefit on the economy and development of Zimbabwe in terms of educational/training provision; the cascaded effect results in better value and cheaper provision of water and energy to an increasing demographic.

10.6 E-commerce in Contemporary Zimbabwe

Zimbabwe has not as yet fully embraced the concept of e-commerce. The culture of online purchasing has not taken hold in the country. Local courier services are expensive and delivery to remote areas is rare. Furthermore, there has been very little marketing of the virtues of e-commerce in the country. While the rest of southern Africa has been carried on the wave of e-commerce, the development of the phenomenon in Zimbabwe has been slow, hampered mainly by reluctance and skepticism toward the emerging trend that has totally transformed the face of global commerce. Although it is not a new phenomenon to Zimbabwean business, having been in existence for a number of years now, the "e-revolution" seems to be taking slightly longer than anticipated to sink its roots in Zimbabwe.

A host of electronic and Internet-based services have found their way onto the local market. Among them are electronic banking, electronic money transfers, electronic bill payments, and electronic product purchases, while the cellphone "craze" has also taken the country by storm. As the level of data connectivity increases, technological innovations will continue to open doors and provide new opportunities for e-commerce platforms. The proliferation of mobile phones is a factor that can make e-commerce the next big thing in Zimbabwe.

Zimbabwe's e-commerce sector is gaining momentum. One example of a Zimbabwean e-commerce platform is "Zimall," an online shopping mall which sells goods via the Internet from local suppliers. Customers can order goods, which are then delivered using Internet-hosted payments such as "PayNow." The platform leverages "EcoCash" and "TeleCash," two popular mobile-money transfer tools in Zimbabwe.

A significant obstacle to setting up an online store in Zimbabwe has been sorting out how to receive payments since everyone has a credit card; there are now local payment gateways, "PayNow" and "Pay4" applications, that give the customers the ability to pay online. Mobile e-commerce consumers are able to transfer money throughout mobile applications, but the volumes are not high due to connectivity problems as well as low Internet penetration ratios. Zimbabwe has shown recent

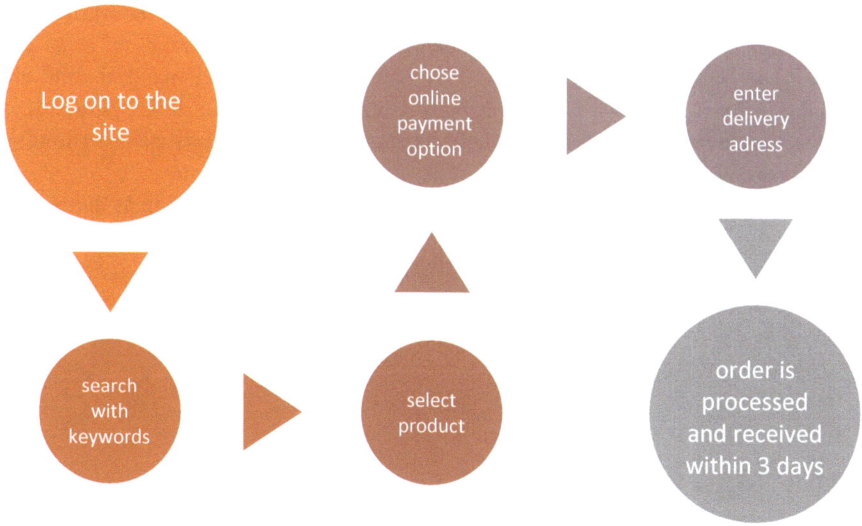

Fig. 10.4 Simplified e-commerce process flow

growth in both total revenue and operating costs across all sectors; mobile networks showed an increase of 26.2% in the first quarter of 2020 (POTRAZ 2020).

Figure 10.4 outlines a simplified e-commerce process flow which customers who wish to purchase goods online go through.

10.6.1 E-commerce Provides the Future of Zimbabwe

E-commerce awards the opportunity for low carbon emissions by creating a paperless economy and resource efficiency by making use of "natural" renewable energy such as solar, wind, or other stored renewable energy in production and transportation of goods and services. Throughout production and in packaging of goods, suppliers are encouraged to make use of recyclable or reusable materials. The use of local individuals in the processes involved gives a sense of social inclusivity through employing the community; in turn this embraces the notion of sustainable development and is an effective implementation of the green economy in Zimbabwe and beyond. There is scope for increased incentives offered to the community for enhancing recycling/reuse schemes.

Manufacturing capacity surplus for Zimbabwe may be easily created through the use of sustainable solar energy, organic plant and animal matter to enable a self-supporting system of Earth materials, and biogeochemical cycles ; Bonnett et al. 2017). There is a hypothesis that deterioration in natural possessions can be managed: that requirements of future generations will be met as long as there is no reduction in economic output. Such an approach is referred to as weak sustainability, whereby natural and other assets are declining, while that of the factory-made capital grows. From such a perspective, natural capital and the "services" provided by nature can be substituted by manufactured capital materials and are valuable only as long as they contribute to economic growth and welfare. It is increasingly evident, however, that below certain stock levels (critical thresholds), natural capital is non-substitutable.

Neo-classical economic theory () has traditionally held that as long as there is no decline in economic growth, substitutes for exhaustible resources can always be found. Thus, in theory there are no resource constraints to economic growth. As human beings slowly but eventually discovered, even if resource limits are relative and can be overcome, the capacity of the planetary ecosystem to absorb the output of economic growth is limited.

10.6.2 Big Data in Zimbabwe

Big data is a term that was coined in the late 1990s with reference to the massive upsurge in data usage on the information superhighway for the purposes of transacting. Big data has five main features: volume, velocity, value, veracity, and variety.

E-commerce makes use of big data which is complex, varied with a dynamic that has to be captured. For instance, the daily transaction data load for "Taobao" (a Chinese application) is 10TB. This is indicative of the workload involved with the use of e-commerce platforms: "eBay" deals use more than 100TB of data daily, which is more than the transaction data of 24 hours for the stock exchange Nasdaq on a prime day; Amazon deals with 398 orders per second. In Zimbabwe the volumes of data are lower because of the population's access to online services. Thus, in order to implement e-commerce seamlessly, the country's telecommunication industry has to be upgraded or outreached in order to raise capacity of managing large amounts of data effectively and with optimum Internet speed. Efficiency is also required to cut down duplication or wasting of signal resources and the efforts of input users. Efficiency may be optimized with the use of cloud computing systems (Qin et al. 2019).

10.6.3 Use of Data in E-commerce

The uses of data in e-commerce fall under the following main categories:

- Market trend prediction through the use of big data captured by the e-commerce platforms. The data kept since the inception of the e-commerce platform will be used to aid the prediction of trends.
- Sales pattern recognition as to how the various stakeholders routinely behave on the e-commerce platform.

- Research data provision for various fields of study.
- Policy formulation as data can be used to track national income and expenses through e-commerce transactions.

The benefits of each of the above uses are vast. Intelligent systems may be created rapidly in order to increase the wealth of the Zimbabwean; however, a major barrier to the use of e-commerce is the lack of willingness of communities, businesses, or individuals to make full disclosure of their activities. It is hoped that the incline of green and ethical business structures will encourage transparency in business owners, though, as the case study in Sect. 10.5 shows, this is only one of the problems which stem the embedded environmental sustainability versus growth of the economy conflict problem (Kamah et al. 2021).

10.7 The Benefits of E-commerce in Different Sectors of the Economy

The current section discusses benefits of e-commerce from the context of the main sectors operating in Zimbabwe.

10.7.1 Agriculture

The principal focus of e-commerce should be in agriculture in Zimbabwe as the sector holds the majority of activity. Zimbabwe has a total of 39 million hectares of land of which 33 million hectares are used for agricultural purposes (FAO 2012). Workers in agriculture-related sectors constitute 60% of the whole country's labor force, and the sector contributes over 40% of the nation's exports.

Zimbabwe has a savannah climate that makes it possible to have a thriving agricultural sector where cereals and grains can be grown. Zimbabwe produces crops such as maize, sorghum, millet, barley, and soya beans (Phiri et al. 2019). Other cash crops that are grown include tobacco and cotton (Chingosho et al. 2020). E-commerce platforms will be useful in marketing products and engaging with clients beyond national boarders which benefits the stakeholders.

Zimbabwe is home to several economically valuable plant species including medicinal plants due to its forest preservation (Maroyi 2013). Rich biodiversity presents an opportunity for online trade of plant extracts, native fruits, natural pesticides, wild mushrooms, tourism services, and trade in goods including eucalyptus, bamboo, and spices (Welford and Briton 2008). Significance of the demand for Zimbabwe's natural products and services will be jumpstarted as e-commerce use spreads, though training and institutional development are also required to facilitate (Kamah et al. 2021).

10.7.2 Manufacturing Sector

Manufacturing in Zimbabwe can be transformed with the use of e-commerce. E-commerce has been instrumental in enabling local manufacturers to purchase raw materials online from affordable suppliers in other countries at lower cost (Ozlen and Mekic 2014). The ability to purchase goods online eliminates travel cost and ensures a direct supply of goods to the end user. There has been a steady increase in the number of manufacturers who depend on international suppliers from China and South Africa as the cost of production is comparatively low, reduced by around 30%. This has assisted local SMEs to increase their profit margins.

10.7.3 Trade Industry

Through the use of e-commerce platforms, the balance of trade in Zimbabwe has been positively impacted. Zimbabwe is an open economy, trading with other economies beyond its borders though has suffered a trade deficit trend for the past decade. In 2017, the country imported goods worth $5472 billion, while exports were only $4353 billion. The Internet has reduced the world into a global village with no boundaries in which businesses in different countries can communicate easily and exchange goods and services. B2B e-commerce enables companies to provide services to companies in other countries, which increases revenue flow for local businesses (Xiao et al. 2019). Local goods and services can be marketed to international markets, thus increasing foreign exchange that the SME earns, increasing the exports of the country. An increase in exports positively impacts the balance of trade in Zimbabwe and increases the GDP in the nation. Export increase will lead to economic growth as it also attracts investment.

10.7.4 Mining Sector

The mining sector stands to benefit from e-commerce in Zimbabwe. The country has a vast array of natural resources in the form of minerals. The country has abundant deposits of gold, copper, precious stones, diamonds, and iron (Malinga 2018). These minerals can be sold for higher prices in international markets, but traditional methods of marketing products make it too expensive for small-scale miners to market their products at competitive prices. By embracing e-commerce, small-scale miners can communicate directly with international players who offer a fair market price. Where small-scale farmers increase their export sales of agricultural products and indeed of minerals, there is an improvement in the overall GDP of the country and a guarantee in economic growth, though the environmental cost of mining and associated risks require vigilant monitoring.

10.7.5 Energy

Zimbabwe has had problems in power generation and distribution which has had a negative effect on domestic industry. In the government national budget presentation of 2019, the responsible ministry reported that the country requires electricity supply of 1800MW annually, but the current capacity was at 400MW (Ncube 2019). The deficit in energy supply has reduced the country's industrial capacity by approximately 40% and increased borrowing, debt, and imports (Ncube 2018). Learning renewable technologies and the spectrum of possibilities for power generation remains the first hurdle to be overcome (Makonese 2014). However, through the use of e-commerce the country can increase its national power supply by accessing alternative sources of energy such as solar and biogas.

Entrepreneurial success utilizing e-commerce has already been heralded in this area: "Oxygen," a small startup company led by Simbarashe Mhuriro, developed a rooftop solar initiative in 2017 with the mixed support of government and investors (Nsehe 2017). As a developing country, Zimbabwe can learn and purchase new technologies from other countries that have developed lean energies (UNDP 2020).

10.7.6 Environmental Sector

The environment has been on the receiving end of human's reckless activities, despite communities' planned behavior; this is especially evident in Zimbabwe where minimal resources are channeled toward environmental conservation efforts (Mawere 2013). E-commerce will go a long way in strengthening environmental conservation efforts, reducing required paperwork and procedures, and lessening deforestation rates. Further trading online will mean less nonrenewable energy consumption and lower gas emissions, and a reduction in the need for physical infrastructure means less land is used for commercial purposes.

10.8 Resource Efficiency Benefits of E-commerce in Zimbabwe

E-commerce provides readily available data and trends on key economic areas; this data aids in the efficient allocation of the scarce resources available as data exhibits which resources are required in a timely manner. This presents several opportunities to the Zimbabwean people: the use of big data provides research data for various research within different sectors, a road map for efficient implementation of economic policies, and national planning. Although there are problems with data interpretation and data redundancy, these problems should resolve themselves with appreciation of the trends within larger data volumes (Kabanda 2019; Resnyansky 2019). The use of e-commerce saves resources as one does not need to have a brick and cement store to function. The use of an online store saves the shop owners money for building as well as time in stocking and maintaining goods, thus simultaneously eliminating insurance costs and rent, giving clients worldwide unlimited access to the products 24 h, 7 days a week. The use of e-commerce enables global marketing and the lowest cost outlay. The ease and convenience of doing business are what make platforms attractive, as they eliminate travel costs when purchasing goods and services which can be done in the comfort of one's home or anywhere in the world. E-commerce platforms enable the Zimbabwean to be easily accessed by the outside world. The instrument of e-commerce operates the use of online transactions as a key significant contributor to the curbing of environmental degradation (Panayotou 1994).

10.8.1 Community Inclusivity and Integration

In every plan or policy, success depends on its stakeholders; e-commerce as a road map comprehends the necessity of community inclusivity. In Zimbabwe the implementation of e-commerce and the green concept will provide jobs for the community while improving the spending power and health of many families (Batani et al. 2015; Mawere and Murero 2019). E-commerce is also an integration tool with the outside world and a one-stop-shop for goods, services, and investment, entailing an array of opportunities for community integration. The Zimbabwe stakeholder community both locally and globally can have access to the local goods and services by a click of a button.

Dialogue within e-commerce seeks to encompass opportunities such as low carbon emission, resource efficiency, and social inclusivity which sustain the notion of sustainable development in a green economy. Concern over sustainability has increased in both academia and industry. Environmental sustainability has become the new form of corporate social responsibility. Zimbabwe received US$3 million funding to enable the setting up of green economy structures to be implemented through community integration from the Green Climate Fund (Zvirevo 2019).

Intelligent driven platforms have benefits derived from their implementation. Projected benefits include creation of jobs in the green sector, an improved standard of living necessitated by an increase in earnings for employees, and most importantly a significant reduction in greenhouse gas emissions making a safe and clean environment. While systems can improve income and reduce poverty, the enhanced efficiency implications across sectors come with significant and potentially irreversible social, environmental, and economic costs – exhibited, for example, through the spread of mining/extraction (Kaseke et al. 2015). An estimated 60% of the world's ecosystem services were found to be degraded or used unsustainably (Sutton et al. 2016). Heralding e-commerce must be associated with risk assessments of industrial diversity (Robaina et al. 2020; Zhao et al. 2019).

10.9 Conclusion

In conclusion, the use of e-commerce as a tool to implement the green concept is fundamental to the improvement of Zimbabwe economically and socially. The acceleration provided is a key feature of a green economy that will enable Zimbabwe to achieve the SDGs. E-commerce ensures that the three pillars of a green economy, low carbon emission, resource efficiency, and community inclusivity, are a reality in the country. Sustainability is the end goal of all activities within the country as there is need to ensure that the country not only provides for the current generation but for generations to come.

With the implementation of green economy concepts, there should be shifts in opinions, behavior, know-how, and familiarity competencies. The first step to alteration to achieve sustainable development is an acknowledgment of the need to progress from a short-term to a long-term opinion of the economy, community, and environment. Zimbabwe is a nation endowed with a wide array of natural resources. Effective use of natural possessions is extensively regarded as the key to revealing economic potential. E-commerce working hand in hand with the implementation of green concepts is considered imperative. With e-commerce and green concept cooperation, several opportunities for community integration enhancing the community residents' quality of life are presented. Subsequently sustainable development is brought to the fore.

References

Akinyele D, Belikov J, Levron Y (2017) Battery storage technologies for electrical applications: impact in stand-alone photovoltaic systems. Energies. https://doi.org/10.3390/en10111760

Batani J, Denhere PT, Mawere T (2015) The role of ecommerce in resuscitating the economy of Zimbabwe. https://www.researchgate.net/publication/280529713_The_Role_of_e-commerce_in_Resuscitating_the_Economy_of_Zimbabwe. Accessed 5 Mar 2021

Bonnett SAF, Maxfield PJ, Hill AA et al (2017) Biogeochemistry in the scales. In: Furze JN, Swing K, Gupta AK et al (eds) Mathematical advances towards sustainable environmental systems. Springer, Cham. https://doi.org/10.1007/978-3-319-43901-3_7

Booth O (2018) Gogo Olive changing the lives of Zimbabwean Women. Available via the Borgen Project. https://borgenproject.org/gogo-olive-zimbabwean-women/. Accessed 4 May 2021

Chingosho R, Dare C, van Walbeek C (2020) Tobacco farming and current debt status among smallholder farmers in Manicaland province in Zimbabwe. Tob Control. https://doi.org/10.1136/tobaccocontrol-2020-055825

Conservation International (2010) Zimbabwe. https://www.nestle.com/sites/default/files/asset-library/documents/creating%20shared%20value/responsible-sourcing/deforestation-guide-zimbabwe.pdf. Accessed 5 Mar 2021

FAO (2012) FAO in Zimbabwe. Available via Food and Agriculture Organisation of the United Nations. www.fao.org/Zimbabwe/fao-inZimbabwe/zimbabwe-at-a-glance/en. Accessed 4 Mar 2021

Fonteneau R, Ernst D (2017) On the dynamics of the deployment of renewable energy production capacities. In: Furze JN, Swing K, Gupta AK et al (eds) Mathematical advances towards sustainable environmental systems. Springer, Cham. https://doi.org/10.1007/978-3-319-43901-3_3

Gyimah-Brempong K (2002) Corruption, economic growth, and income inequality in Africa. Econ Gov 3:183–209
Kabanda G (2019) An evaluation of big data analytics projects and the project predictive analytics approach. Orient J Comput Sci Technol. https://doi.org/10.13005/ojcst12.04.01
Kamah M, Riti JS, Bin P (2021) Inclusive growth and environmental sustainability: the role of institutional quality in sub-Saharan Africa. Environ Sci Pollut Res. https://doi.org/10.1007/s11356-021-13125-z
Kaplan S, Sawhney M (2000) E-hubs: the new B2B marketplaces. Available via Harvard Business Review. https://hbr.org/2000/05/e-hubs-the-new-b2b-marketplaces. Accessed 7 Mar 2021
Kaseke K, Chaminuka L, Musingafi MCC (2015) Mining and minerals revenue distribution in Zimbabwe: learning from our surroundings and past mistakes. J Econ Sustain Dev 6(6):90–96. https://core.ac.uk/download/pdf/234646936.pdf. Accessed 5 Mar 2021
Khan AG (2016). Electronic commerce: a study on benefits and challenges in an emerging economy. Glob J Manag Business Res. https://globaljournals.org/GJMBR_Volume16/3-Electronic-Commerce-A-Study.pdf. Accessed 7 Mar 2021
Makonese T (2014) Renewable energy in Zimbabwe. In: Paper presented at the 2016 International Conference on the Domestic Use of Energy (DUE), Cape Town, South Africa. https://doi.org/10.1109/DUE.2016.7466713
Malinga W (2018) 'From an agro-based to a mineral resources-dependent economy': a critical review of the contribution of mineral resources to the economic development of Zimbabwe. Forum Dev Stud 45(1):71–95. https://doi.org/10.1080/08039410.2017.1378711
Manimekalai P, Harikumar R, Raghavan S (2013) An overview of batteries for photovoltaic (PV) systems. Int J Comput Appl 82(12):28–32
Maroyi A (2013) Traditional use of medicinal plants in south-central Zimbabwe: review and perspectives. J Ethnobiol Ethnomed. https://doi.org/10.1186/1746-4269-9-31
Mawere M (2013) A critical review of environmental conservation in Zimbabwe. Africa Spectr 48(2):84–97. http://www.jstor.org/stable/24589098. Accessed 5 Mar 2021
Mawere GE, Mutero T (2019) Effects of the prevailing Zimbabwean economic situation on e-commerce. Int J Eng Res Technol. https://doi.org/10.17577/ijertv8is010039
Ncube M (2018) Austerity for prosperity. Zimbabwe 2019 Budget Speech. Available via Ministry of Finance and Economic Development. https://www.zimtreasury.gov.zw/index.php?option=com_phocadownload&view=category&id=11&Itemid=760. Accessed 4 Mar 2021
Ncube M (2019) Gearing for Higher Productivity Growth and Job creation. Zimbabwe 2020 Budget Speech. Available via Ministry of Finance and Economic Development. https://t792ae.c2.acecdn.net/wp-content/uploads/2019/11/2020-National-Budget-Speech.pdf. Accessed 7 Mar 2021
Nguyen CP, Su TD, Bui TD et al (2021) Financial development and energy poverty: global evidence. Environ Sci Pollut Res. https://doi.org/10.1007/s11356-021-13038-x
Nsehe M (2017) Meet the 32 year old entrepreneur who is building Zimbabwe's largest solar company. Available via Forbes. https://www.forbes.com/sites/mfonobongnsehe/2017/12/21/meet-the-32-year-old-entrepreneur-who-is-building-zimbabwes-largest-solar-company/?sh=5ca773eb63f7. Accessed 5 Mar 2021
Oláh J, Kitukutha N, Haddad H et al (2019) Achieving sustainable e-commerce in environmental, social and economic dimensions by taking possible trade-offs. Sustainability. https://doi.org/10.3390/su11010089
Ozlen MK, Mekic E (2014) Perceived benefits of e-commerce among manufacturing and merchandising companies. Int J Acad Res Econ Manag Sci. https://doi.org/10.6007/IJAREMS/v3-i2/676
Panayotou T (1994) Economic instruments for environmental management and sustainable development. Available via United Nations Environment Programme. https://core.ac.uk/download/pdf/48031478.pdf. Accessed 5 Mar 2021

Pasura D (2010) Competing meanings of the diaspora: the case of Zimbabweans in Britain. J Ethn Migr Stud 36(9):1445–1461. https://doi.org/10.1080/1369183X.2010.498670

Phiri K, Dube T, Moyo P et al (2019) Small grains "resistance"? Making sense of Zimbabwean smallholder farmers' cropping choices and patterns within a climate change context. Cogent Social Sci. https://doi.org/10.1080/23311886.2019.1622485

POTRAZ (2020) Postal and Telecommunications Regulatory Authority Of Zimbabwe (POTRAZ) Abridged Postal And Telecommunications Sector Performance Report First Quarter 2020. Available via POTRAZ. http://www.potraz.gov.zw/wp-content/uploads/2020/06/Sector_Performance_1stQ2020.pdf. Accessed 4 Mar 2021

Qin H, Yu M, Lai Y et al (2019) Cloud resource orchestration optimisation based on ARIMA. Int J Simul Process Model 14(5):420–430

Robaina M, Villar J, Pereira ET (2020) The determinants for a circular economy in Europe. Environ Sci Pollut Res 27:12566–12578. https://doi.org/10.1007/s11356-020-07847-9

Resnyansky L (2019) Conceptual frameworks for social and cultural Big Data analytics: answering the epistemological challenge. Big Data Soc. https://doi.org/10.1177/2053951718823815

Sheng F, Flomenhoft G, Downs TJ et al (2011) Is the concept of a green economy a useful way of framing policy discussions and policymaking to promote sustainable development? Nat Res Forum 35(1):63–72. https://doi.org/10.1111/j.1477-8947.2011.01347.x

Sanguino R, Barroso A, Fernández-Rodríguez S et al (2020) Current trends in economy, sustainable development, and energy: a circular economy view. Environ Sci Pollut Res 27:1–7. https://doi.org/10.1007/s11356-019-07074-x

Sutton PC, Anderson SJ, Costanza R et al (2016) The ecological economics of land degradation: impacts on ecosystem service values. Ecol Econ 129:182–192. https://doi.org/10.1016/j.ecolecon.2016.06.016

Uddin SMF, Khan MN (2016) Exploring green purchasing behaviour of young urban consumers: empirical evidences from India. South Asian J Glob Business Res 5(1):85–103. https://doi.org/10.1108/SAJGBR-12-2014-0083

UN (2021) The 17 Goals. United Nations Department of Economic and Social Affairs Sustainable Development. Available via https://sdgs.un.org/goals. Accessed 16 Mar 2021

UNDESA (2019) Growing at a slower pace, world population is expected to reach 9.7 billion in 2050 and could peak at nearly 11 billion around 2100. Available via United Nations Department of Economic and Social Affairs. https://www.un.org/development/desa/en/news/population/world-population-prospects-2019.html. Accessed 7 Mar 2021

UNDP (2020) 2030 Agenda for Sustainable Development – Renewable energy. Available via United Nations Development Programme. https://www.undp.org/content/undp/en/home/2030-agenda-for-sustainable-development/planet/sustainable-energy/renewable-energy.html. Accessed 4 Mar 2021

UNFCCC (2015) Adoption of the Paris Agreement. Available via United Nations Framework Convention on Climate Change. https://unfccc.int/resource/docs/2015/cop21/eng/l09r01.pdf. Accessed 7 Mar 2021

Welford L, Briton GL (2008) Bridging the gap: Phytotrade Africa's experience of the certification of natural products. For Trees Livelihoods 18:69–79

WHO (2011) Guidelines for drinking-water quality, fourth edition. World Health Organization. ISBN: 978 92 4 1548151

Xiao L, Guo F, Yu F et al (2019) The effects of online shopping context cues on consumers' purchase intention for cross-border e-commerce sustainability. Sustainability. https://doi.org/10.3390/su11102777

Zhao S, Jiang Y, Wang S (2019) Innovation stages, knowledge spillover, and green economy development: moderating role of absorptive capacity and environmental regulation. Environ Sci Pollut Res. https://doi.org/10.1007/s11356-019-05777-9

Zvirevo W (2019) Green financing in Zimbabwe – case of the Infrastructure Development Bank of Zimbabwe. https://www.idbz.co.zw/file/874/download?token=Afk2CRkK Accessed 5 Mar 2021

Chapter 11
Sustainable Consensus in an Uncertain Environment

Mohamed El Alaoui and Saeid Eslamian

Abstract It is obvious that climate change has become a serious threat to humankind. Unfortunately, even though the scientists are near unanimity—which is rarely achieved in other fields—it is not seen as such by the masses. This chapter investigates the climate change consensus achieved between scientists, politicians, and the public. It maintains that the political aspect is the weak link, since courageous political decisions that benefit future generations can cause great dissatisfaction, among those who are potential voters and sponsors. Furthermore, the lack of a global understanding of what is sustainable impedes potential political agreements. This study clarifies that the public perception on the scientific consensus is largely inferior to the reality; even in highly educated populations, due to policies, educational programs, misinformation, and cultural difference. The study also shows that measuring consensus leads to a false debate about the value reached. Fuzzy logic is used to capture humans' perceptions and proposes achieving the consensus by minimizing weighted incoherencies.

Keywords Climate change · Coherence · Consensus · Policy · Fuzzy logic · TOPSIS

11.1 Introduction

Human influence on Earth is so strong that some scientists suggest that it creates—with some discords on the starting point (Subramanian 2019)—a geological revolution of human origin, capable of marking the lithosphere and procreating a period

M. El Alaoui (✉)
Department of Production and Industrial Engineering, ENSAM-Meknès, Université Moulay Ismaïl, Meknes, Morocco

École supérieure d'Architecture de Casablanca, Casablanca, Morocco
e-mail: mohamedelalaoui208@gmail.com

S. Eslamian
Department of Water Engineering, College of Agriculture, Isfahan University of Technology, Isfahan, Iran
e-mail: saeid@iut.ac.ir

J. N. Furze et al. (eds.), *Earth Systems Protection and Sustainability*,
https://doi.org/10.1007/978-3-030-85829-2_11

called the Anthropocene (Ellis 2018). Frighteningly enough, this period is marked by great accelerations of processes, since many indicators represent an exponential character (Ellis 2018).

Whilst lobbyists cast doubt about the veracity of climate change (Brulle 2018), opposing studies (Colston and Vadjunec 2015) and misinformation (Legates et al. 2015), any respectful study on the future impacts of climate change, will recognize that optimistic scenarios are worrying enough (Arvesen et al. 2011; Rosenzweig 2013). However, consensual decisions reconciling scientists and politicians are still to be found (van der Sluijs et al. 2010) and clarified to the masses (van der Linden et al. 2015; van der Linden et al. 2014).

Unfortunately, even if the scientific consensus on climate change is undeniable (Oreskes 2018), the definition of consensus, as paradoxical as it may sound, is not consensual (El Alaoui et al. 2019a). How can we define consensus? (Briggs 2013) And how can it be measured? (Herrera-Viedma et al. 2014) Furthermore, decisions to be taken are uncertain at several levels (Bakshi 2019). First, they are relative to future events that can only be predicted (Poff et al. 2016). Second, the decision-makers (DM) responsible for making these predictions, no matter how cautious they may be, will never be able to proclaim perfect knowledge, exempt from errors and misjudgments. In addition, information itself can be unavailable or irrelevant. Other classifications of uncertainty causes may be found in Li et al. (2013). Due to the historical development of uncertainty theories, probability was considered to be synonymous to uncertainty (Klir 1989; Zheng et al. 2013). Recently, fuzzy logic imposes itself dogmatically (Bier 1990), especially in decision-making (Özkan and Türkşen 2014), why would it be the best choice?

Sustainability is an ambiguous and vague word, it is ambiguous since it can accept many interpretations (Klir 1987) due to context variation and understanding (Braat 1991; Shearman 1990). It is also vague since there is no clear boundary between what is sustainable and what is not. Fuzzy logic is a powerful tool that can model such concepts. This is especially true when classical yes or no questions are particularly difficult to pose and answer. A review of fuzzy logic use to assess sustainability was proposed by Dumane et al. (2019). Nevertheless, how can we be sure of the veracity of climate change? The succinct proof: human activities cause greenhouse gas (GHG) emissions; GHG warm the planet. To the persistent skeptics advancing poorly documented information, the impact is clearly amplified in arctic regions (Smol and Douglas 2007).

The sequel of this work will treat consensus on climate change through three levels: scientific, political, and public opinion. Section 11.2 will be devoted to the scientific consensus, without which any further discussion is of no value. Section 11.3 treats the political size. It discusses the discordances within and between countries, with an emphasis on different interpretations of what is and what is not sustainable in each region. Section 11.4 points to the dazzling gap between the scientific consensus—which is near unanimity— and how it is perceived by the public. The section also explores some solutions to raise public awareness. Section 11.5, without pretending to be exhaustive; presents the required fuzzy logic background to model human perceptions. Section 11.6 shows the absurdity of the conflict surrounding the exact level of consensus reached and presents an iterative algorithm to achieve it. Section 11.7 contains the proposed approach. Section 11.8 concludes the chapter.

11.2 Scientific Consensus

Sciences rest on a number of axioms, which are by definition admitted without demonstration. According to Planck, scientific thought must relate to something, yet no uniformly acceptable vision of the world has been defined. Hence, it is impossible to place science on fixed and inclusive content (Planck 1949). Furthermore, any system of axioms is incomplete (Gödel 1931); thus, the results obtained are rarely unanimous. This is particularly true while studying the set of physical phenomena involved in climate change.

After turning down ambitious expectations, let us examine how climate change is happening. The heat emitted by the sun is either reflected or absorbed. A portion of the heat absorbed is fixed in the earth atmosphere by GHGs as carbon dioxide, water vapor, methane, and ozone. It is true that significant variations occurred in the past at a geological time scale. However, since the industrial revolution, GHGs concentrations are growing rapidly (Gunaratna 2018). Do we have to eliminate all GHGs? No, if we did, life would be impossible due to freezing temperatures. Coversely, over-concentration will cause extremely high temperatures. In fact, the greenhouse effect explains why Venus' temperature is hotter than Mercury's, despite Venus being farther away from the Sun and receiving just the quarter of the energy received by Mercury (Grego 2007).

While the actual U.S. government is clearly climate skeptic, the National Research Council published since 1979, a report predicting temperature raise between 2.6 and 4.1°C by 2030–2050. It stressed the fact that all the models used predict that the temperature is only going up, and it becomes urgent to react (Charney et al. 1979). In the same year, a second report was prepared for the U.S. department of energy on long-term impact of carbon dioxide on climate (MacDonald et al. 1979). A review of precursor reports and studies published by the National Academy of Science was presented in (Nierenberg et al. 2010) including a report with an evocative title "Changing Climate" (Anon 1983) that estimates a temperature raise between 1.5 and 4.5°C.

Meanwhile, worldwide, the United Nations Environment Program (UNEP) and the World Meteorological Organization (WMO) created the Intergovernmental Panel on Climate Change (IPCC) in 1988. In its first published report released on 1990 (Houghton et al. 1990), the IPCC presents a first international consensus predicting a rise in the average temperature of the globe in the period 1990–2030 (Frame and Stone 2013). Unfortunately, from a climatic and fortunately from a modeling perspective, the forecasts seem accurate given the relative accumulated data. Even though, the results were slightly readjusted later.

Several studies attempted to measure consensus level on climate change. Between 1993 and 2003, 928 abstracts of papers in refereed scientific journals were published (Oreskes 2004). 75% of them explicitly or implicitly accepted the consensus position, while the remaining 25% abstained without rejection of consensus. In 2009, Doran and Zimmerman (2009) pertinently mentioned that scientists were asked to respond to two question: "1. When compared with pre-1800s levels, do you think that mean global temperatures have generally risen, fallen, or remained relatively constant?" and "2. Do you think human activity is a significant contributing factor in changing mean global temperatures?" 96.2% answered "risen" to the first question and "yes" to the second. In Bray (2010) and Bray and von Storch (2007), it is no longer a question of the existence of consensus, but of its dimensions. The authors maintained that even if it does not reach unanimity, asymptotic consensus was continually growing in the studied period.

The scientific consensus on humans involvement in global warming is estimated at 97% among climate scientists (Cook et al. 2013), while some scientists doubt the number (Tol 2016), it remains in the 90–100% interval (Cook et al. 2016). However, it is not a question of numbers, with 1% less, our judgment will remain intact; the question can be more disturbing when passing the symbolic barrier of 50%. Furthermore, recent publications advocates 100% (Powell 2019). A historical development of consensus on climate change was discussed in Cook et al. (2019).

While skepticism was barely accepted in the 1990s (Michaels and Knappenberger 1996) due to abundant uncertainties, it becomes—scientifically speaking—evident that climate change is occurring and something must be done urgently. The Darfur conflict was partially due to environmental reasons. The persistent discourse is only on the possible repercussions: mass migration, policy, and others (Salehyan 2008).

11.3 Political Consensus

Despite the undeniable scientific consensus on climate change, policy makers still have large disagreements (Walther et al. 2005). We must recognize that a courageous political decision will create a lot of dissatisfaction, who are potential voters and sponsors, for results promised in a future way beyond the electoral mandate.

11.3.1 Between Countries

The negotiation process is tough and complex in its nature. This is particularly true for global climate negotiations involving more than 180 countries and countless observers and participants in the United Nations Framework Convention on Climate Change (UNFCCC). Under the constant threat of VETO's right, it is preferable to seek a consensus instead of a vote that will stumble, in all likelihood, over a VETO (Depledge 2013). Furthermore, the UNFCCC's Conference Of the Parties (COP) experienced a dramatic increase in sessions, subgroups, and so forth, making the discussions even harder (Vihma 2015).

To clarify each state responsibility, they were linked to historical GHGs emissions. However, a consensus is far from being achieved on how past emissions contribute on present and future warming (Friman 2016). Thereby, developed countries keep sending the ball back and forth between each other preventing any definitive agreement. Ironically, they warn developing countries to follow the same path, which will curb their development and further widen the gap between industrialized and developing countries (Gupta 1997). Furthermore, the arctic region may contain undiscovered reserves of oil and natural gas that might be easily accessible after iceberg melting. The latter also eases shipping through the arctic seas with influent beneficiary nations including Russia, Canada, Norway, and the United States (Gunaratna 2018). This may explain why the United States weakened G8 summit plans (Wallis 2006). Even worse, there is no guarantee that all the worst polluters will honor concluded accords (Gunaratna 2018).

11.3.2 Local or Regional Level

Going beyond these issues, how can we measure countries commitment to sustainability issues while, they face fundamentally opposed issues? Whereas the Finnish Ministry of Environment includes reindeer and gray seals numbers in addition to the breaking date of icy rivers to measure sustainability (Wallis 2006). These indicators are only applicable in few countries. Leaving a wide-open space for regional (Bogaert et al. 2018; Langan and Price 2016; Maurer et al. 2009; Rodrigues 2015) and national thinking and applications (Matsuura and Baba 2016; Sapountzaki and Wassenhoven 2005; Sinclair and Smith 1999; Xu et al. 2014). With some vision adopted to specific sectors and industries such as agriculture (Donini et al. 2016; Drewnowski 2018; Eakin et al. 2017); aquaculture (Caffey et al. 2000); mineral industry (Segura-Salazar and Tavares 2018); urban planning (Gil Solá et al. 2018); nano-remediation (Corsi et al. 2018); health (Reddock 2017; von Schirnding 2002); transportation (Keseru et al. 2016); wood and wood product (McCown 2014).

While the Bertrand's report on sustainability definition is largely accepted, "development that meets the needs of the present without compromising the ability of the future generations to meet their own needs" (World Commission On Environment and Development 1987), a practical definition is still to be achieved. In fact, what will the needs of future generations look like? Have we already condemned them? Can they find their own path independently?

When health and safety is fiercely attacked by politicians, employers, and media which are described as barriers to growth and development (Kay 2008), it is hard to convince people that sustainability and economy can go hand in hand. However, its relation to economy (Luckhurst 2018) and food security (Berry et al. 2015) can be easily proved.

According to the Organization for Economic Cooperation and Development (OECD), an indicator is "a parameter, or a value derived from parameters, which points to, provides information about, or describes the state of a phenomenon/

environment/area, with a significance extending beyond that directly associated with its value" (OECD 2006). Further approaches were proposed in the literature (Hák et al. 2016; Singh et al. 2012) varying from warming (Oppenheimer 2014), the natural step (Upham 2000), to consensus itself (Bender and Simonovic 1997). While some consensual indicators could be found as "recycling" (Figge et al. 2018); that must be distinguished from upcycling (El Alaoui 2020a; Jørgensen and Pedersen 2018); it is extremely hard to fix a comprehensive and inclusive view. Indicators can only be used to draw comparisons with similar situations (Cartwright 2000) and not as an end on themselves (El Alaoui and Ben-azza 2017).

11.4 Mass Consensus

Previously, temperatures were a little warmer or a little colder in each region. Now, it is hotter everywhere. These perceptible fluctuations, accompanied by heat peaks, favor public awareness and support for government action to reduce climate change impacts (Lee et al. 2018). However, are humans willing to endure radical change affecting their diet (Landzettel and Bommelt 2019); transport modes (Chapman 2007); and daily life? (Ouest and Lafon 2016). These questions are all the more persistent since the recent COVID crisis proved that some people hardly respect temporary confinement, even if they know possible repercussions on their lives. Public participation in such efforts remains poorly theorized and understudied (Sarzynski 2015). Ironically, the focus should be on how to fructify citizens participation (Doelle and Sinclair 2006). Public perception on the scientific consensus is largely inferior to what it is in reality (Kerr and Wilson 2018a, b); even in highly educated populations (Díaz Estévez et al. 2014); due to policies (Bolsen and Druckman 2018), educational programs (Colston and Vadjunec 2015), misinformation (Bedford and Cook 2013; Legates et al. 2015), culture (Kunkle and Monroe 2019), and other reasons (Carlton et al. 2015). Thus, it becomes obvious that climate change causes and consequences should be clarified to the public (Maibach et al. 2014). Especially when the message is fiercely attacked (Pearce et al. 2015), from scientists in other fields (Carlton et al. 2015). Hence, the focus now should be on efficient solutions to convey the right information (Cortés and Eugenia 2014; van der Linden et al. 2014).

Worse yet, true scientists are confronted on TV sets by dissenters, which make the impression, especially if it is a one-on-one discussion; on a fifty–fifty chance of scientific consensus (Oreskes 2018), we must not forget that we do not have the same awareness to environmental issues nor the same scientific background. Hence, to sensitize a large audience, satirical programs do the job, which against documentaries and specialized shows, can reach a public with no real motivation to seek such information (Brewer and McKnight 2017). A pertinent approach was proposed by Pepermans and Maeseele, since the political disagreements are far from being handled, the solution could be achieved through eliminating the political aspect, because after all, it is not a political issue, it is a moral one (Pepermans and Maeseele 2014).

While public awareness is paramount to achieve concrete results (Madumere 2017); not all scientific information can be debated in public, without a minimum investigation by experts (Stanhill 2007). Premature public involvement increased by media over simplification can be harmful. In fact, the ozone hole—over-mediated and ultimately recovered relatively easily—resulted in a false belief, suggesting the possibility—or even the ability—that any climate process is reversible.

11.5 Required Fuzzy Logic Background

Fuzzy logic permits dealing quantitatively with imprecisions (Zadeh 1965). It can use linguistic variables that ease capturing human perceptions (El Alaoui et al. 2016). Let X be a universal set and F a fuzzy subset in X, F is defined as follows (El Alaoui 2018, 2020b; Skalna et al. 2015):

$$\tilde{F} = \left\{ \left\langle x, \mu_{\tilde{F}}\left(x\right) \right\rangle | x \in X \right\}. \tag{11.1}$$

Where $\mu_F(x)$ is the degree of membership of x in F in the unity interval

$$\mu_F : X \to \left[0,1\right]. \tag{11.2}$$

A fuzzy set is convex if and only if: $\mu_F(\lambda x_1 + (1 - \lambda)x_2) \geq \min(\mu_F(x_1), \mu_F(x_2))$ for all $x_1, x_2 \in X$ and $\lambda \in [0, 1]$. A fuzzy set F is normalized if $\sup(\mu_F) = 1$ otherwise it is called sub-normal.

A Fuzzy Number (FN) is a convex normalized fuzzy set of the real line R^1 whose membership function is piecewise continuous. A FN N is called positive, denoted by $N > 0$, if its memberships function $\mu_N(x) = 0$ for all $x < 0$.

The frequently used notation is based on Trapezoidal Fuzzy Numbers (TFN) (Grzegorzewski and Mrówka 2005). A positive TFN $\tilde{A}$ is denoted by 4-tuple (a^1, a^2, a^3, a^4) as follows:

$$\mu_{\tilde{A}}\left(x\right) = \begin{cases} \dfrac{x - a^1}{a^2 - a^1}, & a^1 \leq x \leq a^2 \\ 1, & a^2 \leq x \leq a^3 \\ \dfrac{x - a^4}{a^3 - a^4} & a^3 \leq x \leq a^4 \\ 0, & otherwise \end{cases} \tag{11.3}$$

If $a^2 = a^3$ then the TFN is called a Triangular FN (TrFN); if $a^1 = a^2 = a^3 = a^4$ then we have a classical crisp number.

For any two positive TFN $\tilde{A}\left(a^1,a^2,a^3,a^4\right)$ and $\tilde{B}\left(b^1,b^2,b^3,b^4\right)$ the fuzzy addition, subtraction, multiplication, and multiplication by a scalar are, respectively, defined as follows:

$$\tilde{A}\oplus\tilde{B}=\left(a^1+b^1,a^2+b^2,a^3+b^3,a^4+b^4\right) \tag{11.4}$$

$$\tilde{A}-\tilde{B}=\left(a^1-b^4,a^2-b^3,a^3-b^2,a^4-b^1\right) \tag{11.5}$$

$$\tilde{A}\otimes\tilde{B}=\left(a^1*b^1,a^2*b^2,a^3*b^3,a^4*b^4\right) \tag{11.6}$$

$$\beta\otimes\tilde{A}=\left(\beta*a^1,\beta*a^2,\beta*a^3,\beta*a^4\right) \tag{11.7}$$

The distance between two TFNs $\tilde{A}\left(a^1,a^2,a^3,a^4\right)$ and $\tilde{B}\left(b^1,b^2,b^3,b^4\right)$ can be computed as follows:

$$D\left(\tilde{A},\tilde{B}\right)=\sqrt{\frac{1}{4}\left[\left(a^1-b^1\right)^2+\left(a^2-b^2\right)^2+\left(a^3-b^3\right)^2+\left(a^4-b^4\right)^2\right]} \tag{11.8}$$

That becomes for TrFN $\tilde{A}\left(a^1,a^2,a^3\right)$ and $\tilde{B}\left(b^1,b^2,b^3\right)$ as follows:

$$D\left(\tilde{A},\tilde{B}\right)=\sqrt{\frac{1}{3}\left[\left(a^1-b^1\right)^2+\left(a^2-b^2\right)^2+\left(a^3-b^3\right)^2\right]} \tag{11.9}$$

11.6 Consensus

Do we need an agreement or a consensus? According to Amanda Machin, we only need to agree to disagree, since disagreement is a necessary step before achieving climate change actions (Machin 2013). We must not forget that it is not a number issue; 97% consensus is near agreement, while 10% consensus is far from being a consensus. The heap paradox, formulated in the 4th-century BC by Eubulide, can be seen in two ways: in the first, everyone can say that a grain of sand cannot be a pile. Adding a second grain does not make a heap either. By continuing in the same way by adding each time a grain, it is obvious that the final result will be a pile, since there seems to be no valid reason to stop at a given point rather than another (Black 1963). The second way— the decreasing one—starts from a pile of sand and stipulates that by removing a grain of sand, the result would always be a pile. The question that can be deduced is: what is the number of grains that classify a pile. This shows the absurdity of the conflict surrounding the exact level of consensus reached.

Ideally, consensus would be synonymous with unanimity. Such a situation is rarely or hardly attainable. In practice, it is the aggregation of opinions into one that best represents that of everyone (El Alaoui et al. 2019a). It is not a question of majority either. Suppose a situation where 10 persons would have to choose between two alternatives $A1$ and $A2$, supposing in addition than the votes for, against, and neutral were as follows (4, 0, 6) for the 1^{s+t} and (6, 3, 1) for the 2nd. If we are satisfied with the votes, for more people voted for the 2nd. However, the latter is far from being the object of consensus. The goal is to reach consent, not necessarily the agreement (Herrera-Viedma et al. 2014).

Several methods have been proposed to achieve consensus in a fuzzy context (Ding et al. 2017; El Alaoui et al. 2018a, b; Lee 2002; Li et al. 2019; Xu et al. 2016). The consensus procedure adopted from Lee (2002) and used in (El Alaoui et al. 2018c, 2019c) can be summarized as follows:

Let $\tilde{R}_k$ be the DM' expressed fuzzy opinions. The consensus is achieved through minimizing the following function:

$$\min_{M\times IR^4} \sum_{i=1}^{K} w_k^t * \left(c - S\left(\tilde{R}_k, \tilde{R}\right)\right), \tag{11.10}$$

where $M = \left\{ \begin{array}{c} W = \left(w_1, w_2, \ldots, w_K\right), w_k \geq 0, \\ \sum_{k=1}^{K} w_k = 1, \end{array} \right\}$

t is a positive integer $t > 1$, c is a real number $c > 1$, $\tilde{R}_k\left(r_k^1, r_k^2, r_k^3, r_k^4\right)$ are the individual opinions expressed by each DM. $\tilde{R}\left(r^1, r^2, r^3, r^4\right)$ is the aggregated consensus $S\left(\tilde{R}_k, \tilde{R}\right)$ the similarity between the ith decision and the consensus.

The similarity used was computed as a distance inverse as follows (El Alaoui and El Yassini 2020):

$$S\left(\tilde{R}_k, \tilde{R}\right) = 1 - \frac{1}{4u^p}\left(D\left(\tilde{R}_k, \tilde{R}\right)\right)^p \tag{11.11}$$

where $u = \max_{k,q}\left(r_k^q\right) - \min_{k,q}\left(r_k^q\right)$

In (El Alaoui et al. 2019d; El Alaoui 2021), it was proved experimentally that changing the normalizing term u by the width of the interval used can fasten the convergence.

This can be solved by the following algorithm:

Algorithm 11.1

Step 1: each DMs express his opinion by a FN (TFN or TrFN) $\tilde{R}_k$

Step$_K$ 2: fix the initial weights $W^{(0)}\left(w_1^{(0)}, \ldots, w_K^{(0)}\right)$ verifying $0 \leq w_k^{(0)} \leq 1$ and $\sum_{k=1} w_k^{(0)} = 1$. The iterations will be marked by $l = 0, 1, \ldots$

Step 3: compute

$$\tilde{x}^{(l+1)} = \frac{\sum_{k=1}^{K} w_k^{(l)q} \otimes \tilde{R}_k}{\sum_{k=1}^{K} w_k^{(l)q}} \tag{11.12}$$

Step 4: compute

$$w_j^{(l+1)} = \frac{\left(1/\left(c - S\left(\tilde{R}_k, \tilde{R}^{(l+1)}\right)\right)\right)^{1/(q-1)}}{\sum_{j=1}^{m}\left(1/\left(c - S\left(\tilde{R}_k, \tilde{R}^{(l+1)}\right)\right)\right)^{1/(q-1)}} \tag{11.13}$$

Step 5: if $\|W^{(l+1)} - W^{(l)}\| \leq \varepsilon$ stop, else set $l = l + 1$ and go to step 3.

11.7 Proposed Method and Illustrative Example

Sustainability requires merging different aspects, economic, social, environmental, that might be conflictual and extremely difficult to compare. Technique for Order Preference by Similarity to Ideal Solution (TOPSIS) is a powerful tool in such situations (El Alaoui et al. 2019b); it consists on choosing the solution at the same time, the closest to the best solution is called the Positive Ideal Solution (PIS), and the furthest from the worst solution is called the Negative Ideal Solution (NIS) (Hwang and Yoon 1981).

The problem can be formulated as follows:

K DMs $k = (1, \ldots, K)$, evaluate n alternatives $A = (A_1, A_2, \ldots, A_n)$, according to m criteria $C = (C_1, C_2, \ldots, C_m)$.

Each DM will assess criteria weight $\tilde{W} = \left(\tilde{w}_{jk}; j = 1,\ldots,m; k = 1,\ldots,K\right)$ and alternative in accordance to criteria of evaluation $R = \left(\tilde{R}_{ijk}; i = 1,\ldots,n; j = 1,\ldots,m; k = 1,\ldots,K\right)$

While obtaining the consensual collective evaluation will occur according to Algorithm 11.1. The rest will follow a classical TOPSIS algorithm as follows (El Alaoui et al. 2019b):

Step 1: each DM assesses criteria importance by linguistic variables.
Step 2: each DM assesses alternatives with respect to each criterion.
Step 3: convert the linguistic variables to TFN.
Step 4: construct the collective assessment for each alternative in accordance to each criterion $\tilde{x}_{ij}$ and the collective assessment for criteria weights $\tilde{w}_j$ using Algorithm 11.1.
Step 5: compute the weighted decision for each decision

$$\tilde{G}_{ij} = \tilde{x}_{ij} \otimes \tilde{w}_j ; i = 1,\ldots,n; j = 1,\ldots,m \tag{11.14}$$

Step 6: determine the PIS $\tilde{S}_j^+$ and the NIS $\tilde{S}_j^-$.
Step 7: compute the distance to each ideal solution

$$D_i^- = \sum_{j=1}^{m} D\left(\tilde{G}_{ij},\tilde{S}_j^-\right) \; i = 1,\ldots,n \tag{11.15}$$

$$D_i^+ = \sum_{j=1}^{m} D\left(\tilde{G}_{ij},\tilde{S}_j^+\right) \; i = 1,\ldots,n \tag{11.16}$$

Step 8: compute the closeness coefficients

$$CC_i = \frac{D_i^-}{D_i^- + D_i^+} \tag{11.17}$$

where D_i^- is the distance between the alternative assessment and the NIS and D_i^+ the distance between the alternative assessment and the PIS

Step 9: rank alternatives according to the closeness coefficients.

In the example treated in El Alaoui et al. (2019b, d), and Tan et al. (2010), three Agriculture Supply Chains (ASC) were evaluated to choose the most sustainable one. The selection process was held in accordance to four criteria (reliability, effectiveness cost, and asset management) by three DMs. While the final ranking obtained by all methods is similar, the sorting capacity in the methods proposed in (El Alaoui et al. 2019b, d)—taking into account parameter tuning—is clearly superior. Since, they best fit TOPSIS logic, making the best alternative both closer to the PIS and furthest from the NIS preventing any doubts. They also can prevent rank reversal due to the choice of ideal solutions.

11.8 Summary and Conclusions

Previously, human activities were considered insignificant compared to geological processes. Since the industrial revolution, GHG emissions, tree cutting, and so forth are among others that are producing a noticeable geological impact creating a new geological era called the Anthropocene. The era is by no means a danger to our planet, since it recovered from more violent phenomena in the past, but it is for humankind. Is there unanimity around the subject since science advances through unanimity or even consensus!? In this work, a review of climate change consensus is given in accordance with three levels: the scientific, the political, and public opinions. It is has been shown that the scientific level is near unanimity, while other perceptions of consensus are biased by misinformation, political interests, and so on. Are there uncertainties around? Yes there are, but not on the subject veracity, rather on possible repercussions. Fuzzy logic, as a powerful tool to represent uncertainty, was used to capture human's opinion in order to derive a consensus.

References

Anon (1983) Changing climate: report of the Carbon Dioxide Assessment Committee. National Academies Press, Washington, DC

Arvesen A, Bright RM, Hertwich EG (2011) Considering only first-order effects? How simplifications lead to unrealistic technology optimism in climate change mitigation. Energy Policy 39(11):7448–7454. https://doi.org/10.1016/j.enpol.2011.09.013

Bakshi BR (2019) Sustainable engineering: principles and practice. Cambridge University Press

Bedford D, Cook J (2013) Agnotology, scientific consensus, and the teaching and learning of climate change: a response to Legates, Soon and Briggs. Sci Edu 22(8):2019–2030. https://doi.org/10.1007/s11191-013-9608-3

Bender MJ, Simonovic SP (1997) Consensus as the measure of sustainability. Hydrol Sci J 42(4):493–500. https://doi.org/10.1080/02626669709492049

Berry EM, Dernini S, Burlingame B et al (2015) Food security and sustainability: can one exist without the other? Public Health Nutr 18(13):2293–2302. https://doi.org/10.1017/S136898001500021X

Bier VM (1990) Is fuzzy set theory more general than probability theory? In: Proceedings of ISUMA '90, the first international symposium on Uncertainty Modeling and Analysis, College Park, MD, 1990

Black M (1963) Reasoning with loose concepts. Dialogue 2(1):1–12. https://doi.org/10.1017/S001221730004083X

Bogaert P, van Oers H, Oyen HV (2018) Towards a sustainable EU health information system infrastructure: a consensus driven approach. Health Policy 122(12):1340–1347. https://doi.org/10.1016/j.healthpol.2018.10.009

Bolsen T, Druckman JN (2018) Do partisanship and politicization undermine the impact of a scientific consensus message about climate change? Group Processes Intergr Relat 21(3):389–402. https://doi.org/10.1177/1368430217737855

Braat L (1991) The predictive meaning of sustainability indicators. In: Kuik O, Verbruggen H (eds) In search of indicators of sustainable development, environment & management. Springer, Dordrecht

Bray D (2010) The scientific consensus of climate change revisited. Environ Sci Policy 13(5):340–350. https://doi.org/10.1016/j.envsci.2010.04.001

Bray D, von Storch H (2007) The perspectives of climate scientists on global climate change. GKSS, Geesthacht

Brewer PR, McKnight J (2017) "A statistically representative climate change debate": satirical television news, scientific consensus, and public perceptions of global warming. Atl J Commun 25(3):166–180. https://doi.org/10.1080/15456870.2017.1324453

Briggs B (2013) Introduction to consensus. Vervante

Brulle RJ (2018) The climate lobby: a sectoral analysis of lobbying spending on climate change in the USA, 2000 to 2016. Clim Change 149(3):289–303. https://doi.org/10.1007/s10584-018-2241-z

Caffey RH, Kazmierczak RF, Avault JW (2000) Developing consensus indicators of sustainability for southeastern United States Aquaculture. Available via SSRN. https://papers.ssrn.com/sol3/papers.cfm?abstract_id=242312. Accessed 20 May 2021

Carlton JS, Perry-Hill R, Huber M et al (2015) The climate change consensus extends beyond climate scientists. Environ Res Lett 10(9):094025. https://doi.org/10.1088/1748-9326/10/9/094025
Cartwright LE (2000) Selecting local sustainable development indicators: does consensus exist in their choice and purpose? Plan Pract Res 15(1–2):65–78. https://doi.org/10.1080/713691874
Chapman L (2007) Transport and climate change: a review. J Trans Geogr 15(5):354–367. https://doi.org/10.1016/j.jtrangeo.2006.11.008
Charney JG, Arakawa AD Baker J et al (1979) Carbon dioxide and climate: a scientific assessment. National Academy of Sciences. https://www.bnl.gov/envsci/schwartz/charney_report1979.pdf. Accessed 20 May 2021
Colston NM, Vadjunec JM (2015) A critical political ecology of consensus: on 'teaching both sides' of climate change controversies. Geoforum 65:255–265. https://doi.org/10.1016/j.geoforum.2015.08.006
Cook J, Nuccitelli D, Green SA et al (2013) Quantifying the consensus on anthropogenic global warming in the scientific literature. Environ Res Lett 8(2):024024. https://doi.org/10.1088/1748-9326/8/2/024024
Cook J, Oreskes N, Doran PT et al (2016) Consensus on consensus: a synthesis of consensus estimates on human-caused global warming. Environ Res Lett 11(4):048002. https://doi.org/10.1088/1748-9326/11/4/048002
Cook J, van der Linden S, Maibach E et al (2019) The consensus handbook. Center for Climate Change Communication. George Mason University
Corsi I, Winther-Nielsen M, Sethi R et al (2018) Ecofriendly nanotechnologies and nanomaterials for environmental applications: key issue and consensus recommendations for sustainable and ecosafe nanoremediation. Ecotoxicol Environ Saf 154:237–244. https://doi.org/10.1016/j.ecoenv.2018.02.037
Cortés G, Eugenia M (2014) Relaying the message of climate change: the IPCC experts comment on media coverage of scientific consensus. Interact Stud Commun Cult 5(1):41–50. https://doi.org/10.1386/iscc.5.1.41_1
Depledge J (2013) The organization of global negotiations: constructing the climate change regime. Routledge
Díaz Estévez M, de Frutos García RA, Peña Moya J (2014) Communication of the scientific consensus on climate change to the citizenship: knowledge and perception of Young University students from five countries regarding the media treatment of global warming. Interact Stud Commun Cult 5(1):51–70. https://doi.org/10.1386/iscc.5.1.51_1
Ding J, Xu Z, Liao H (2017) Consensus-reaching methods for hesitant fuzzy multiple criteria group decision making with hesitant fuzzy decision making matrices. Front Inf Technol Electron Eng 18(11):1679–1692. https://doi.org/10.1631/FITEE.1601546
Doelle M, Sinclair AJ (2006) Time for a new approach to public participation in EA: promoting cooperation and consensus for sustainability. Environ Impact Assess Rev 26(2):185–205. https://doi.org/10.1016/j.eiar.2005.07.013
Donini LM, Dernini S, Lairon D et al (2016) A consensus proposal for nutritional indicators to assess the sustainability of a healthy diet: the mediterranean diet as a case study. Front Nutr 3:37. https://doi.org/10.3389/fnut.2016.00037
Doran PT, Zimmerman MK (2009) Examining the scientific consensus on climate change. Eos, Trans Am Geophys Union 90(3):22–23. https://doi.org/10.1029/2009EO030002
Drewnowski A (2018) The Chicago consensus on sustainable food systems science. Front Nutr 4. https://doi.org/10.3389/fnut.2017.00074
Dumane PR, Sarate AD, Chavan SS (2019) Sustainability assessment by use of fuzzy logic—a review. In: Iyer B, Nalbalwar S, Pathak N (eds) Computing, communication and signal processing, Advances in intelligent systems and computing, vol 810. Springer, Singapore. https://doi.org/10.1007/978-981-13-1513-8_38
Eakin H, Connors JP, Wharton C et al (2017) Identifying attributes of food system sustainability: emerging themes and consensus. Agric Human Values 34(3):757–773. https://doi.org/10.1007/s10460-016-9754-8

El Alaoui M (2021) Fuzzy TOPSIS: logic, approach and case studies. CRC Press. 216p. ISBN: 978-0-367-76748-8. https://doi.org/10.1201/9781003168416
El Alaoui M (2018) SMART grid evaluation using fuzzy numbers and TOPSIS. IOP Conf Ser Mater Sci Eng 353(1):012019. https://doi.org/10.1088/1757-899X/353/1/012019
El Alaoui M (2020a) A fuzzy multiplicative performance indicator to measure circular economy efficiency. Int J Math Eng Manag Sci 5(6):1118–1127
El Alaoui M (2020b) Fuzzy goal programming for biodiesel production. Int J Green Energy 17(10):1–8. https://doi.org/10.1080/15435075.2020.1779075
El Alaoui M, Ben-azza H (2017) Aggregation of performance indicators for supply chain and fuzzy logic extensions applied to green supply chain. In: 2017 International Colloquium on Logistics and Supply Chain Management (LOGISTIQUA), Rabat, Morocco, 2017
El Alaoui M, El Yassini K (2020) Fuzzy similarity relations in decision making. In: Handbook of research on emerging applications of fuzzy algebraic structures, pp 369–385. https://doi.org/10.4018/978-1-7998-0190-0.ch020
El Alaoui M, Ben-azza H, Zahi A (2016) New multi-criteria decision-making based on fuzzy similarity, distance and ranking. In: Proceedings of the third international Afro-European Conference for Industrial Advancement — AECIA 2016, Advances in Intelligent Systems and Computing. Springer, Cham
El Alaoui M, Ben-azza H, El Yassini K (2018a) Optimal weighting method for fuzzy opinions. In: Proceedings of the International Conference on Industrial Engineering and Operations Management Paris, France, July 26–27, 2018
El Alaoui M, Ben-azza H, El Yassini K (2018b) Optimal weighting method for interval-valued intuitionistic fuzzy opinions. Notes Intuitionistic Fuzzy Sets 24(3):106–110. https://doi.org/10.7546/nifs.2018.24.3.106-110
El Alaoui M, El Yassini K, Ben-azza H (2018c) Enhancing MOOCs peer reviews validity and reliability by a fuzzy coherence measure. In: SCA '18: proceedings of the 3rd international conference on Smart City Applications, October 2018. https://doi.org/10.1145/3286606.3286834
El Alaoui M, Ben-azza H, El Yassini K (2019a) Achieving consensus in interval valued intuitionistic fuzzy environment. Procedia Comput Sci 148:218–225. https://doi.org/10.1016/j.procs.2019.01.064
El Alaoui M, Ben-azza H, El Yassini K (2019b) Fuzzy TOPSIS with Coherent measure: applied to a closed loop agriculture supply chain. In: Ezziyyani M (ed), Advanced Intelligent Systems for Sustainable Development (AI2SD'2018), Advances in intelligent systems and computing. Springer
El Alaoui M, El Yassini K, Ben-azza H (2019c) Peer assessment improvement using fuzzy logic. In: Ahmed MB, Boudhir AA, Younes A (eds) Innovations in Smart Cities applications edition 2, Lecture notes in intelligent transportation and infrastructure. Springer
El Alaoui M, El Yassini K, Ben-azza H (2019d) Type 2 fuzzy TOPSIS for agriculture MCDM problems. Int J Sust Agric Manag Inf 5(2/3):112–130. https://doi.org/10.1504/IJSAMI.2019.10023177
Ellis EC (2018) Anthropocene: a very short introduction. Oxford University Press, Oxford/New York
Figge F, Thorpe AS, Givry P et al (2018) Longevity and circularity as indicators of eco-efficient resource use in the circular economy. Ecol Econ 150:297–306. https://doi.org/10.1016/j.ecolecon.2018.04.030
Frame DJ, Stone DA (2013) Assessment of the first consensus prediction on climate change. Nat Clim Change 3(4):357–359. https://doi.org/10.1038/nclimate1763
Friman M (2016) Consensus rationales in negotiating historical responsibility for climate change. Int Environ Agreem Politics Law Econ 16(2):285–305. https://doi.org/10.1007/s10784-014-9258-1
Gil Solá A, Vilhelmson B, Larsson A (2018) Understanding sustainable accessibility in urban planning: themes of consensus, themes of tension. J Transp Geogr 70:1–10. https://doi.org/10.1016/j.jtrangeo.2018.05.010

Gödel K (1931) Über formal unentscheidbare Sätze der Principia Mathematica und verwandter Systeme I. Monatshefte für Mathematik und Physik 38(1):173–198. https://doi.org/10.1007/BF01700692

Grego P (2007) Venus and mercury, and how to observe them. Springer

Grzegorzewski P, Mrówka E (2005) Trapezoidal approximations of fuzzy numbers. Fuzzy Sets Syst 153(1):115–135. https://doi.org/10.1016/j.fss.2004.02.015

Gunaratna KL (2018) Climate change: the scientific consensus. In: Gunaratna KL (ed) Towards equitable progress: essays from a South Asian perspective, South Asia economic and policy studies. Springer, Singapore

Gupta J (1997) The climate change convention and developing countries: from conflict to consensus? Springer, Dordrecht

Hák T, Janoušková S, Moldan B (2016) Sustainable development goals: a need for relevant indicators. Ecol Indic 60:565–573. https://doi.org/10.1016/j.ecolind.2015.08.003

Herrera-Viedma E, Cabrerizo FJ, Kacprzyk J et al (2014) A review of soft consensus models in a fuzzy environment. Inf Fusion 17:4–13. https://doi.org/10.1016/j.inffus.2013.04.002

Houghton JT, Jenkins GJ, Ephraums JJ (1990) Climate change: the IPCC scientific assessment. Cambridge University Press, Cambridge

Hwang CL, Yoon K (1981) Multiple attribute decision making: methods and applications a state-of-the-art survey. Springer, Berlin/Heidelberg

Jørgensen S, Pedersen LJT (2018) The circular rather than the linear economy. In: RESTART sustainable business model innovation. Palgrave studies in sustainable business in association with future earth. Palgrave Macmillan, Cham. https://doi.org/10.1007/978-3-319-91971-3_8

Kay S (2008) Social consensus under attack. Int Union Rights 15(1):24–25

Kerr JR, Wilson MS (2018a) Changes in perceived scientific consensus shift beliefs about climate change and GM food safety. PLOS ONE 13(7):e0200295. https://doi.org/10.1371/journal.pone.0200295

Kerr JR, Wilson MS (2018b) Perceptions of scientific consensus do not predict later beliefs about the reality of climate change: a test of the gateway belief model using cross-lagged panel analysis. J Environ Psychol 59:107–110. https://doi.org/10.1016/j.jenvp.2018.08.012

Keseru I, Bulckaen J, Macharis C et al (2016) Sustainable consensus? The NISTO evaluation framework to appraise sustainability and stakeholder preferences for mobility projects. Transp Res Procedia 14:906–915. https://doi.org/10.1016/j.trpro.2016.05.070

Klir GJ (1987) Where do we stand on measures of uncertainty, ambiguity, fuzziness, and the like? Fuzzy Sets Syst 24(2):141–160. https://doi.org/10.1016/0165-0114(87)90087-X

Klir GJ (1989) Is There More to Uncertainty Than Some Probability Theorists Might Have Us Believe? Int J Gen Syst 15(4):347–378. https://doi.org/10.1080/03081078908935057

Kunkle KA, Monroe MC (2019) Cultural cognition and climate change education in the U.S.: why consensus is not enough. Environ Educ Res 25(5):633–655. https://doi.org/10.1080/13504622.2018.1465893

Landzettel M, Bommelt W (2019) La fin de l'alimentation: Comment le changement climatique va bouleverser ce que nous mangeons. Vuibert

Langan M, Price S (2016) The EU and 'pro-poor' contributions to sustainable development in the post-2015 consensus. Third World Thematics A TWQ J 1(4):431–436. https://doi.org/10.1080/23802014.2016.1327797

Lee HS (2002) Optimal consensus of fuzzy opinions under group decision making environment. Fuzzy Sets Syst 132(3):303–315. https://doi.org/10.1016/S0165-0114(02)00056-8

Lee GE, Loveridge S, Winkler JA (2018) The influence of an extreme warm spell on public support for government involvement in climate change adaptation. Ann Am Assoc Geogr 108(3):718–738. https://doi.org/10.1080/24694452.2017.1375888

Legates DR, Soon W, Briggs WM et al (2015) Climate consensus and 'misinformation': a rejoinder to agnotology, scientific consensus, and the teaching and learning of climate change. Sci Educ 24(3):299–318. https://doi.org/10.1007/s11191-013-9647-9

Li Y, Chen J, Feng L (2013) Dealing with uncertainty: a survey of theories and practices. IEEE Trans Knowl Data Eng 25(11):2463–2482. https://doi.org/10.1109/TKDE.2012.179

Li YL, Wang R, Chin KS (2019) New failure mode and effect analysis approach considering consensus under interval-valued intuitionistic fuzzy environment. Soft Comput. https://doi.org/10.1007/s00500-018-03706-5

Luckhurst J (2018) The shifting global economic architecture: decentralizing authority in contemporary global governance. Palgrave Macmillan

MacDonald GH, Abarbanel H, Carruthers P et al (1979) The Long term impact of atmospheric carbon dioxide on climate. JASON technical report JSR-78-07. SRI International, Arlington

Machin A (2013) Negotiating climate change: radical democracy and the illusion of consensus. Zed Books

Madumere N (2017) Public enlightenment and participation – a major contribution in mitigating climate change. Int J Sust Built Environ 6(1):9–15. https://doi.org/10.1016/j.ijsbe.2016.10.003

Maibach E, Myers T, Leiserowitz A (2014) Climate scientists need to set the record straight: there is a scientific consensus that human-caused climate change is happening. Earth's Future 2(5):295–298. https://doi.org/10.1002/2013EF000226

Matsuura M, Baba K (2016) Consensus building for long-term sustainability in the non-North American context: reflecting on a stakeholder process in Japan. Negot Confl Manag Res 9(3):256–268. https://doi.org/10.1111/ncmr.12076

Maurer EP, Adam JC, Wood AW (2009) Climate model based consensus on the hydrologic impacts of climate change to the Rio Lempa Basin of Central America. Hydrol Earth Syst Sci 13(2):183–194. https://doi.org/10.5194/hess-13-183-2009

McCown CA (2014) The development of consensus-based standards for wood preservatives/protectants and treated wood products. In: Deterioration and protection of sustainable biomaterials. vol 1158, ACS symposium series. American Chemical Society

Michaels PJ, Knappenberger PC (1996) The United Nations intergovernmental panel on climate change and the scientific "consensus" on global warming. Energy Environ 7(4):333–348. https://doi.org/10.1177/0958305X9600700403

Nierenberg N, Tschinkel WR, Tschinkel VJ (2010) Early climate change consensus at the national academy: the origins and making of changing climate. Hist Stud Nat Sci 40(3):318–349. https://doi.org/10.1525/hsns.2010.40.3.318

OECD (2006) Environment at a glance OECD environmental indicators: OECD environmental indicators. OECD Publishing

Oppenheimer M (2014) Warming goal: still the best indicator. Nature 514:434–434. https://doi.org/10.1038/514434a

Oreskes N (2004) The scientific consensus on climate change. Science 306(5702):1686–1686. https://doi.org/10.1126/science.1103618

Oreskes N (2018) The scientific consensus on climate change: how do we know we're not wrong? In: Lloyd EA, Winsberg E (eds) Climate modelling: philosophical and conceptual issues. Palgrave Macmillan. https://doi.org/10.1007/978-3-319-65058-6_2

Ouest, JS, Lafon C (2016) Notre quotidien en 2050: Comment le réchauffement climatique bouleverse notre futur. Journal Sud Ouest

Özkan İ, Türkşen IB (2014) Uncertainty and fuzzy decisions. In: Banerjee S, Erçetin SS, Tekin A (eds) Chaos theory in politics, understanding complex systems. Springer, Dordrecht. https://doi.org/10.1007/978-94-017-8691-1_2

Pearce W, Brown B, Nerlich B et al (2015) Communicating climate change: conduits, content, and consensus. Wiley Interdiscip Rev Clim Change 6(6):613–626. https://doi.org/10.1002/wcc.366

Pepermans Y, Maeseele P (2014) Democratic debate and mediated discourses on climate change: from consensus to de/politicization. Environ Commun 8(2):216–232. https://doi.org/10.1080/17524032.2014.906482

Planck M (1949) The meaning and limits of exact science. Science 110(2857):319–327. https://doi.org/10.1126/science.110.2857.319

Poff NL, Brown CM, Grantham TE et al (2016) Sustainable water management under future uncertainty with eco-engineering decision scaling. Nat Clim Change 6(1):25–34. https://doi.org/10.1038/nclimate2765

Powell J (2019) Scientists reach 100% consensus on anthropogenic global warming: bulletin of science. Technol Soc. https://doi.org/10.1177/0270467619886266

Reddock J (2017) Seeking consensus on universal health coverage indicators in the sustainable development goals. J Health Serv Res Policy. https://doi.org/10.1177/1355819617704676

Rodrigues MG (2015) Bringing local voices to the global negotiation table: norm dissemination and consensus building on tropical forests and climate change. New Glob Stud 9(2):125–157

Rosenzweig C (2013) Handbook of climate change and agroecosystems: global and regional aspects and implications. World Sci

Salehyan I (2008) From climate change to conflict? No consensus yet. J Peace Res 45(3):315–326. https://doi.org/10.1177/0022343308088812

Sapountzaki K, Wassenhoven L (2005) Consensus building and sustainability: some lessons from an adverse local experience in Greece. Environ Dev Sustain 7(4):433–452. https://doi.org/10.1007/s10668-004-2376-0

Sarzynski A (2015) Public participation, civic capacity, and climate change adaptation in cities. Urban Clim 14:52–67. https://doi.org/10.1016/j.uclim.2015.08.002

Segura-Salazar J, Tavares LM (2018) Sustainability in the minerals industry: seeking a consensus on its meaning. Sustainability 10(5):1429. https://doi.org/10.3390/su10051429

Shearman R (1990) The meaning and ethics of sustainability. Environ Manage 14(1):1. https://doi.org/10.1007/BF02394014

Sinclair AJ, Smith DL (1999) Policy review the model forest program in Canada: building consensus on sustainable forest management? Soc Nat Resour 12(2):121–138. https://doi.org/10.1080/089419299279795

Singh RK, Murty HR, Gupta SK, Dikshit AK (2012) An overview of sustainability assessment methodologies. Ecol Indic 15(1):281–299. https://doi.org/10.1016/j.ecolind.2011.01.007

Skalna I, Rębiasz B, Gaweł B et al (2015) Advances in fuzzy decision making – theory and practice. Springer, Cham

Smol JP, Douglas MSV (2007) From controversy to consensus: making the case for recent climate change in the arctic using lake sediments. Front Ecol Environ 5(9):466–474. https://doi.org/10.1890/060162

Stanhill G (2007) Climate change: controversy and consensus. Weather 62(5):137–137. https://doi.org/10.1002/wea.32

Subramanian M (2019) Anthropocene now: influential panel votes to recognize Earth's new epoch. Nature. https://doi.org/10.1038/d41586-019-01641-5

Tan Y, Cai Z, Qi H (2010) A process-based performance analysis for closed-loop agriculture supply chain. In: 2010 international conference on intelligent system design and engineering application. vol 1, 2010

Tol RSJ (2016) Comment on 'Quantifying the consensus on anthropogenic global warming in the scientific literature'. Environ Res Lett 11(4):048001. https://doi.org/10.1088/1748-9326/11/4/048001

Upham P (2000) Scientific consensus on sustainability: the case of the natural step. Sustain Dev 8(4):180–190. https://doi.org/10.1002/1099-1719(200011)8:4<180::AID-SD143>3.0.CO;2-R

van der Linden SL, Leiserowitz AA, Feinberg GD et al (2014) How to communicate the scientific consensus on climate change: plain facts, pie charts or metaphors? Clim Change 126(1):255–262. https://doi.org/10.1007/s10584-014-1190-4

van der Linden SL, Leiserowitz AA, Feinberg GD et al (2015) The scientific consensus on climate change as a gateway belief: experimental evidence. PLOS ONE 10(2):e0118489. https://doi.org/10.1371/journal.pone.0118489

van der Sluijs JP, van Est R, Riphagen M (2010) Beyond consensus: reflections from a democratic perspective on the interaction between climate politics and science. Curr Opin Environ Sustain 2(5):409–415. https://doi.org/10.1016/j.cosust.2010.10.003

Vihma A (2015) Climate of consensus: managing decision making in the un climate change negotiations. Rev Eur Comp Int Environ Law 24(1):58–68. https://doi.org/10.1111/reel.12093

von Schirnding YE (2002) Health-and-environment indicators in the context of sustainable development. Can J Public Health 93(1):S9–S15. https://doi.org/10.1007/BF03405112

Wallis AM (2006) Sustainability indicators: is there consensus among stakeholders? Int J Environ Sustain Dev 5(3):287–296. https://doi.org/10.1504/IJESD.2006.010898

Walther GR, Hughes L, Vitousek P et al (2005) Consensus on climate change. Trends Ecol Evol 20(12):648–649. https://doi.org/10.1016/j.tree.2005.10.008

World Commission On Environment and Development (1987) Our Common Fut, 1st edn. Oxford University Press

Xu W, Khoshroo N, Bjornlund H et al (2014) Effects of 'grain for green' reforestation program on rural sustainability in China: an AHP approach to peasant consensus of public land use policies. Stoch Environ Res Risk Assess 28(4):867–880. https://doi.org/10.1007/s00477-012-0592-y

Xu Y, Sun H, Wang H (2016) Optimal consensus models for group decision making under linguistic preference relations. Int Trans Oper Res 23(6):1201–1228. https://doi.org/10.1111/itor.12154

Zadeh LA (1965) Fuzzy sets. Inf Control 8(3):338–353. https://doi.org/10.1016/S0019-9958(65)90241-X

Zheng K, Lou HH, Huang Y (2013) Sustainability under severe uncertainty: a probability-bounds-analysis-based approach. In: Jawahir IS, Sikdar SK, Huang Y (eds) Treatise on sustainability science and engineering. Springer, Dordrecht

Chapter 12
Robot Path-Planning Research Applications in Static and Dynamic Environments

Firas A. Raheem, Safanah M. Raafat, and Shaymaa M. Mahdi

Abstract The purpose of path planning is primarily to find a set of continuous motions for moving robots from an initial point and ending at the goal point in Cartesian or configuration space. Solving the robot arm path-planning problem can be considered as one of the most significant aspects of robot navigation to guarantee the best path free of obstacle collisions before tracking. The complexity of the robot environment is one of the most important characteristics of robots, especially in environments that contain static and/or dynamic obstacles such as robots, human operators, and other moving objects. The problem of path planning is subjected to the constraints of finding the possible path. The constructed path satisfies specific optimization criteria if the environment is known or has a property of intelligence if the environment is unknown. Many recent path-planning approaches have optimization and intelligent properties when the robot environment is partially unknown. In addition to intelligent methods that are used for planning a successful path, recent advances in path-planning methodologies involve using both heuristic and meta-heuristic with artificial intelligence concepts with relevance across various field types with great application in calculated optimization of sustainability and resource use.

Keywords Robot arm path planning · Robot environment · Optimization · Intelligent methods · Metaheuristic · Dynamic · Static

F. A. Raheem (✉) · S. M. Raafat · S. M. Mahdi
Automation and Robotics Research Unit, Control and Systems Engineering Department, University of Technology-Iraq, Baghdad, Iraq
e-mail: firas.a.raheem@uotechnology.edu.iq; safanah.m.raafat@uotechnology.edu.iq; shaymaa.m.mahdi@uotechnology.edu.iq

J. N. Furze et al. (eds.), *Earth Systems Protection and Sustainability*,
https://doi.org/10.1007/978-3-030-85829-2_12

12.1 Introduction to Path Planning

In any robot workspace, the solution of the motion control problem includes firstly finding a suitable path which can be indicated as computing the movement of any robot joint continuously, which enables a collision-free robot motion with all located obstacles in its environment from the start point or configuration to the goal point or goal configuration at any chosen point of its motion path.

Planning a path requires a map depicting the robot and its surrounding environment, so that the robot knows its position on the map and can determine its location using the map to avoid transitory obstacles on its path. How to find the robot location and how to take into consideration the undefined location information will be the core of the rest of this chapter.

In a cluttered and crowded environment with obstacles, path-planning solutions for manipulators and autonomous robots and the determination of optimum path are considered to be attractive research topics. Most research focuses on the working conditions of the robot and the application of path-finding steps in fully known static or dynamic environments. In fractionally known or totally undiscovered, uncharted, unexplored environments, the steps of planning a path are difficult and cannot be applied directly. The robot encounters uncertainties, reflected in how it makes decisions to continue moving through unexplored and explored areas of the robot workspace. These types of planning methods are related to known areas and operate in a work zone where the target point is placed. Moreover, we will discuss the type of data structure required to support outstanding algorithms for path planning.

Before considering path-planning research, it is useful to resolve the problem of solving path-planning issues from a human perspective – how people plan paths.

From the perspective of investigating path-planning issues, it should be noted that the robot path planner must obtain knowledge of the strategy used and the related features that must be considered when adjusting the route-planning algorithm (Hameed 2019).

12.2 Path-Planning Concept with Obstacle Avoidance

Planning a path is a subtask of general motion planning tasks. Finding a collision-free path from the initial configuration to the target configuration is a purely geometric challenge. Obstacle avoidance means that the robot moves in its workspace with its diverse mobility and controllability without colliding with environmental obstacles. The task of planning paths and avoiding obstacles is more difficult for manipulators than for mobile robots; the problem is not only to find the path of the end effector but also to avoid collisions with manipulator links. Path planning requires information of the size of the robot, first and final robot configurations, work environment, and static stationary obstacles.

Planning a trajectory can be considered independently from path planning. Trajectory planning is defined as the problem of determining the path as a function of time, which means determining the velocity and the acceleration of the robot at each point (Hussain 2017).

12.3 Path-Planning Classifications

The solution of robot path-planning task can be divided into five categories: (i) planning type, (ii) planning time, (iii) environment type, (iv) obstacle motion behavior, and (v) according to the robot space used for trajectory computations. Category (i) includes global path planning and local path planning. Through global path planning, the robot knows all the information in the work area prior to planning and can directly determine its path without collision. Global path planning is sometimes referred to as a deliberative method. Planning based on robot sensors' information measured at each sample of motion is referred to as local path planning where the environment is incomplete or unknown and the robot uses feedback from sensors to find the path. Category (ii) includes planning a path depending on the time to execute the robot motion and can be divided into online planning and offline planning. Online real-time planning affects robot movement in systematic environments. Offline planning provides a complete path before the robot's motion. Category (iii) includes path planning subject to the environment type and can be classified into known, totally unknown, and partially known environments. Category (iv) depends on the obstacle motion behavior and can be divided into static stationary obstacles and dynamic or movable obstacles (Hussain 2017). Category (v) divides the robot path and trajectory computations according to Cartesian space, joint space, and configuration space.

In robotics, trajectory planning strategies are computed and combined together according to the different situations of the robot and its environment that match the above classifications (Hameed 2019); these classifications are described in detail in the rest of this section.

12.3.1 Path Planning According to Obstacles

- Static obstacles
- Moving or dynamic obstacles

12.3.2 Path Planning Depends on Environment and Obstacle Types

- Known static environment
- Known dynamic environment
- Partially known static and dynamic environment
- Unknown static environment
- Unknown dynamic environment

12.3.3 Types of Planning Methodologies

- Global planning: This methodology assumes that the total information that represents the robot environment is known and available.
- Local planning: This methodology uses only a part of the global model to find the robotic path and control the motion, which is a disadvantage due to the loss of path optimization cases. The local plan is effectively suffering from local minima.

A preferred key point of the local and global methods is that local planning calculations are not as complex as global calculations. This is especially important when the world model is updated with each sample due to sensor collision data measurements which pose a known problem to autonomous robots.

12.3.4 Path Planning According to Robot Space

- Cartesian space
- Joint space
- Configuration space (C-space)

12.3.5 Path According to Planning Time

- Online planning
- Offline planning

Path-planning schema of classifications and approaches is shown in Fig. 12.1.

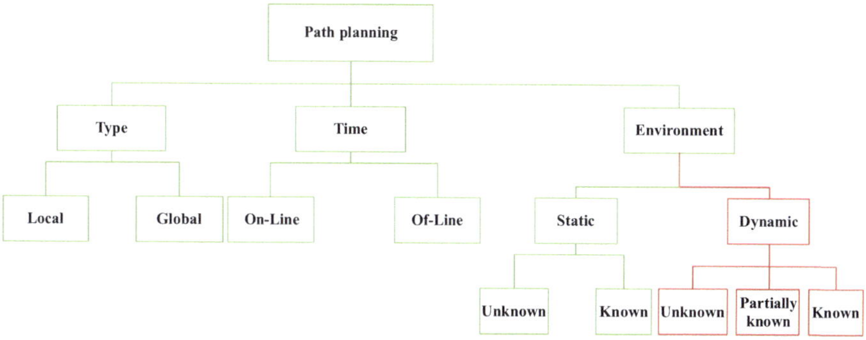

Fig. 12.1 Classifications and approaches of path planning

12.4 Cartesian Space Path Planning

Robot workspace or Cartesian space is the space in which the robot operates. Generally, the work area is the part of the environment in which the robot executes a motion plan. The planning problem in Cartesian workspace is the movement of an end effector of the manipulator on a specific path and specific trajectory. The advantage of path planning in Cartesian workspace is that it is conceptually straightforward, and the required configuration of the end effector is easy to sense and understand. Moreover, it is easy to imagine, define, and determine the motion of the manipulator end effector. Inverse kinematics is applied multiple times at each point along the trajectory to obtain values of robot joints. Commonly, when using Cartesian planning algorithms, the planned path passes through singular configurations of the robot manipulator, or it may override the manipulator's reachable workspace. In practice, in order to kinematically avoid the singularity problem of the manipulator, it is necessary to provide a sufficient number of via points inside the workable area. Cartesian trajectory planning is ordinarily used for precise local motion (Hussain 2017).

12.5 Path Planning Using Configuration Space

The main problem of planning a robotic path is finding a continuous, collision-free path for the robot from start to target configuration. Configuration space (C-space) can be considered an important tool for finding and formulating the path solution. C-space points are computed mathematically, disregarding all singular points so that the planned path represents a unique inverse kinematic solution, without redundancy or undefined states. Another benefit of using C-space for path planning is that a final path can be summarized easily in a unified method of finding the appropriate map that converts the robot's complex geometry to a single C-space point. The main difficulty of using C-space planning is that the intricacy of the path-planning difficulties raises exponentially with increasing number of degrees of freedom (DOF), which leads to increasing the C-space dimensions and the required parameters to specify the robot configuration.

The first contribution to precise C-space path planning was made by Lozano-Perez. Lozano-Perez and Wesley proposed simplification of complex robot geometry shape to a defined point in "configuration space" (C-space). In his path-planning method, two sub-problems were specified: (i) finding the C-space map and (ii) finding the appropriate collision-free path. Configuration space requires intensive computations and is often overlooked by researchers (Hussain 2017), though is a useful tool for finding path solutions.

12.6 Heuristic Methods

12.6.1 A Algorithm*

The A* algorithm is the most commonly used search algorithm for finding the shortest path. It represents an extension of Dijkstra's algorithm. The usage of the heuristic process is what makes the A* different from other graph search algorithms. The heuristic provides an estimated distance from a current node to the goal node, denoted by $h(n)$. The A* takes into account the cost from the start node to the current node, denoted as $g(n)$. The cost function is determined as follows (Raheem and Hussain 2017a, b):

$$f(n) = g(n) + h(n) \tag{12.1}$$

Each C-space node in the total map is represented by a joint angle pair, $n : (\theta_1,\ \theta_2)$. The A* algorithm includes two lists: open list (O_List) and closed list (C_List). It chooses the nodes that have minimum cost functions to create a path from the start node to the goal node.

12.6.2 D Algorithm*

Algorithm D* or "dynamic algorithm A*" behaves similarly to algorithm A*. The cost of the arc changes dynamically during the execution of the algorithm. The D* algorithm uses a Cartesian grid made up of eight-connected nodes to attain the robot position. The D* algorithm can be used to solve the path-planning problem based on the assumption that the robot must move in free space from the start point and continue exploring the area in order to find the minimum-cost path until reaching the destination point. The D* algorithm analyzes the robot's environment and expresses the problem space as a series of cells that indicate the position of the robot with the corresponding arcs or connections (Choset 2007; Nosrati et al. 2012).

Each cell in the D* algorithm contains an estimated cost of the path to the destination point and a back-pointer to one of its neighbor cells as an indicator of the destination direction. The arc's direction (connections) between a pair of neighboring cells is assigned with a positive scalar value of the motion cost (Choset 2007; Nosrati et al. 2012).

Algorithm D* uses a reverse direction search strategy, in which the D* algorithm forces the robot to start searching from the target cell and moves backward to other cells along arcs until it reaches the starting cell by repeatedly selecting a cell from the open list to evaluate and calculate the path cost to the goal. The D* algorithm moves from cell to cell along arcs and repeatedly selects a cell from the open list to evaluate and calculate the cost to the target cell. Finally, eight minimum cost neighboring cells are put on the open list (Raheem et al. 2019).

The D* algorithm sets conditions that specify when all alterations are made, either to find a new path or to continue using an old optimal path that has been chosen previously. D* is computationally efficacious and memory active and can be applied in unrestrained areas. Formulating the conditions for repeating the optimal search path problem in a directed graph includes marking the arcs by the transition cost values to prepare a cost range on the continuum. The arc cost corrections that are normally measured by sensors can be saved for a little time, while the well-known, approximately evaluated and calculated arc costs are the elements that make the environment map. The D* algorithm ensures keeping a closed list to save the path nodes and obstacles nodes and opens an expansion list for further path computations. In the D* algorithm, the arc cost parameter changes dynamically at each iteration. The D* algorithm sense is dynamic, and the path traverse cost can be minimized or maximized dynamically. This strategy can be used for any plan computation, including visibility graphs and grid-cell structures (Hameed 2019; Stentz 1994). The objective function (F) D* can be illustrated by A*:

$$F = g + h \tag{12.2}$$

where the cost estimated from the initial point is g while h is the motion cost to the target point.

Cost changes to neighbors are propagated as shown in Fig. 12.2a, b.

$$\mathrm{N}(\mathrm{x1}, \mathrm{x2}) = 1$$

$$\mathrm{N}(\mathrm{x1}, \mathrm{x3}) = 1.414$$

$$\mathrm{N}(\mathrm{x1}, \mathrm{x4}) = 100000, \text{in case of x4 has an obstacle and x1 is a free cell.}$$

$$\mathrm{N}(\mathrm{x1}, \mathrm{x5}) = 100000.4, \text{in case of x5 has an obstacle and x1 is a free cell}$$

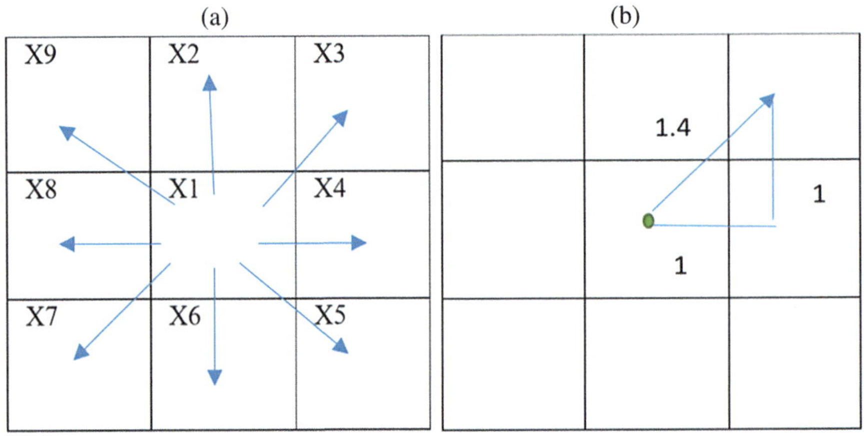

Fig. 12.2 Node expansion and calculation. (**a**) Node expansion, (**b**) node expansion calculation

The initial node has an arc cost equal to (0.0), and the arc cost of the vertical and horizontal nodes can be computed as:

$$\text{From x1 to x2} = \sqrt{(0-0)^2 + (0-1)^2} = 1$$

$$\text{From x1 to x4} = \sqrt{(0-1)^2 + (0-0)^2} = 1$$

In case of a diagonal nodes calculation, the arc cost is:

$$\text{From x1 to x3} = \sqrt{(0-1)^2 + (0-1)^2} = \sqrt{2} = 1.414 \text{ and so on.}$$

12.7 Theory of Particle Swarm Optimization (PSO)

PSO is an optimization method that is considered a stochastic onerous technique akin to the intelligent flying motion of birds. PSO was inspired by natural social and dynamic motion behavior, as well as the communication of fish, insects, or birds. Convergence speed and quality of solutions are advantages of the technique (Raheem and Hameed 2018; Boonyaritdachochai et al. 2010).

The principle of PSO relies on generating a fixed number of particles at random positions in a certain workspace, with velocity of particles specified randomly. Each particle has a memory that stores all the best cell locations that have been visited prior to the current cell location; and stores the improved fitness over time (Poli et al. 2007). In each PSO method iteration, the pos_{ij} and $vel^t_{i,j}$ vectors of particle i

are adjusted for each j dimension to make the particle i toward either the previous best vector $pbest_{ij}$ or toward the best vector that represent the swarm's ($gbest_{ij}$).

$$vel_{i,j}^{t+1} = w \times vel_{i,j}^{t} + c_1 \times r_{1,j} \times \left(pbest_{ij} - pos_{ij} \right) + c_2 \times r_{2,j} \times \left(gbest_{ij} - pos_{ij} \right) \quad (12.3)$$

Each particle (the point that is chosen from D* path) has an updated position according to the equation below, using the new velocity for that particle:

$$pos_{ij}^{t+1} = pos_{ij}^{t} + vel_{ij}^{t+1} \quad (12.4)$$

where the cognitive coefficients $c_1$and c_2 are applicable under the condition ($c_1 + c_2 \leq 4$) and $r_{1,j}$ and $r_{2,j}$ are real numbers that can take a value between [0, 1] randomly, while controlling the momentum of the particle can be done by the weight of inertia w. The velocity vel_{ij} can take a value inside the range [$-vel_{max}$, vel_{max}] to decrease leaving chance of being out of the search space by the particle. Given a space of search that can be identified within the bounds [$-pos_{max}$, pos_{max}], then the vel_{max} value setting is usually $vel_{max} = k * pos_{max}$, where $0.1 \leq k \leq 1.0$.

A weight (w) that represents large inertia has two cases when its value is small, making the global search easier. A large w value simplifies localization (Poli et al. 2007).

12.8 Research Case Studies

12.8.1 Heuristic Path-Planning Enhancement Based on Free Cartesian Space Analysis

In heuristic path planning, when free Cartesian space analysis is known, the shortest path and trajectory planning of the two-link robot arm with 2 DOF in the 2D static known environment can be analyzed. The analysis deals with three main problems. The first concern regards the construction of free Cartesian space by analyzing the inverse kinematic solutions, which guarantees collision-free path planning. The second problem focuses on generating the shortest path that satisfies the aims of motion and applies the D* algorithm. The third problem is the selection of the specified number of intermediate via points and attaining the corresponding smooth trajectory through using fifth-order polynomial equations. Results illustrate that free Cartesian space ensures a collision-free path and trajectory planning (Raheem et al. 2019).

12.8.2 Implementation of D* Algorithm for Path Planning

The modification of the D* algorithm is proposed for solving the problem of generating the shortest collision-free path for a 2-DOF planar robotic arm in a known stationary (static) environment. The primary principle of applying the D* algorithm in proposed robot arm path planning makes use of navigation and environment and analyzes criteria to construct the short path of the arm end effector subject to certain constraints regarding safety and smoothness.

The proposed method is composed of three stages: free Cartesian space analysis followed by path construction and smooth trajectory generation. In the first stage, the proposed system is initialized by selecting a robot arm environment where the D* algorithm plans the shortest path. Subsequently, the analysis of the corresponding free Cartesian space has been made according to the free Cartesian space analysis procedure in order to reveal free space, ensuring a collision-free path planning (Raheem et al. 2019).

In the second stage of path construction, the D* algorithm is applied to generate the shortest path from start to goal points by avoiding all the obstacles. The principle of operation of the proposed D* algorithm is local planning for the shortest path within the free Cartesian space until it reaches the goal point. Moreover, the D* algorithm is initialized by placing the start "current" node on the open list, which is inserted into the currently planned path. Subsequently, the current node is expanded to eight connected neighborhood nodes for determining the next arm movement or the "candidate next node." However, only nodes that belong to the free Cartesian space are used. Ultimately, this iterative process of node expansion and selection is terminated as soon as the goal point is appended to the closed list (Raheem et al. 2019).

In the third stage of the method, fifth-order polynomial equations are applied for smooth trajectory generation based on the generated path of the D* algorithm. Initially, several intermediate points are randomly selected from the shortest path generated by the D* algorithm to be added to the initial and goal points. Then quintic fifth-order polynomial equations are used to add the required smoothness action and guarantee generating a smooth trajectory at a specific time according to equations (Raheem et al. 2019; Sreenivasulu 2012), after transforming the intermediate points to a joint angle using an inverse kinematics function. Consequently, a checking function is applied for checking and testing the generated segments within the free Cartesian space. In the case of being outside the free Cartesian space, the process iteratively selects and generates another point and corresponding trajectory until achieving a possible trajectory whose segments belong to the free Cartesian space. Subsequently, the cost value is calculated for the trajectory according to the following equation:

$$\text{Trajectory} - \text{Cost} = \sum_{i=1}^{k} \sqrt{\left(x_i - x_{i+1}\right)^2 + \left(y_i - y_{i+1}\right)^2} \tag{12.5}$$

where k is the number of the configuration that is equal to the amount of specific time along the trajectory, (x_i, y_i) representing the current coordinate of ith points of the trajectory.

12.8.3 Simulation Result

Environment maps with various difficulties were used to test and verify the proposed method. The dimension limits of the environments from −50 to 50 cm for both x and y contain various shapes of static obstacles. Moreover, both robot arm links have an equal length of 25 cm. In addition, the suggested joint variable limitations are specified as $0 \leq \theta_1 \leq 360$ and $0 \leq \theta_2 \leq 360$. Initially, the procedure of the free Cartesian space analysis has been applied to verify the required analysis for the environments (as shown in Fig. 12.3), in order to discover the free space and ensure finding a collision-free planning path. Accordingly, inverse kinematics functions are applied to construct offline planning both of the free elbow-up solution area and free elbow-down solution area in the Cartesian space (Raheem et al. 2019).

The final smooth trajectory of the robot is shown in Fig. 12.3 which is the result of applying the equations of the quintic polynomial trajectory. This clarifies the difference between the planned path of the D* algorithm and the smoothed generated final trajectory; green points indicate the intermediate via points that have been selected from the planned D* path (Raheem et al. 2019).

Figure 12.4 demonstrates the movements of the arm that follow the shortest path, where the red line indicates the first link of the robot arm and the blue line indicates the second link of the robot arm.

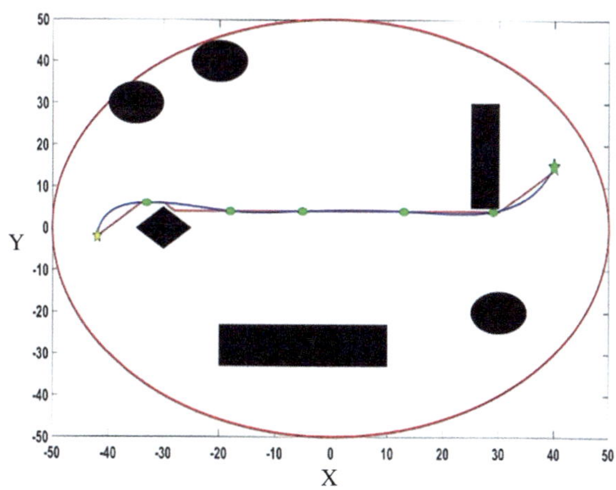

Fig. 12.3 The planned D* path and generated trajectory. (Raheem et al. 2019).

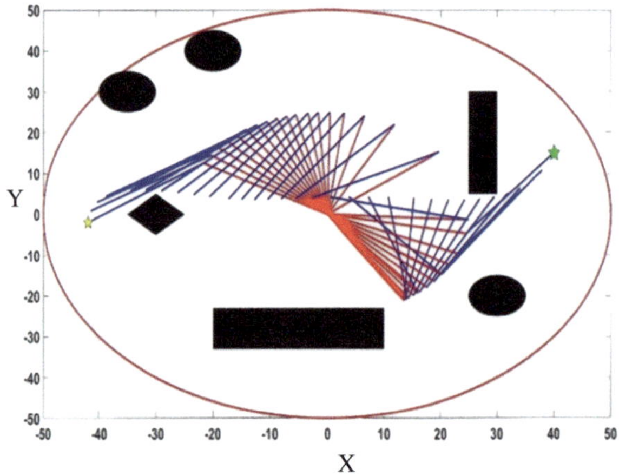

Fig. 12.4 Two-link robot arm movements. (Raheem et al. 2019).

12.9 Artificial Potential Field (APF) Based on PSO for Factor Optimization

An improved and comprehensive hybrid method for robot path planning is proposed, which firstly uses PSO for determining the best value for each factor of the artificial potential field (APF) and makes a progressive improvement in the resultant path in an iterative process until the shortest path is found. A spline equation was used for smoothing the path and producing a final smoothed trajectory. The results clearly show the strength and the efficiency of the hybrid PSO and APF method (Raheem and Badr 2017).

12.9.1 Artificial Potential Field Theory

In a mass point path-planning case, q refers to the position coordinate of the robot moving in a 2D environment. The current position coordinate of the robot is referred to as $q = [x\ y]$, while the obstacle position coordinate is referred to as $q_{obs} = (x_{obs},\ y_{obs})$; similarly the coordinate of the goal position is denoted by $q_{goal} = (x_{goal},\ y_{goal})$. The parabolic shape is the general style of the artificial potential field function, where Fig. 12.5a shows the attractive potential, which increases quadratically with the distance to the target point (Raheem and Badr 2017).

$$U_{att}(q) = \frac{1}{2} k_a\, d^2\left(q, q_{goal}\right) \tag{12.6}$$

where k_a refers to the relative factor that constructs the attractive potential surface, $d(q,\ q_{goal})$, which is the Euclidean distance from the current position of the robot to

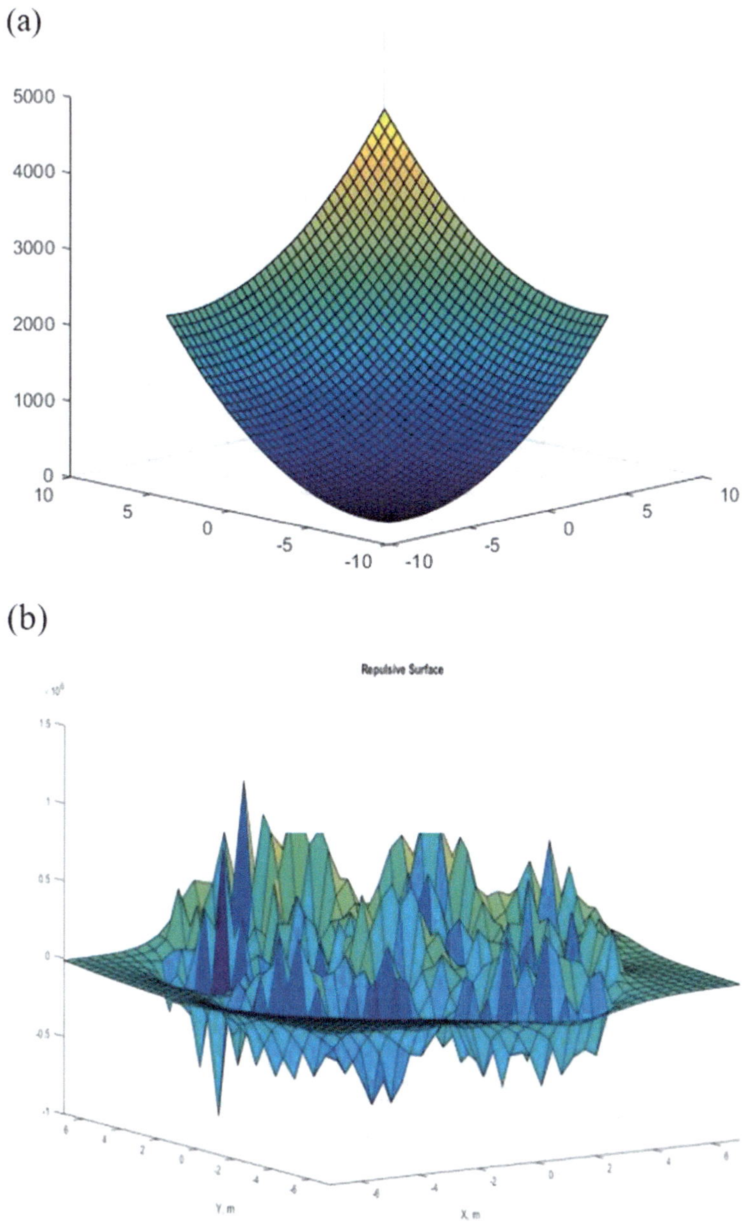

Fig. 12.5 Potential artificial fields, (**a**) attractive surface, (**b**) repulsive surface

the goal point (target) q_{goal}. The attractive force can be calculated as the attractive potential field negative gradient (Raheem and Badr 2017):

$$F_{att}(q) = -\nabla U_{att}(q) = -k_a \, d(q, q_{goal}) \tag{12.7}$$

There is a relative relationship between the repulsive force and the obstacle distance to the robot position. The surface of the potential repulsive was constructed using the repulsive forces produced from all obstacles located within the environment. The repulsive potential function can be represented by Eqs. (12.3) and (12.4), while Fig. 12.5b represents the corresponding repulsive surface (Raheem and Badr 2017).

$$U_{rep}(q) = \sum_i U_{repi}(q) \tag{12.8}$$

where $U_{repi}(q)$ denotes the repulsive potential field which is constructed by the obstacles and i represents obstacle number.

$$U_{rep}(q) = \begin{cases} \frac{1}{2} K_{rep} \left(\frac{1}{d(q, q_{obs})} - \frac{1}{d_0} \right)^2 d(q, q_{goal})^n & if \; d(q, q_{obs}) < d_0 \\ 0 & if \; d(q, q_{obs}) > d_0 \end{cases} \tag{12.9}$$

The variable q represents the current position of the robot, n refers to a real integer number, the obstacle position is q_{obs}, and d_0 refers to a positive number, which represents the distance to the efficacious obstacle; a distance that measured between all obstacles and the robot can be denoted as $d\ (q,\ q_{obs})$, while the repulsive potential surface factor which is an adaptable constant is represented in Eq. (12.9) as K_{rep}. The overall repulsive force has a negative slope due to its nature as explained in Eq. (12.10) (Gue et al. 2013):

$$F_{rep}(q) = -\nabla U_{rep}(q) \begin{cases} K_{rep} \left(\frac{1}{d(q, q_{obs})} - \frac{1}{d_0} \right) \frac{(q, q_{obs})}{d^{3(q, q_{obs})}} & if \; d(q, q_{obs}) < d_0 \\ 0 & if \; d(q, q_{obs}) > d_0 \end{cases} \tag{12.10}$$

The two surfaces, the attractive potential surface U_{att} and the repulsive potential surface U_{rep} , can be combined into the total potential field as shown in Eq. (12.11):

$$U(q) = U_{att}(q) + U_{rep}(q) \tag{12.11}$$

The applied forces toward the robot are produced from the negatively gradient that utilizes a well-known method called a steepest descent to drive the direction of the robot to its desired target value.

$$F(q) = -\nabla U(q) = -\nabla U_{att}(q) - \nabla U_{rep} \tag{12.12}$$

where ∇U represents the gradient vector of U, while the total impact force that affects the robot work can be calculated as the two components' summation, the attractive vector force and the repulsive vector force, F_{att} and F_{rep}, respectively (Li et al. 2000):

$$F(q) = F_{att}(q) + F_{rep}(q) \tag{12.13}$$

12.9.1.1 Proposed Method

The optimization of the artificial potential field factor (APF) uses the PSO technique. Two modifications have been carried out. The force and its direction in the ordinary artificial potential field will be initially modified. The environment can be represented and converted to a grid of points. Each point on the grid has a force that is computed from two sources: the attractive force due to the goal point and the repulsive force due to the obstacle (if the range of influence covers these points). Each point can be affected by two forces, firstly toward the x-axis direction and secondly toward the y-axis direction. Some researchers change these forces to impose the path of the robot according to Eqs. (12.14), (12.15), (12.16), and (12.17) (Raheem and Badr 2017):

$$F_{x\_total} = F_{x\_att} + F_{x\_rep} + F_{y\_rep} \tag{12.14}$$

$$F_{y\_total} = F_{y\_att} + F_{y\_rep} - F_{x\_rep} \tag{12.15}$$

$$F_{x\_total} = F_{x\_att} + F_{x\_rep} - F_{y\_rep} \tag{12.16}$$

$$F_{y\_total} = F_{y\_att} + F_{y\_rep} + F_{x\_rep} \tag{12.17}$$

The results of Eqs. (12.14) and (12.15) (Fig. 12.6) and Eqs. (12.16) and (12.17) (Fig. 12.7) clarify that all forces have one direction usually surrounding the obstacle.

In this work, the third term in Eqs. (12.14) through (12.17) has been removed; another term has been added, which was found by a trial and error tuning process to be appropriate to specific cases; and this was added to each term in the equation making the overall force in the coordinates of x-axis and y-axis going toward the target point and simultaneously ensuring a collision-free path as presented in Eqs. (12.18) and (12.19) and Fig. 12.8:

$$F_{x\_total} = F_{x\_att} + F_{x\_rep} - 0.75 \tag{12.18}$$

$$F_{y\_total} = F_{y\_att} + F_{y\_rep} + 0.5 \tag{12.19}$$

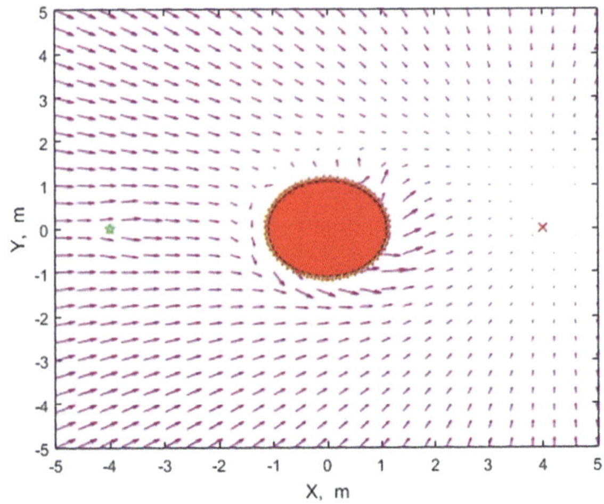

Fig. 12.6 The forces' direction according to Eqs. (12.14) and (12.15). (Raheem and Badr 2017)

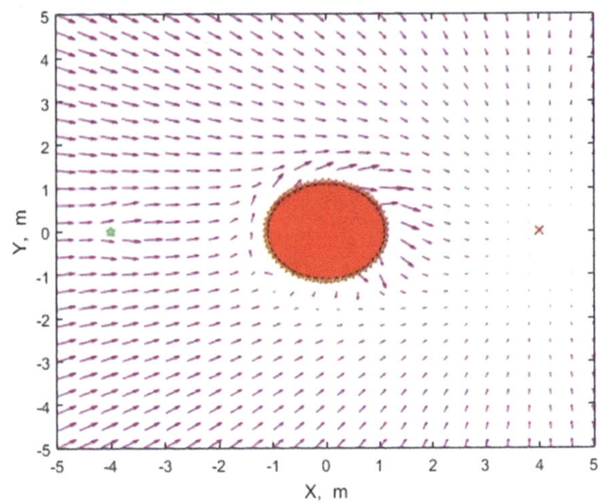

Fig. 12.7 The forces' direction according to Eqs. (12.16) and (12.17). (Raheem and Badr 2017)

The second modification includes focusing on the APF attractive force factor (k_a) and repulsive force factor (k_r). This modification plays a significant function in this proposed work. In case the value of the attractive force is very high, only its effectiveness will be considered. Furthermore, if it has a small value, then its effect can be disregarded. PSO is applied to determine the forces; the repulsive force factor will affect a certain particular area, which indicates the APF influence around obstacles that need to be optimized. The mathematical application of the PSO algorithm is according to Raheem and Badr (2017).

The details of the proposed method include the computation of the factors of both attractive and repulsive forces and can be listed in accordance with the following ten steps:

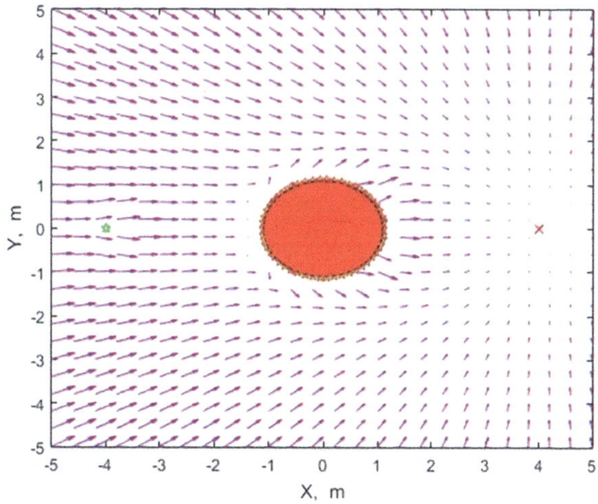

Fig. 12.8 The forces' direction according to Eqs. (12.18) and (12.19). (Raheem and Badr 2017)

1. Initialize the PSO parameters and specify the APF factors (K_a, K_{rep}) and their suitable ranges.
2. Initiate the generation APF factors randomly according to the suggested extents.
3. Perform the algorithm APF method instantly by using the factors that are randomly generated after the construction of the APF, and then find a possible path with the calculations of the cost function.
4. Update the PSO algorithm factors which include the local and the best global factors in accordance with the acceptable path.
5. Repeat the previous three steps, (step 2 to step 4) for the chosen particle number.
6. Update the values of position and velocity according to the mathematical equations below for the following APF factor calculation:

$$v_i(k+1) = w \times v_i(k) + c_1 \times rand \times \left(P_{iatt}(k) - x_{iatt}(k)\right) + c_2 \times rand \times \left(P_{gatt}(k) - x_{iattL}(k)\right) \tag{12.20}$$

$$x_i(k+1) = x_i(k) + v_i(k+1) \tag{12.21}$$

$$v_i(k+1) = w \times v_i(k) + c_1 \times rand \times \left(P_{irep}(k) - x_{irep}(k)\right) + c_2 \times rand \times \left(P_{grep}(k) - x_{irepL}(k)\right) \tag{12.22}$$

$$x_i(k+1) = x_i(k) + v_i(k+1) \tag{12.23}$$

where *rand* is a random real number between [0, 1], P_{iatt} is the best position for the *iatt* particle for finding the attractive factors, x_{ia} is the current existing position, and $x_{iaL}$is the best local position where the letters (*att*) indicate the attractive factors, a similar definition to the Eqs. (12.20) through (12.23) where the letters (*rep*) indicate the repulsive factor.

7. Apply the APF method using the updated values of the repulsive and the attractive factors, then compute the cost objective function.
8. Update the APF values locally and globally to determine the factors' best global values.
9. Repeat the previous three steps (step 6 to step 8) using the updated factors according to the selected population number and the number of iterations.
10. After finding the best value globally for each factor, the related route generated will be the best computed path.

12.9.2 Simulation Result

12.9.2.1 The First Proposed Environment

Figure 12.9 shows the first robot navigation environment using the proposed method of APF-based PSO as a planning algorithm to find an optimal path, while the smoothed path after using the spline method is shown in Fig. 12.10. In this suggested test environment, the robot moves from the start point (2, 4) and reaches the target point (−1, −6), in which seven static obstacles are included in a test environment of dimensions [−7 to 7] for both x and y coordinates.

12.9.2.2 The Second Environment

Figure 12.11 represents the first resultant path of APF-based PSO, but Fig. 12.12 is the final smoothed path after applying the spline equations for smoothness. In this environment, seven stationary (static) obstacles are included, and the starting point

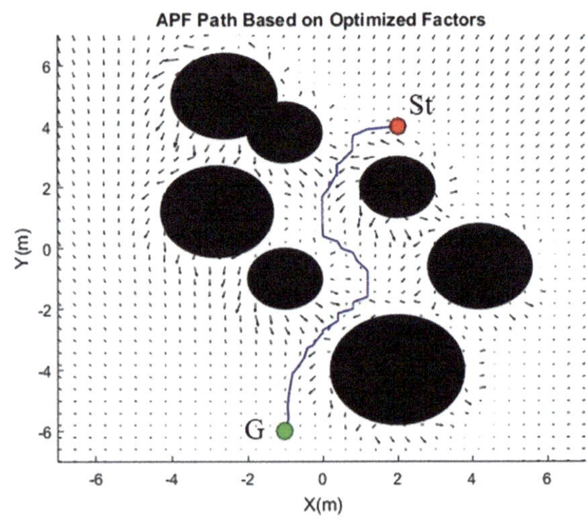

Fig. 12.9 Robot path planning for the first environment after employing the APF method. (Raheem and Badr 2017)

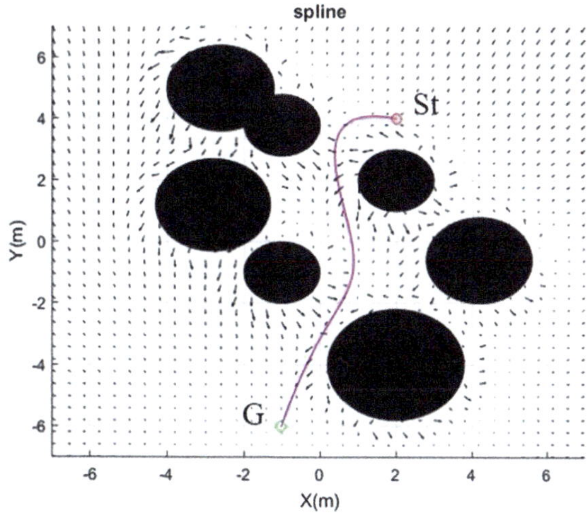

Fig. 12.10 Robot path planning for the first environment after employing the spline method. (Raheem and Badr 2017).

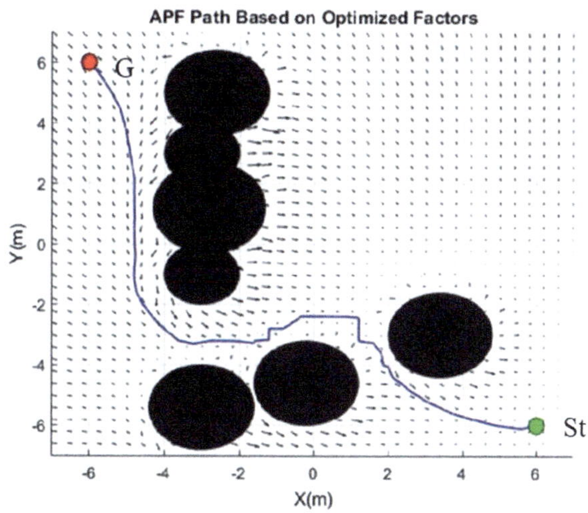

Fig. 12.11 Robot path planning for the second environment after employing the APF method. (Raheem and Badr 2017)

is at (−6,6) while the target point is at (6,−6); the environment has the same dimensions.

12.10 Dynamic Environment Path Planning Using D* Heuristic Method Based on PSO

This method is used for robots to find a safe and short route of planning in a dynamic moving obstacle environment. These moving obstacles can be various objects, people, animals, or other moving robots. Hybrid robotic path-planning methods use the

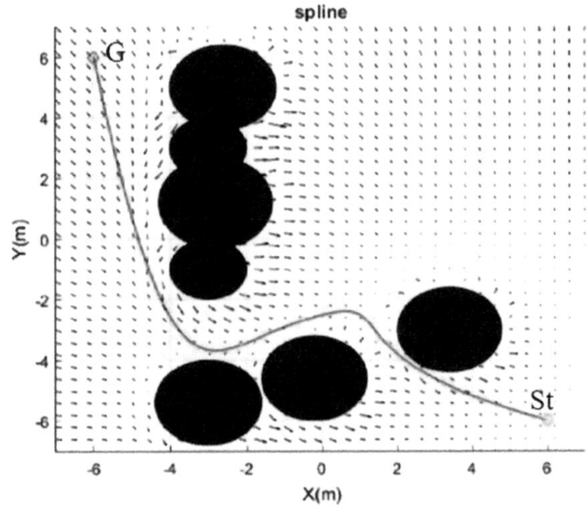

Fig. 12.12 Robot path planning for the second environment after employing the spline method. (Raheem and Badr 2017)

combination of heuristic calculations and an optimization algorithm. The D* heuristic method was used again to determine the shortest path. In addition, the method of particle swarm optimization (PSO) can be applied after the D* algorithm for further improvement of the path and to provide the final optimal path. Computations of this hybrid method in a dynamic environment consider complete changes at the domain of each sample of motion. The effectiveness of this hybrid proposed method was verified through simulation results (Raheem and Hameed 2018).

12.10.1 Proposed Method: Hybridization of D Heuristic Method Based on PSO*

The heuristic D* algorithm is used as an algorithm for path planning, which relies on node expansion in a test environment with dynamic moving objects. The creation of the test environment comprises defining the limitation of the map and starting and ending points in addition to the obstacles' location and size (Raheem and Hameed 2018).

After the creation of an environment, the execution of the algorithm includes standing the robot at the goal node; it starts moving virtually with an initial expansion to its eight connected neighbors. Each node has an initial cost function, and the robot selects the node with the lowest cost function for moving and places it in the D* closed list; further nodes will be placed into the open list. The huge cost-function dynamic obstacles that were sorted from the beginning of the procedure in the closed list and until reaching back to the starting node are taken into account (the D* method is called a reverse searching method). Upon completion of the heuristic D* search, the robot moves from the previously saved starting node in the closed list to

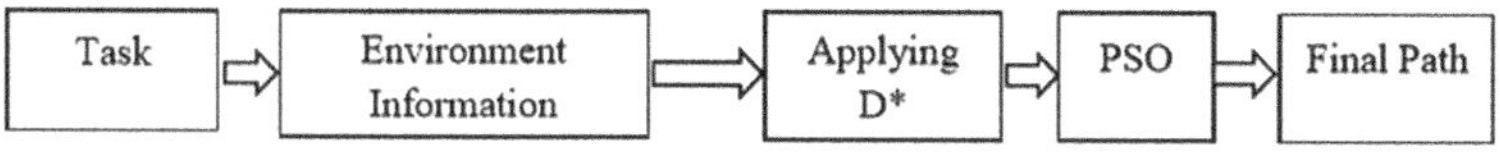

Fig. 12.13 Mass point proposed path-planning stages in a dynamic environment. (Raheem and Hameed 2018)

the goal node averting all the obstacle nodes, current node, and the node of the next step of motion.

The moving obstacles have a constant velocity, while the mass point (a virtual point that represents a robotic object and can be considered as either a mobile robot center or after some modifications as a robot manipulator end effector) velocity is changing with time (*MS* × *Ts*) during navigation and can be determined by (Raheem and Hameed 2018):

$$v_x(t) = \frac{\Delta x(t)}{MS \times T_s} \tag{12.24}$$

$$v_y(t) = \frac{\Delta y(t)}{MS \times T_s} \tag{12.25}$$

where Δx is the change in x-axis distance, Δy is the change in y-axis distance, *MS* refers to the sample of motion, and *Ts* is the time of the sample at 0.01 s intervals.

After determining the path using the heuristic D* method, a path improvement stage requires applying the PSO optimization method for further path enhancement by eliminating the sharp edges and shortening the path. The path produced from the D* algorithm generally takes a stair shape. The stages of the proposed mass point path planning in a known dynamic environment are presented in Fig. 12.13.

12.10.2 Simulation Result

12.10.2.1 Test Environment Number One

A non-interactive path solution dynamic environment has been tested to verify the proposed method of path planning by finding a collision-free and shortest path solution based on the combination of the D* heuristic method and PSO optimization technique among the moving obstacles and the robot. In this environment, many obstacles have different sizes and different behaviors of motion. The environment contains eight obstacles, six moving obstacles with different styles of movement, and two static obstacles. In this critical case, only a few possible paths exist that can maintain the shortest route length (Raheem and Hameed 2018).

The use of the PSO and D* algorithm ensures finding the shortest path avoiding collisions with dynamic and static obstacles, as shown in Fig. 12.14a. The hybrid

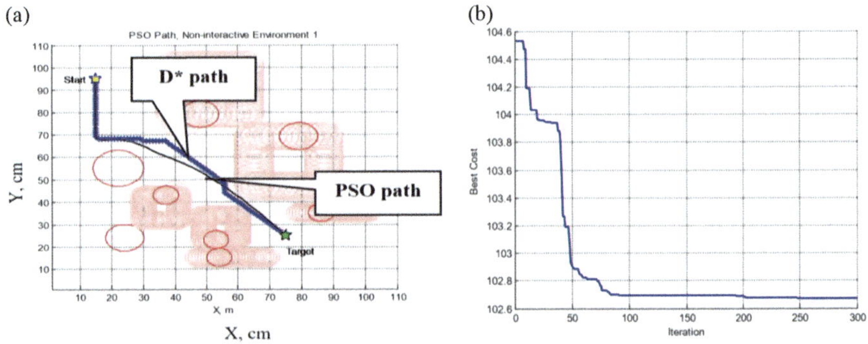

Fig. 12.14 Non-interactive path planning using D* algorithm based on PSO: (a) shows D* and D*-based PSO paths, (b) shows the curve of the path cost function using PSO

approach of the D*-based PSO enhanced the path to 102.6672 cm, as presented in Fig. 12.14b. The cost function reaches a 3.9013 cm minimum value after 300 motion samples (Raheem and Hameed 2018).

12.10.2.2 Test Environment Number Two

Non-interactive path solutions in a known dynamic environment that has been tested to verify the proposed method of path planning by finding a collision-free and shortest path solution based on the combination of the D* heuristic method and PSO optimization technique are provided here. In this environment, many dynamic obstacles have different sizes and different and difficult behaviors of motion. The environment contains six moving obstacles with different sizes and difficult styles of movement.

The use of the PSO and D* algorithm ensures finding the shortest path avoiding collisions with dynamic obstacles, as shown in Fig. 12.15a. The hybrid approach of the D*-based PSO enhanced the path to 106.6039 cm, as presented in Fig. 12.15b. The cost function reaches a 4.0651 cm minimum value after 300 motion samples (Raheem and Hameed 2018).

12.11 Interactive Path Solution Using Heuristic D* Method and PSO in Known Dynamic Environment

A new methodology is presented here for finding an interactive robot path-planning solution in a fully known dynamic environment. This methodology includes the use of the PSO technique together with an improved and modified D* heuristic method which is applied to the total analysis of the free Cartesian space at each sample of motion. The essential D* method has been modified to meet the requirements of the

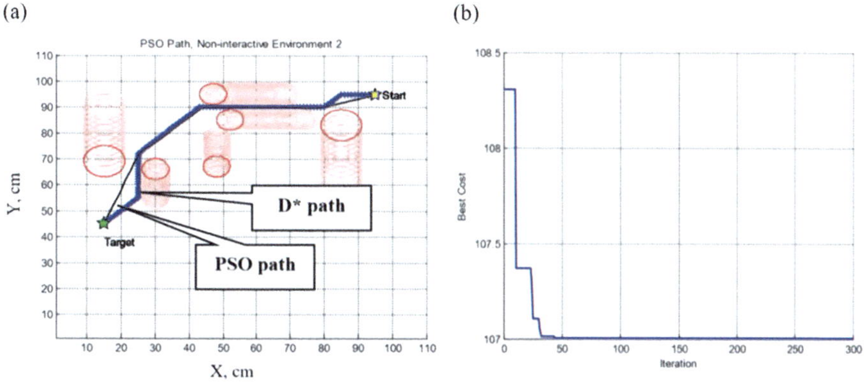

Fig. 12.15 Non-interactive path planning using D* algorithm based on PSO: (**a**) shows D* and D*-based PSO paths, (**b**) shows the curve of the path cost function using PSO

known dynamic environment by considering the addition of two cases, a motion stop and a return backward case. These cases are not covered in the original theory of the D* heuristic method (Raheem and Hameed 2019).

12.11.1 Proposed Method: Hybridization of Modified D Heuristic Method and PSO*

After applying the D* heuristic method and studying the final path solution which was a non-interactive path in a well-known environment that contains movable obstacles, an enhancement by finding an interactive path solution method has been proposed for finding an enhanced path by taking into consideration the full dynamic behavior of the obstacles, including spatial and temporal information. Finding the robot interactive path solution by applying the heuristic D* method in case of known obstacle time and position dynamic environment requires a map with all task information and limitations, start and goal positions, and the position, shape, and size of the obstacles. This method was used with a 2-DOF planar robotic arm as presented in Fig. 12.16. The interactive path solution includes the free Cartesian space calculations (FCS) in each motion sample of the robot and from node to node. In this case, the motion range of the robot can be described accurately. It is more intelligent to analyze and calculate multiple and changeable FCS maps for each motion sample. A new calculated FCS map will indicate the feasible area for the robot to move. Although this is a time-consuming process, it produces an accurate decision for the right motion step with high robot motion accuracy. Since this calculation process is an offline search method that starts after constructing the environment, the method will start when the robot is located at the goal node first and then expanded initially to its eight connected neighbors. Each node has an initial value of the cost objective

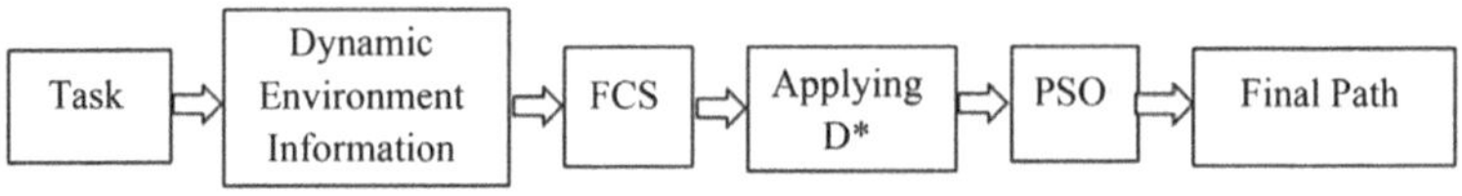

Fig. 12.16 Proposed interactive path-planning method in a dynamic environment for 2-DOF planar robotic arm

function, where the robot selects the node that has the minimum cost to continue moving on and then save it in the closed list, at the same time further remaining nodes will be saved in the open list. In the same way, keep in mind the obstacles that have big or huge values of the cost function, sorting these values and storing the initial starting positions in the closed list. At each sample of motion, the D* closed list is updated, to know the new locations of the obstacles, and the closed list will be dynamically changing at each motion sample and storing the nodes representing the obstacle at the current motion sample. The process will complete an offline iterative process only in case of reaching the start node. In case of search completed, the robot moves interactively from the initial node considering the closed list immediately to have required information about the free nodes prior to move on toward the goal node. The heuristic D* function $f = g + h$, where f is the objective function and g is the estimated cost function value from the start point while h is the cost to the goal (Raheem and Hameed 2019).

12.11.2 Simulation Results

12.11.2.1 First Environment

As a robot example, a 2-DOF planar robotic arm is used here. The arm length is assumed to be equal to half of the environment length, and each robot-link length is equal to 50 cm. The arm joints can theoretically rotate 360 degrees. Figure 12.17 presents the path of the D* algorithm for the 2-DOF planar robotic arm, where part (a) shows the end effector located at the start point at the beginning of the planning process. Part (b) shows the arm configuration after eight samples of motion. The arm at motion sample 15 was returned backward to avoid collision as shown in part (c). Parts (d) and (e) show the motion planning of the arm after 21 and 36 samples. At the sample of motion number 38, the arm was returned backward to avoid collision with a moving obstacle, while two samples later, the arm safely reached the target point as shown in parts (g) and (h) (Raheem and Hameed 2019).

Since this is an offline interactive path solution approach, optimizing the D* by applying the PSO algorithm has been done offline to remove the sharp edges and to shorten the D* path. Figure 12.18a shows the blue line of the D* path, and the black line is the final D*-based PSO path solution. Moreover, Fig. 12.18b explains that the length of the D*-based PSO path solution was reduced by 3.2846 cm compared with the original D* path (Raheem and Hameed 2019).

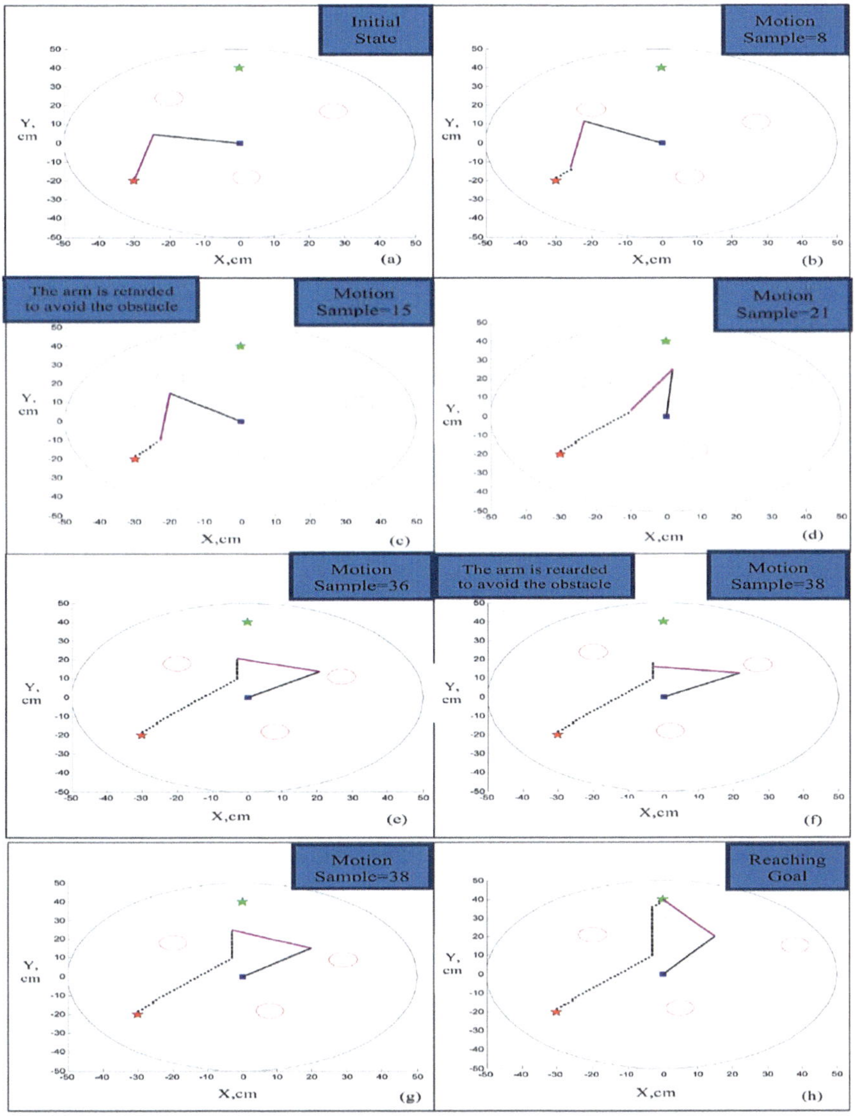

Fig. 12.17 2-DOF planar robot path results after applying the D* algorithm. (Raheem and Hameed 2019)

The motion style of this path solution is clearly featured as an interactive path movement, where the robot must make a decision under certain circumstances and may be stopped or returned backward in such a way that the stopping state and backward movement reduce the time and ensure getting a shorter collision-free path.

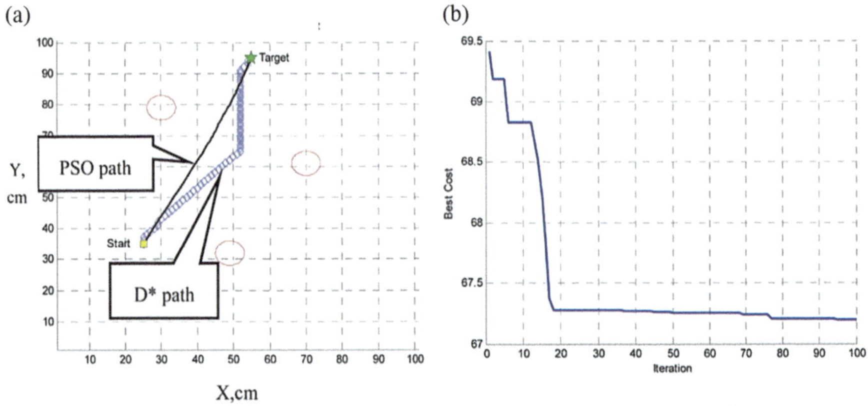

Fig. 12.18 Interactive path planning using D* algorithm based on PSO: (**a**) final D* and D*-based PSO paths and (**b**) cost function of the path

12.11.2.2 Second Environment

For further testing, this method was proposed and verified by a second more complicated environment setup where four dynamic obstacles move closer to the links of the robot and two of these obstacles have a square path behavior of motion. The robot can make an appropriate decision for choosing the next points freely and intelligently within an allowable area. Figure 12.19 presents the D* path of the 2-DOF planar robotic arm, where part (a) indicates the start of planning (the robotic end effector located on the initial starting point), part (b) presents the robot motion at the end of 11 samples, part (c) presents the robot in a return backward state at sample 21 of motion, parts (d) and (e) present the robot motion configuration after 33 and 42 samples of motion, respectively, and parts (g) and (h) clarify the robot motion after 46 samples of motion; the robot was returned backward to avoid collision with the closest moving obstacle before reaching the target point of destination safely (Raheem and Hameed 2019).

Since the location and time behavior of the moving obstacles are known, finding a path solution as a process can be performed offline, so PSO plays an important role in completing the process and finding the final applicable best, shortest, and optimal path with removing the sharp edges of the heuristic D* original path. Figure 12.20a shows both of these paths, D* path and D* based on PSO path, where the blue line indicates the path of D*, while the black line indicates the D* based on PSO path. Figure 12.20b plots the cost progress of the objective function optimization and verifies that the last definitive optimized path utilizes heuristic D* based on optimization technique (PSO); the length of the path was shorter than the original D* path by 4.3724 cm (Raheem and Hameed 2019).

In comparison, Cubero (2007) tried to use the heuristic D* method to solve the same problem for a mobile robot path-planning case. The proposed method introduced here differs because of its new characteristic of the modification added to the

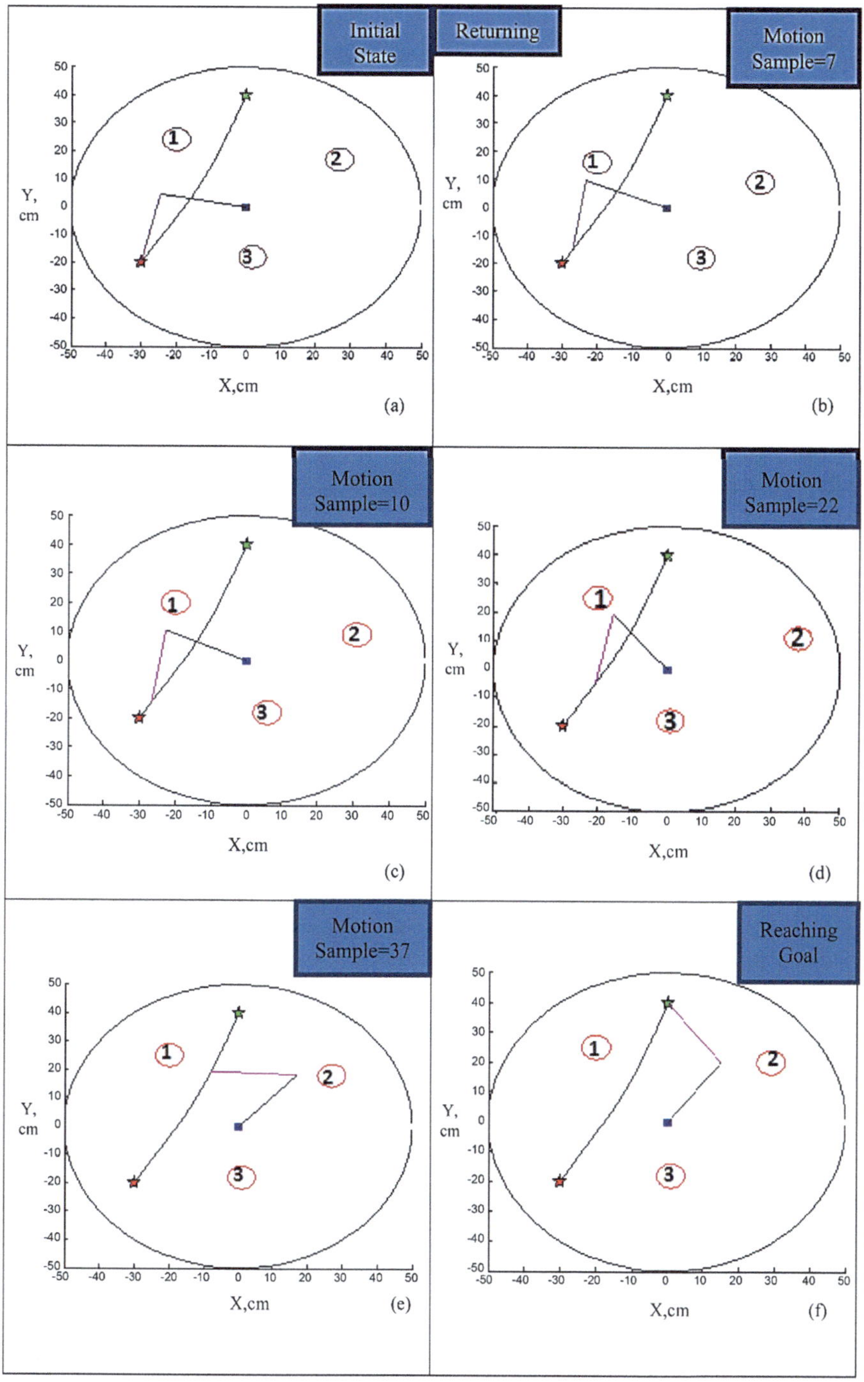

Fig. 12.19 The two-link robot arm result for the D* path for the second environment. (Raheem and Hameed 2019)

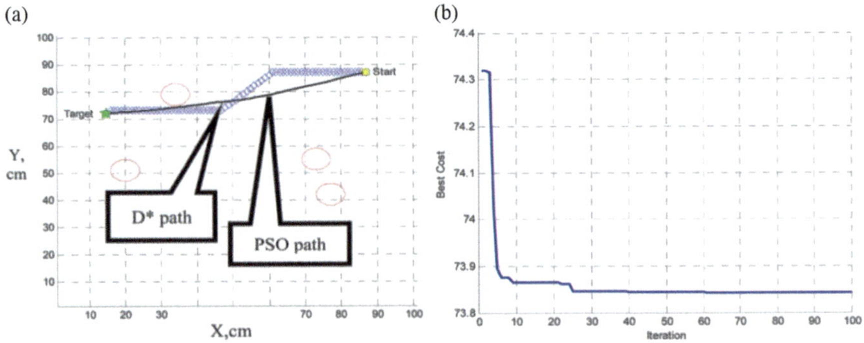

Fig. 12.20 (**a**) Final D* and PSO path and (**b**) the path cost function for the second environment

D* heuristic method which is suggested in our work. This modification ensures the safety, interactivity, and finesses of the final PSO-optimized path.

12.12 Heuristic A* Path Solution Based on C-Space Analysis

In this method, a configuration space (C-space)-based path-planning approach for a 2-DOF planar robotic arm was presented. The analysis of the constructed C-space map makes the process of finding a feasible, safe, and successful path solution much easier. After constructing and analyzing the C-space map for the planar robot which comprises all the possible collision situations between the obstacles and the robot links and joints from the base till the end effector, the heuristic A* method has been applied to find an optimal path in the C-space map that represents a successful heuristic path in the Cartesian workspace. A modification on the original heuristic A* method equations and calculations has been carried out to make this method applicable in the C-space. Moreover, further modifications are required to apply the A*-based C-space methodology for robotic manipulators with high degrees of freedom (more than 2 DOF). The method was verified by simulation results and proved the overall C-space map construction accuracy. These results were efficient and successful and guarantee a collision-free path from the start point to the target point by priorly eliminating all the probabilities of collisions with any robot point that appeared clearly in the C-space map (Raheem and Hussain 2017a, b).

12.12.1 C-Space Derivation

For a robotic arm, C-space is the space that can be expressed by the joint's variable parameters, whether it is a linear positioning joint or a rotational positioning joint. C-space map can be divided into two main areas, C-obstacle area which indicates the robot configurations in which collisions between the robot links and obstacles definitely occur and C-free area which represents the free allowable area for planning the path (Raheem and Hussain 2017a, b; Pan and Mancha 2015). The obstacle(s) can be represented geometrically and indicated as B in the equations below, so the C-space of the obstacles (C_{obs}) is as follows (Althoefer 1996):

$$C_{obs} = \{q : A(q) \cap B \neq \varnothing\} \tag{12.26}$$

while the C-space free (C-free) area can be referred to as below (Pan and Mancha 2015):

$$C_{free} = \{q : A(q) \cap B = \varnothing\} \tag{12.27}$$

where C_{obs} is C-space obstacles, q is robot configuration, and $A(q)$ are the set of points that are included in an area confined by configuration q of the robot. The overall C-space map can be represented as (Pan and Mancha 2015):

$$C_{space} = C_{obs} \cup C_{free} \tag{12.28}$$

The mathematical study of the surface properties, which do not change if this surface has deformations, such as bending or stretching, is called the topology of space. The topology for robot arms with two revolute joints can be described as (Lynch and Park 2017):

$$S^1 * S^1 = T^2 \tag{12.29}$$

The above equation means that the C-space for two revolute joint robot arms is a torus (with no joint limits for the joints), where S^1 is the circle description and $T^2$is the two-dimensional torus surface (Lynch and Park 2017). Each joint angle pair corresponds with a unique point on the torus, as shown in Fig. 12.21.

To find the mapping of C-space, basic shapes of obstacle(s) will be detailed.

12.12.1.1 Point Obstacle C-Space Construction

Mathematically, the simplest and essential for the beginning of C-space calculations which can be located in a robot workspace is the point obstacle. Herein, a modified derivation for C-space analysis will be discussed, in which collision checking is based on geometrical analysis. In the case of a point obstacle and according to the calculations of all probabilities of collisions, it can be noted that the collision area with the first link contains straight lines, while the collision between the point obstacle and the second link appears as a curved shape. As the point obstacle approaches the end effector of the planar robotic arm, the points of the C-space obstacle will be reduced, as presented in Fig. 12.22 (Raheem and Hussain 2017a, b).

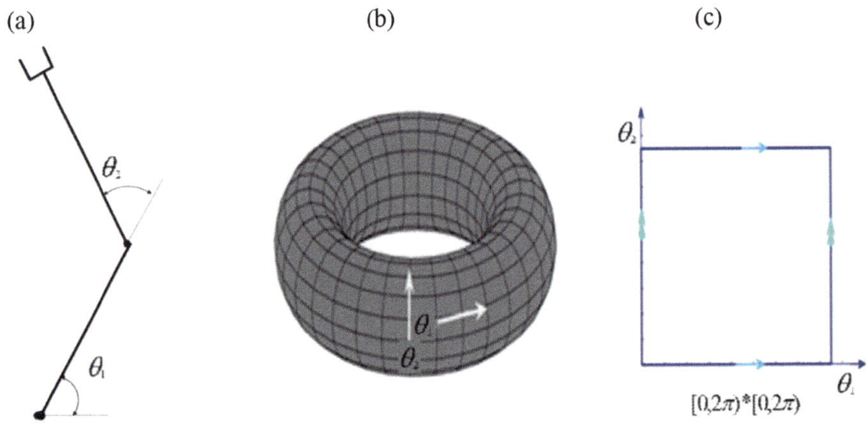

Fig. 12.21 Topological representation of the two-dimensional C-space. (**a**) 2R robot arm; (**b**) 2Torus; (**c**) sample representation. (Lynch and Park 2017)

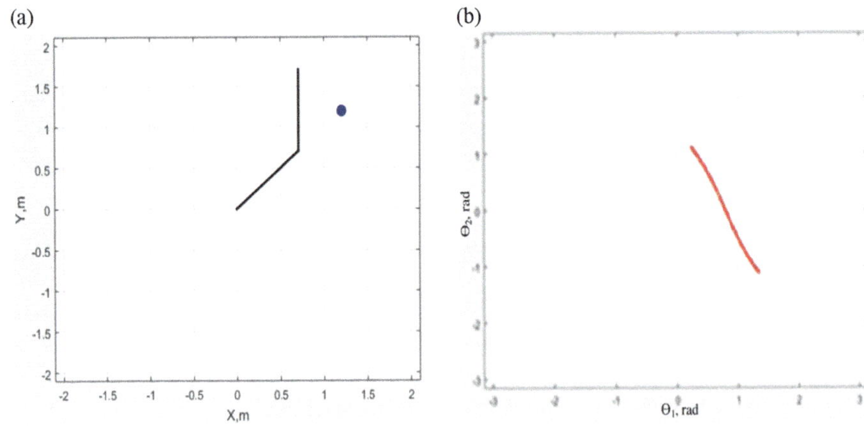

Fig. 12.22 C-space mass point analysis for (**a**) point obstacle collides with the second link only; (**b**) C-space curve where the curve points decrease as the point is nearest to the end effector

12.12.1.2 Line Obstacle C-Space Construction

Robot links can be handled as two lines, regardless of their width. Here, the derivation of C-space analysis comprises two analyses for collision testing; the same C-space curved map of the point obstacle can be applicable for each line(s) point. Thus, these compact form curves will represent the C-space of the line or polygonal obstacles, as illustrated in Fig. 12.23 (Raheem and Hussain 2017a, b).

A C-space map for any 2D shape formed by three or more straight lines will be considered a C-space map for a polygon. In case of a triangle (three lines), square, rectangle, rhombic (four lines), or more than four lines, the same method of

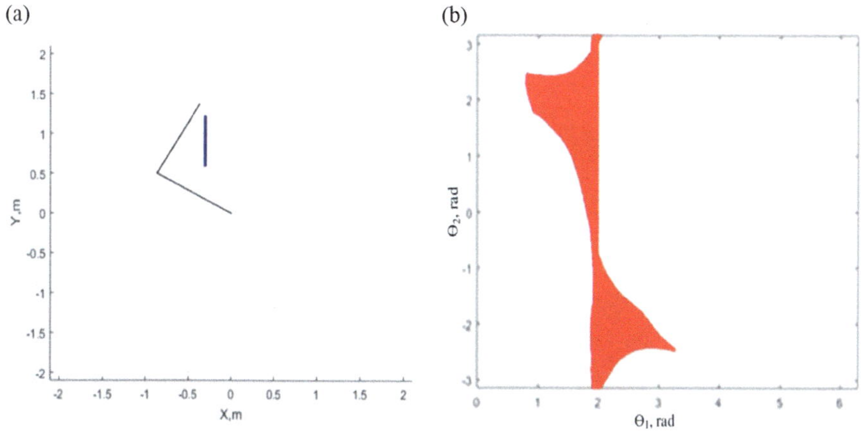

Fig. 12.23 C-space line obstacle analysis for (**a**) robot arm with line obstacle in the workspace, (**b**) corresponding C-space map

calculations and analysis for constructing the C-space map will be applied here as the method used for line obstacle (Bloch 2015).

In case the obstacle in the Cartesian workspace approaches closer to the first link, then the number of points on the C-space map increases. Besides that, the constructed C_{free} space will be on both sides of the constructed C_{obs} space. For that reason, finding a feasible path solution from one side of the C-space map to the other side requires choosing a suitable range of changing θ_1 for better unification of both regions which leads to finding a C_{free} path planning much easier, as presented in Fig. 12.24 (Raheem and Hussain 2017a, b).

12.12.1.3 Circle Obstacle C-Space Construction

A circle obstacle can be defined by a radius (rad) and center (x_c, y_c); the collision can be analyzed and tested by the derivation of C-space analysis.

Two main parts of edges of the C-space map constructed for a circular obstacle are presented in Fig. 12.25. Part (a) includes the collision between the circle obstacle circumference and the first link, which denote the vertical straight lines with all the probabilities of θ_2 values, while part (b) comprises the outer lower and outer upper curved shapes denoting the collision between the circle obstacle circumference and the second link (Raheem and Hussain 2017a, b).

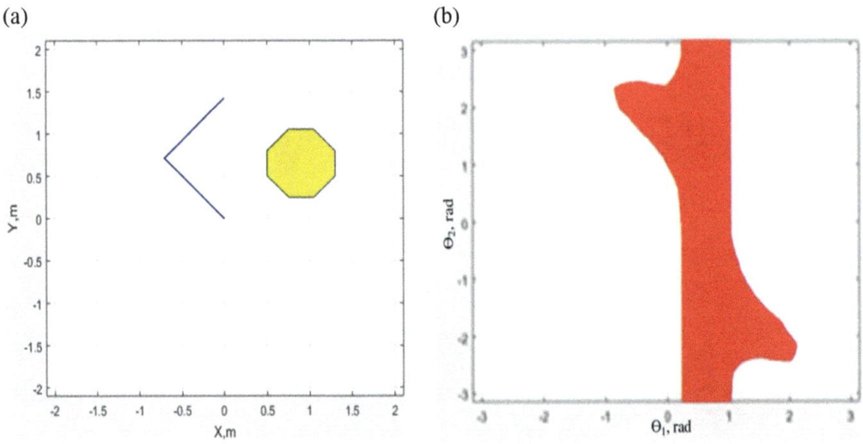

Fig. 12.24 C-space polygon obstacle analysis for (**a**) 2-DOF planar robotic arm Cartesian workspace contains octagonal obstacle. (**b**) Equivalent constructed C-space map

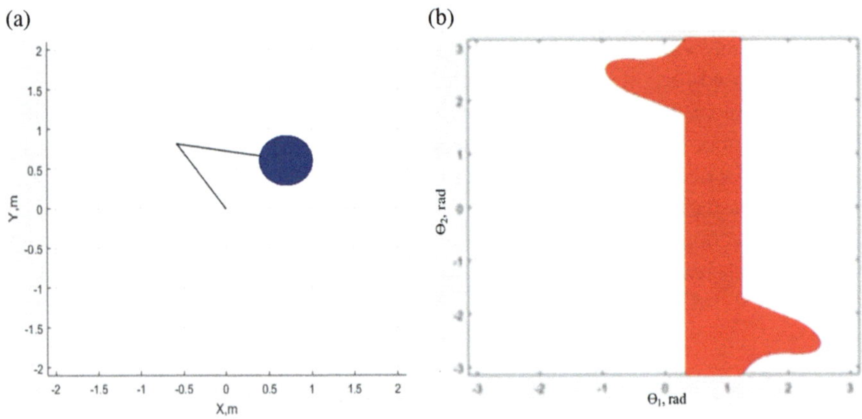

Fig. 12.25 C-space circle obstacle analysis for (**a**) 2-DOF planar robotic arm Cartesian workspace contains circle obstacle. (**b**) Equivalent constructed C-space map

12.12.2 Results of Applying A* Algorithm on Modified C-Space

Two environments have different obstacle shapes and tasks. First, we apply all the steps of C-space analysis and then perform the construction to get the exact map, which includes the collision and free areas. Second, the start and goal configuration, which are represented by the dot and the star shapes, respectively, have been computed by inverse kinematic equations. For the given test environment, the start and goal point are represented by the dot and the star shapes, respectively, as shown in Figs. 12.26 and 12.27 (Raheem and Hussain 2017a, b).

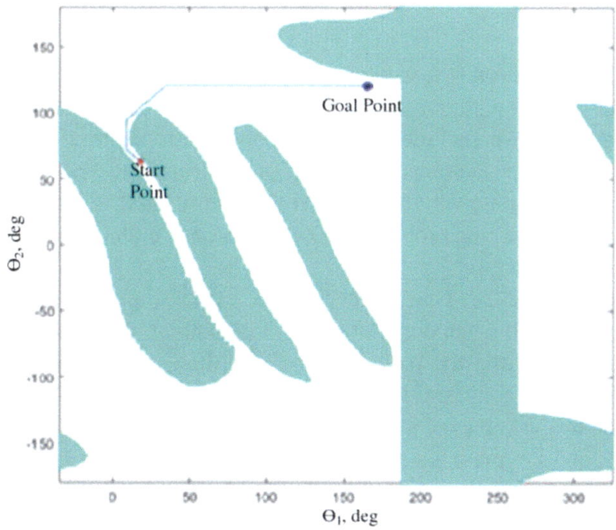

Fig. 12.26 A* path solution in the constructed C-space map according to the given environment task

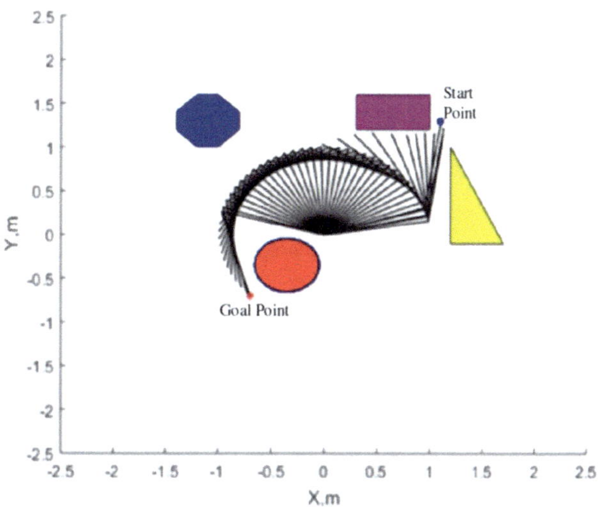

Fig. 12.27 2-DOF planar robotic motion from start to goal point for given environment task

12.13 Summary and Conclusion

Usually, finding the shortest feasible path is considered an optimization problem to determine the shortest path length from the start to the target points without colliding with obstacles. We included different robot path-planning applications and methodologies of their solutions depending on environment type, robot space, and the required calculations; we also showed the theory used to find a suitable path solution. The path-planning problem has been solved in both Cartesian and configuration spaces.

The analysis of the free Cartesian space for a 2-DOF planar robotic arm represents a significant task when complete information about the static robot arm environment is available and the path is globally offline, planned prior to the robot motion. While in the configuration space, a specific configuration of the robot can be represented by the values of joint angles as a point in the C-space map. Moreover, heuristic, A*, and D* methods are used to compute the best path solution in both known static and known dynamic environment.

The particle swarm optimization (PSO) technique has been used to make the smoothest, shortest path. The artificial potential field (APF) was coupled with PSO to tune the factors of the field construction to find the best mass point path solution. The results of these path-planning applications are simulated, analyzed, and compared to prove the effectiveness of the solutions. The techniques and solutions presented in this chapter have numerous "real-life" applications as the use of robotics increases to ensure unnecessary resource loss is not incurred in both static and dynamic environments in chemical, biological, geological, and temporal fields.

References

Althoefer K (1996) Neuro-fuzzy motion planning for robotic manipulators. Department of Electronic and Electrical Engineering, King's College London, University of London

Bloch ED (2015) Polygons, polyhedra, patterns & beyond. http://faculty.bard.edu/bloch/math107_notes.pdf. Accessed 22 May 2021

Boonyaritdachochai P, Boonchuray C, Ongsakul W (2010) Optimal congestion management in an electricity market using particle swarm optimization with time-varying acceleration coefficients. Comput Math Appl 60:1068–1077. https://doi.org/10.1016/j.camwa.2010.03.064

Choset H (2007) Robotic motion planning: A* and D* search. Carnegie Mellon University School of Computer Science: The Robotics Institute. Available via https://www.cs.cmu.edu/~motionplanning/lecture/AppH-astar-dstar_howie.pdf. Accessed 22 May 2021

Cubero S (ed) (2007) Industrial robotics theory, modelling, and control. Verlag Robert Mayer-Scholz

Gue J, Gao Y, Cui G (2013) Path planning of mobile robot base on improved potential field. Inf Technol J 12(11):2188–2194. https://doi.org/10.3923/itj.2013.2188.2194

Hameed UI (2019) Development of path planning algorithm for robotic manipulator in dynamic environment. MSc thesis. University of Technology-Iraq, Baghdad, Iraq

Hussain AA (2017) New approaches for c-space map construction for heuristic path solutions based on PSO technique. MSc thesis. University of Technology-Iraq, Baghdad, Iraq

Li G, Yamashita A, Tamura HAY (2000) An efficient improved artificial potential field based regression search method for robot path planning. In: Proceedings of the 2000 IEEE international conference on robotics & automation, San Francisco, CA
Lynch KM, Park FC (2017) Introduction to robotics: mechanics, planning and control. Cambridge University Press
Nosrati M, Karimi R, Hasanvand H (2012) Investigation of the * (Star) search algorithms: characteristics, methods, and approaches. World Appl Program 2(4):251–256
Pan J, Mancha D (2015) Efficient configuration space construction and optimization for motion planning. Engineering Sciences Press
Poli R, Kennedy J, Blackwell T (2007) Particle swarm optimization. Swarm Intell 1:33–57. https://doi.org/10.1007/s11721-007-0002-0
Raheem FA, Badr MM (2017) Development of modified path planning algorithm using artificial potential field (APF) based on PSO for factors optimization. Am Sci Res J Eng Technol Sci 37(1):316–328
Raheem FA, Hameed UI (2018) Path planning algorithm using D* Heuristic method based on PSO in dynamic environment. Am Sci Res J Eng Technol Sci 49(1):257–271
Raheem FA, Hameed UI (2019) Interactive heuristic D* path planning solution based on PSO for two-link robotic arm in dynamic environment. World J Eng Technol 7:80–99
Raheem FA, Hussain AA (2017a) Applying A* path planning algorithm based on modified C-space analysis. Al-Khwarizmi Eng J 13(4):124–136
Raheem FA, Hussain AA (2017b) Applying A* path planning algorithm based on modified C-Space analysis. Al-Khwarizmi Eng J 13(4):124–136
Raheem FA, Sadiq AT, Abbas NF (2019) Robot arm free cartesian space analysis for heuristic path planning enhancement. Int J Mech Mech Eng IJMME-IJENS 19(01):29–42
Sreenivasulu R (2012) Simulation of Desired end point trajectory for A 2-Dof planar manipulator. Int J Adv Sci Tech Res 5:688–696
Stentz A (1994) The D* algorithm for real-time planning of optimal traverses. CML-RI-TR-94-37. The Robotics Institute, Carnegie Mellon University, Pittsburgh

Index

J. N. Furze et al. (eds.), *Earth Systems Protection and Sustainability*,
https://doi.org/10.1007/978-3-030-85829-2

Printed by Printforce, the Netherlands